D0789366

CARRIER GRADE VOICE OVER IP

McGraw-Hill Telecommunications

Carrier Grade Voice over IP

Daniel Collins

McGraw-Hill
New York   San Francisco   Washington, D.C.
Auckland   Bogotá   Caracas   Lisbon   London
Madrid   Mexico City   Milan   Montreal   New Delhi
San Juan   Singapore   Sydney   Tokyo   Toronto

McGraw-Hill

A Division of The McGraw·Hill Companies

Copyright © 2001 by The McGraw-Hill Companies, Inc. All rights reserved.
Printed in the United States of America. Except as permitted under the United
States Copyright Act of 1976, no part of this publication may be reproduced or
distributed in any form or by any means, or stored in a data base or retrieval
system, without the prior written permission of the publisher.

2 3 4 5 6 7 8 9 0   DOC / DOC   0 6 5 4 3 2 1

ISBN 0-07-136326-2

The sponsoring editor for this book was Stephen S. Chapman and the production
supervisor was Pamela A. Pelton. It was set in Century Schoolbook by D&G Limited,
LLC.

Printed and bound by R. R. Donnelley & Sons Company.

McGraw-Hill books are available at special quantity discounts to use as premiums
and sales promotions, or for use in corporate training programs. For more information, please write to the Director of Special Sales, Professional Publishing, McGraw-
Hill, Two Penn Plaza, New York, NY 10121-2298. Or contact your local bookstore.

Information contained in this work has been obtained by The McGraw-Hill
Companies, Inc. ("McGraw-Hill") from sources believed to be reliable. However,
neither McGraw-Hill nor its authors guarantee the accuracy or completeness of
any information published herein and neither McGraw-Hill nor its authors shall
be responsible for any errors, omissions, or damages arising out of use of this
information. This work is published with the understanding that McGraw-Hill
and its authors are supplying information but are not attempting to render engineering or other professional services. If such services are required, the assistance of an appropriate professional should be sought.

This book is printed on recycled, acid-free paper containing a minimum of
50 percent recycled de-inked fiber.

To my wife, Ann, my inspiration. The trip to Galveston was worth it.

To my wonderful parents: If everyone had parents like mine, the world would be a much better place. Happy birthday, Dad.

CONTENTS

Contents

Contents

Contents

PREFACE

As I write this book, the United States is enjoying the longest period of economic growth in history and many other countries are experiencing similar economic expansion. There may be many reasons for this growth, but most people agree that technology is a major contributing factor—if not the primary driver. The technology in question is the information technology that allows us to communicate faster and share information of all kinds, thereby increasing productivity.

Perhaps the most important component of information technology is the *Internet Protocol* (IP) suite. This is the set of protocols that enables file sharing, shared printers, e-mail, the World Wide Web, streaming audio and video, instant messaging and numerous other applications. These are applications that have revolutionized the way we work and have given us new ways to communicate. Now, IP has taken a step further and is starting to drastically change something that has existed for over a century—telephony.

Unlike the other applications previously mentioned, telephony and real-time applications in general present new problems. Perhaps the biggest is simply the fact that telephony already exists. It is one thing to introduce a new capability such as the Web, when nothing has gone before to compare it with. It is quite another to take something that already exists and do it in a different way. The new way has to be a lot better. There are many, and myself among them, who believe that *Voice over IP* (VoIP) is better than traditional telephony. VoIP is better because it allows for regular voice conversation to be integrated with all of the data applications that IP already supports. It is better because it provides the user with advanced features that the user can quickly and easily customize and use—no more confusing star-codes and hook-flashes. It is better because it enables greater competition in the telecommunications industry—something that benefits all users.

OK, so VoIP is so much better. However, until recently, it had been worse. The quality was not as good, and the systems did not scale very well. There is little benefit to having fancy new services if the voice quality is not as good as the existing service or if only a limited number of users can be supported. In other words, VoIP has to be at least as good in every respect as what is already there and must be better in a number of areas. That day is at hand. Thanks to the technical expertise and dedicated efforts of many individuals, VoIP technology offers high quality, scalability and reliability, coupled with support for a wide range of new features and capabilities.

The overall aim of this book is to describe the technical solutions that enable VoIP, with an emphasis on how VoIP networks can be made carrier-grade. I attempt to ask and answer the following questions regarding VoIP: What is it? How does it work? How can it be made to work just as well as what is there already? Why is it better? How can we benefit from it?

This book is largely a technical work and as such, it can serve as a useful reference for those in technical disciplines within companies that develop or plan to develop VoIP solutions and within companies that plan to offer VoIP solutions to customers. I have attempted to present the technical material in a manner that will also be meaningful and useful to those who do not work with technical details on a daily basis. Interested individuals in all areas of the telecommunications and information technology industries should find that this book provides a useful introduction to VoIP and a practical explanation of this new and exciting technology.

At the end of the day, VoIP is about communication, which in turn, is about presenting information in a way that can be understood by others. Protocols are those sets of rules governing how information is structured and exchanged between parties, be those parties computers, telephones, other devices or even people. Consequently, this book deals with a number of protocols in some detail. There is no point in technology for technology's sake, however. Therefore, I have tried to present the various VoIP-related protocols and technical solutions in sufficient detail for them to be understood, but without losing sight of the fact that they exist for a real reason, that reason being the improvement of people's lives through better communication between people. I hope that I have succeeded.

Finally, I wish to thank those organizations and individuals that that have made material available to me for inclusion in this book. First, the talented and dedicated individuals who make up the *Internet Engineering Task Force* (IETF). Without them, the Internet would not be the success that it is. Second, the *International Telecommunication Union* (ITU), an organization that has granted me permission to use ITU material in several places in this book. Third, Ralph Hayon at Congruency and Jim Hourihan at Pingtel. Finally, I wish to thank Mike Gallagher at Synacom, for giving me the book-writing urge in the first place.

—*Daniel Collins*

CHAPTER 1

Introduction

Introduction

The two terms that make up the title of this book—carrier grade and *Voice over IP* (VoIP)—mean many things to different people, and each term might have several meanings depending on the context. Furthermore, many people might consider the two terms mutually exclusive, making the phrase "carrier grade VoIP" an oxymoron. That situation might have been the case until recently. However, the excitement that VoIP has generated and the resources that have been applied to developing technical solutions for VoIP mean that VoIP has become a serious alternative for voice communications. In other words, VoIP is a serious alternative to the circuit-switched telephony that we have known for decades. Not only can VoIP technology now offer straightforward telephony services, but it can also offer a lot more—and with the same quality of service that we have become familiar with in traditional telephony.

The overall objective of this book is to examine the technical solutions that make VoIP a reality and to illustrate the manner in which VoIP can offer every feature that traditional circuit switching provides, plus a lot more. This first chapter introduces the reader to VoIP in general and examines some of the reasons why this topic is so hot. Thereafter, we present a brief overview of technical challenges and solutions. You can consider this chapter a glimpse of the topics that are covered in more depth in later chapters. First, however, we need to understand what exactly we mean when we talk about terms such as VoIP and carrier grade.

What Does Carrier Grade Mean?

Perhaps the easiest way to describe the term carrier grade is to think of the last time you picked up your residential phone and failed to obtain a dial tone. With the exception of extraordinary circumstances, such as an earthquake or a careless backhoe operator inadvertently cutting through some cables, most people cannot remember such a time. Basically, the phone always works.

Carrier grade means extremely high reliability. In fact, the requirement for reliability in commercial telephony networks is 99.999 percent. This figure corresponds to a down time of about five minutes per year and is known as five-nines reliability.

Carrier grade refers to the capability to support hundreds of thousands (if not millions) of subscribers. For example, AT&T, the world's largest telecommunications carrier, serves more than 80 million customers.[1] Serving such large numbers of customers also requires the capability to carry hundreds of thousands of simultaneous phone calls. For example, in the last three months of 1999, the domestic Bell Atlantic network supported more than 46 billion minutes of use.[2]

Carrier grade means that when you dial a number, you get through to the number that you dialed. Also, when you finish dialing, the phone at the other end starts ringing within two to three seconds. Also, when someone answers the phone and a conversation takes place, the speech quality is high and no perceptible echoes, noticeable delays, or annoying noises are heard on the line.

These standards are tough to meet. In terms of the network, these requirements translate to systems that are fully redundant; in some cases, self-healing, highly scalable, and manageable. They mean compliance with numerous technical specifications to ensure interoperability with other networks. These requirements also imply a highly skilled network-maintenance organization with technicians on duty around the clock who are ready to respond to a network problem in an instant.

What Does VoIP Mean?

VoIP is simply the transport of voice traffic by using the *Internet Protocol* (IP), and this definition is hardly surprising. You should note, however, that VoIP does not automatically imply Voice over the Internet.

The Internet is a collection of interconnected networks, all of which use IP. Anyone and everyone can use the connections between these networks for a wide range of applications, from electronic mail (e-mail) to file transfer to electronic commerce (e-commerce). As we will see later, one of the greatest challenges of VoIP is voice quality, and one of the keys to acceptable voice quality is bandwidth. If we are to ensure that sufficient bandwidth is available to enable high-quality voice, then we need to control and prioritize access to the available bandwidth. Currently, that regulation does not occur over the Internet. In fact, each Internet user is at the mercy of all other users. One person transferring a huge file might cause other users' transactions to proceed more slowly. As a result, voice quality over the Internet today might vary from acceptable to atrocious.

---

[1] www.att.com/network

[2] www.bellatlantic.com/invest/financial/quarterly/4q99_telecomop.htm

All is not lost, however. The Internet is changing rapidly in terms of its size, number of users, and technology being used. Great effort is being applied to the development of technology for the next generation of the Internet. Therefore, it is possible (and even likely) that high-quality voice over the Internet might become the norm rather than the exception. That day is still not at hand, however. Consequently, while the term Carrier Grade VoIP might not be an oxymoron, the term Carrier Grade Voice over the Internet might well be. Internet telephony might be considered a subset of, or a special case of, VoIP (a.k.a., IP telephony). Within this book, however, when we refer to VoIP, we generally exclude Internet telephony. We will see that high voice quality over IP networks requires the use of managed networks, *Quality of Service* (QOS) solutions, and service-level agreements between providers. While the Internet might include all of those things in the future, we have not yet arrived at that point.

In the meantime, a significant opportunity exists to deploy VoIP over networks in which access and bandwidth are better managed. One such environment is that of next-generation telecommunications companies—those companies who are building telephone networks that use VoIP from the outset and who are posing a significant challenge to incumbent carriers.

A Little About IP

IP is a packet-based protocol, which means that traffic is broken into small packets that are sent individually to their destination. In the absence of special technical solutions, the route that each packet takes to its destination is determined independently at each network node on a packet-by-packet basis. IP is not the only packet-based protocol in existence; however, it is by far the most successful. The explosive growth of the Internet proves this statement.

IP itself provides no guarantees. For example, different packets can take different routes from the origin to the destination, leading to the possibility of packets arriving at a destination in a different order than they were originally sent. Not only can packets arrive out of sequence, but some packets might not arrive at all or might be severely delayed. To combat these shortcomings, other protocols have been developed to operate in conjunction with IP, in order to ensure that packets are delivered to their ultimate destination in the correct sequence and without loss. The most notable of these protocols is the *Transmission Control Protocol* (TCP). TCP includes functions for retransmission of packets that might have been lost or delayed and for the assembly of packets in the correct order at the destination. TCP is so widespread that the term TCP/IP is almost as common as IP and is sometimes used synonymously with IP. While the mechanisms that TCP uses are appropriate and successful for data transfer (e.g., file transfers and e-mail), they are not appropriate for the delivery of voice traffic.

Most data traffic is asynchronous and extremely error sensitive. For example, we usually do not care if an e-mail message takes 10 seconds or 30 seconds to reach its destination, but it is critical for every bit to be received correctly. On the other hand, voice traffic is synchronous in nature and is a little more tolerant of errors. When someone speaks, for example, the listener should hear the words almost immediately, while it might not be as critical that every millisecond of speech is heard.

Given that IP provides no guarantees regarding the efficient transport of data packets, one wonders why IP would even be considered as a means for transporting voice—particularly with the stringent delay requirements that voice imposes. One also wonders how VoIP can be made to match the quality, reliability, and scalability of traditional networks. After all, if VoIP is to be a successful competitor to traditional telephony technology, then it must meet all of the requirements that are met by traditional telephony and must offer new and attractive capabilities beyond traditional telephony. VoIP must also accomplish these goals at a lower cost.

Why VoIP?

We should break this question into two parts. First, why worry about carrying voice at all when there is a large and lucrative market for data services? The second question is, assuming that a carrier intends to compete in the voice market, "Why should IP be chosen as the transport mechanism?"

Why Carry Voice?

IP and the Internet have led to many new and exciting services. Whereas in the past it might have taken weeks to access certain types of information, you can now find out practically anything in an instant. You can now communicate at the touch of a button and share information between colleagues, friends, and family. Many people now shop online, and the variety of goods that are available is growing rapidly. Just to demonstrate the variety of services that are available over the Internet, at this writing, a man in

Dallas who is known as "dot-com guy" is spending a year without leaving his home and is surviving completely by using the Internet.

In other words, people are using the Internet in new and exciting ways. The Internet also means significant revenue opportunities and has led to the creation of companies and enterprises that are capitalizing on these opportunities. However, the revenues generated by e-commerce are miniscule compared to the revenues generated by telephone companies who carry voice. Voice is still the killer application.

To illustrate this point, let's look at some real examples. In 1999, Amazon.com generated approximately $1.6 billion in revenues but still failed to make a profit.[3] Yahoo!, one of the best-known companies in the IP world, generated a net income of $61.1 million in 1999 from revenues of $588.6 million.[4] By comparison, BellSouth generated a net income of almost $3.5 billion from revenues of more than $25 billion, with most of its revenues from domestic wireline operations.[5] Meanwhile, Bell Atlantic generated a net income of more than $4.7 billion from revenues of more than $33 billion, again with a large portion of the revenue from regular telephony.[6] To further emphasize this point, you can now purchase calling cards at practically any convenience store, and there are more and more new voice carriers appearing almost daily. Voice is big business.

Why Use IP for Voice?

Traditional telephony carriers use circuit switching for carrying voice traffic. Given that these carriers generate large revenues and profits, why should anyone consider using a packet technology such as IP to transport voice instead of using circuit switching? Circuit switching was designed for voice from the outset and does its job extremely well. So, why try to fix something that is clearly not broken?

The fact that circuit switching was designed for voice is both its strength and its weakness. Without a doubt, circuit switching carries voice well, although one could argue that circuit switching is an expensive solution. Circuit switching is not particularly suitable for much else, however. These

---

[3]Amazon.com Annual Report, 1999

[4]Yahoo!, Inc. Annual Report, 1999

[5]Bell South Annual Report, 1999

[6]Bell Atlantic Annual Report, 1999

days, people want to talk as much as ever, but they also want to communicate in a myriad of other ways—through e-mail, instant messaging, video, the World Wide Web, etc. Circuit switching really does not qualify as a suitable technology for this new world of multimedia communications. Furthermore, circuit-switching equipment has a number of attributes, including vendors and operators, that are less than ideal. IP offers not only another technology choice, but also a new way of doing business (as we will describe shortly).

IP is an attractive choice for voice transport for many reasons, including the following:

- Lower equipment cost
- Integration of voice and data applications
- Lower bandwidth requirements
- The widespread availability of IP

Lower Equipment Cost

Although there are thousands of standards and technical specifications for circuit-switched telephony, the systems themselves are generally proprietary in nature. A switch from a given supplier will use proprietary hardware and a proprietary operating system, and the applications will use proprietary software. A given switch supplier might design and produce everything including processors, memory cards, cabling, power supplies, the operating system, and the application software—perhaps even the racks to hold the equipment. The opportunities for third parties to develop new software applications for these systems are extremely limited. The systems generally also require extensive training to operate and manage. Consequently, when an operator chooses to implement systems from a given vendor, it is not unusual to choose just a single vendor for the whole network (at least, at the beginning). If the operator wishes to replace operating equipment with equipment from a new vendor, the cost and effort involved is huge. The result is that the equipment vendor, once chosen, is in a position to generate revenue not only from the sale of the equipment—but also from training, support, and feature development. In this environment, proprietary systems are the order of the day.

This world is much like the mainframe computer business of several years ago. In those days, computers were large, monolithic systems. The system vendor was responsible for all of the components of the system, plus installation, training, and support. The choice of vendor was limited, and once a system from a particular vendor was installed, changing to another vendor was extremely difficult. Since those days, however, the computer world has changed drastically from the mainframe model. Although many mainframe systems are still in place, many of the big vendors of mainframe systems no longer exist. Those vendors who do still exist do so in quite a different form.

The IP world is different from the monolithic systems of mainframe computers and circuit-switching technology. Though some specialized hardware systems exist, much of the hardware is standard computer equipment and is mass-produced. This situation offers a greater choice to the purchaser and the opportunity to benefit from the large volumes that are produced. Commonly, the operating system is less tightly coupled to the hardware, and the application software is quite separate again. This situation also enables a greater range of choices for the purchaser of the system and the capability to contract with a separate company to implement unique features. Furthermore, IP systems tend to use a distributed client-server architecture rather than large monolithic systems, which means that starting small and growing as demand dictates is easier. In addition, this type of architecture means that there are companies that make only portions of the network solution, enabling the customer to pick those companies that are best in different areas and creating a solution that is optimum in all respects. Figure 1-1 gives a pictorial comparison between the mainframe approach of circuit-switched systems and that of packet switching technologies. In the words of John Chambers, Cisco Systems CEO, "When a horizontal business model meets a vertical business model, horizontal wins every time."

Figure 1-1
Architecture and business model comparison

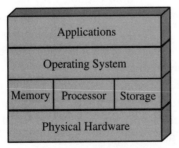

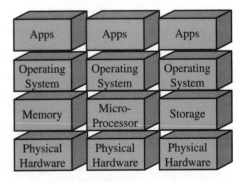

Not only are IP architectures more open and competition friendly, but the same characteristics also apply to IP standards. IP standards are more open and flexible than telephony standards, enabling the implementation of unique features so that a provider can offer new features more quickly. Also, a provider will be better able to customize service offerings. New features can be developed and deployed in a few months, which is a big difference from the traditional telephony world—where new features can take 18 months to two years.

One could argue that an *Intelligent Network* (IN) architecture in telephony also offers great flexibility and rapid feature development. Although that is what IN has promised from the beginning, it does not match the openness and flexibility of IP solutions. With the exception of a few highly successful services, such as toll-free calling, IN has not had the enormous success that was originally envisioned. Of course, those IN services that are successful happen to be extremely important and also need to be provided by an IP-based solution. As we will see in Chapter 7, "VoIP and SS7," VoIP networks can interoperate with *Signaling System 7* (SS7) and can take advantage of IN services that are built on SS7. Hence, IP can take advantage of what is already there and can do even more.

All of these characteristics mean greater competition, greater flexibility on the part of the network operator, and a significantly lower cost. Furthermore, the packet-based systems obey Moore's Law, which states that processing power doubles roughly every 18 months, which is a much faster pace of development than traditional circuit-switched systems. Figure 1-2 shows a comparison between the two. We can clearly see that circuit-switching equipment lags behind packet and frame technologies where speed of development is concerned.

Of course, all of the major manufacturers of traditional telephony switches now also produce VoIP solutions, but a quick search on the Internet will reveal a large number of new companies also producing VoIP products. The range of choices is large; the equipment cost is drastically lower than that of circuit switching products; and the pace of development is incredibly fast.

Voice/Data Integration and Advanced Services

IP is the standard for data transactions—everything from e-mail to Web browsing to e-commerce. When we combine these capabilities with voice transport, we can easily imagine advanced features that can be based on

Figure 1-2
Cost performance
comparison

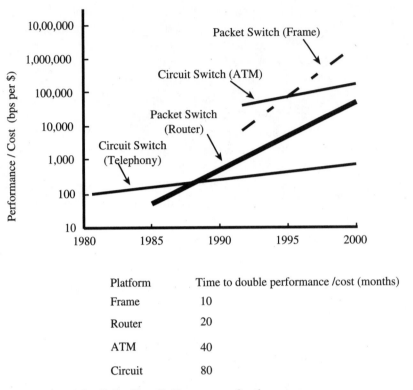

Platform	Time to double performance /cost (months)
Frame	10
Router	20
ATM	40
Circuit	80

Source: Peter J. Sevcik, President, NetForecast, peter@netforecast.com,
"Why Circuit Switching is Doomed," *Business Communications Review*, Sep 1997.

the integration of the two. One example is an application in which the receiver of an e-mail can call the sender simply by clicking a header within the e-mail. Another example is a button on a Web page that a user can click to speak to a customer-service representative for assistance with navigating the site or conducting a transaction, as depicted in Figure 1-3. When the user clicks that button, the call is directed to the most appropriate customer-service representative based on what the user was attempting to do just before deciding to seek help.

Today, shopping online is becoming increasingly popular but is still at the stage where it is most useful when the shopper knows what he or she wants to buy and knows how much he or she is willing to pay. Also, waiting a few days for delivery must not be a problem. We are still not at the stage where browsing through an online store is particularly simple. That day is quickly

Figure 1-3
Click to talk
application

Figure 1-4
Example application:
Web collaboration

approaching, however. Furthermore, Web collaboration already exists—where one user can navigate the Web and another user at another location can view the same Web page. In that case, why not shop with a friend? You can both see the same Web page and share opinions before purchasing, as depicted in Figure 1-4.

Yet another example is video conferencing. Not only does this feature include person-to-person video telephony, but it can also include shared white-board sessions. Design teams can come together to discuss issues and sketch solutions while sitting hundreds or thousands of miles away from each other. With IP multicasting, virtual auditorium services are available —something of great interest for distance learning in particular.

These types of applications are not just wishful thinking. Many such applications exist today, and more are coming. Later in this chapter, we describe a number of examples of existing VoIP solutions, including IP-based *Private-Branch Exchanges* (PBXs), IP-based call centers, IP-based voice mail, and straightforward telephony.

Lower Bandwidth Requirements

Circuit-switched telephone networks transport voice at a rate of 64kbps. This rate is based on the Nyquist Theorem, which states that it is necessary to sample an analog signal at twice the maximum frequency of the signal in order to fully capture the signal. Typical human speech has a bandwidth of somewhat less than 4,000Hz. Therefore, when digitizing human speech, a telephone system takes 8,000 samples per second as directed by the Nyquist Theorem, and each sample is represented with 8 bits. This amount leads to a bandwidth requirement of 64kbps in each direction for a standard telephone call. This mechanism is standardized in *International Telecommunications Union* (ITU) Recommendation G.711.

While G.711 is the standard for telephony, many other coding schemes are available. These schemes use more sophisticated coding algorithms that enable speech to be transmitted at speeds such as 32kbps, 16kbps, 8kbps, 6.3kbps, or 5.3kbps. Furthermore, some of these coding techniques utilize silence suppression, such that traffic is passed over the network only when something is being said. Given that most conversations involve one person listening and one person talking at a given time, the bandwidth saved by not transmitting silence can be significant. We discuss various speech-coding schemes in detail in Chapter 3, "Speech-Coding Techniques," and all of those schemes described can be used in VoIP systems. Therefore, VoIP offers significant advantages over circuit-switched telephony from a bandwidth-efficiency point of view.

You should note, however, that the availability of these coding schemes is not something that is unique to VoIP. Strictly speaking, these various coders could also be deployed in traditional telephony networks. They are not generally implemented in circuit-switched networks, however, because it would require an enormous investment. Currently, you can place a call from any telephone to any other telephone in the world. Taking advantage of more advanced coding schemes would require these coding techniques to be implemented for practically every telephone switch in the world. Alternatively, it would require existing networks to support the capability for the two ends of a conversation to negotiate the coding scheme to use. At the bare minimum, it would require calls to be routed from origination to destination based on the coding choice of the originator, so that if different coders are used at either end, the call could be passed though some interworking device. None of these capabilities are supported in traditional telephony networks. As a result, traditional telephony uses G.711 and is unlikely to ever change.

On the other hand, VoIP solutions are far more recent than traditional telephony. They generally support more efficient coding schemes than G.711, and the standards written for VoIP enable the two ends of a call to negotiate which coding scheme to use. Thus, VoIP is in a position to significantly reduce bandwidth requirements to as little as one-eighth of what is used in the circuit-switched world. Given that transmission capacity can account for a large percentage of a carrier's initial investment and a large portion of a carrier's operational costs, these bandwidth savings can mean a big difference to the bottom line.

The Widespread Availability of IP

IP is practically everywhere. Every personal computer produced today supports IP. IP is used in corporate *Local-Area Networks* (LANs) and *Wide-Area Networks* (WANs), dial-up Internet access, etc. IP applications now even reside within hand-held computers and various wireless devices. As a result, there is widespread availability of IP expertise and numerous application-development companies. This factor alone makes IP a suitable choice for transporting voice (or any other digital media stream, for that matter).

Of course, VoIP is not the only packet-based solution available to a commercial carrier. In fact, *Voice over Frame Relay* (VoFR) and *Voice over Asynchronous Transfer Mode* (VoATM) are powerful alternatives. One of the great disadvantages of such solutions, however, is that they do not have the same ubiquitous presence as IP. While many users currently use IP, whether they know it or not, the same cannot be said for Frame Relay or ATM. IP is already at the desktop, and it is hard to imagine that the same characteristic will ever apply to ATM or to Frame Relay.

To some extent, however, that factor might not matter to a next-generation telecommunications company if the main objective is to simply capture a market share in the long-distance business. Such a business could be easily served by other technologies besides IP. To ignore IP, however, would be to decline possible opportunities that are offered by voice and data integration. Ignoring IP would limit choices and opportunities for advanced applications and would reduce the possible benefits from the myriad of advances that are being made almost daily in the IP community. That is not to say that ATM and Frame Relay do not have a role to play. They certainly do, but their role lies in the core of the network as an underlying network transport for IP (and not at the user's end).

The VoIP Market

Clearly from previous descriptions, VoIP can offer cost savings, new services, and a new way of doing business—both for carriers and for network operators. Also, we can see that voice is an important service but not the *only* important service. Rather, voice is part of a suite of services that, when packaged together, can offer exciting new capabilities. All of this information translates into serious revenue opportunities. While the business is currently in its infancy and the revenues have yet to be seen, the future is certainly bright. Figure 1-5 gives one projection of the market for Voice over Packet.

Figure 1-5
Voice over packet revenue projection

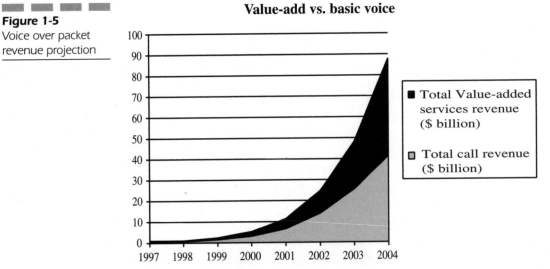

Value-add vs. basic voice

■ Total Value-added services revenue ($ billion)

□ Total call revenue ($ billion)

Source: Gartner/Dataquest/Bhawani Shankar /"Voice on the Net"/March 2000

Figure 1-6

Projected breakdown
of VoIP-related
revenue

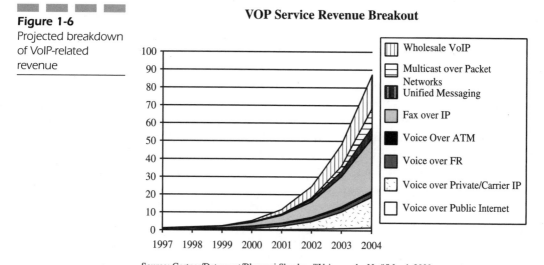

VOP Service Revenue Breakout

Wholesale VoIP

Multicast over Packet
Networks

Unified Messaging

Fax over IP

Voice Over ATM

Voice over FR

Voice over Private/Carrier IP

Voice over Public Internet

Source: Gartner/Dataquest/Bhawani Shankar /"Voice on the Net"/March 2000

You will find it interesting to note that Figure 1-5 refers to Voice over Packet, as opposed to VoIP. You might be curious to know what the breakdown is—in particular, how IP fares against other technologies (such as ATM or Frame Relay). Figure 1-6 provides that breakdown and shows that IP is expected to be dominant, indicating that ATM and Frame Relay are expected to comprise a relatively small percentage of the market. Perhaps the most interesting factor is that facsimile (fax) service is expected to be big over IP networks—almost as big a revenue generator as voice.

VoIP Challenges

To offer a credible alternative to traditional circuit-switched telephony, VoIP must offer the same reliability and voice quality. In other words, the five-nines reliability requirement must be met, and speech must be of toll quality. Toll quality relates to a *Mean Opinion Score* (MOS) of 4.0 or better on the following scale, as specified in ITU-T Recommendation P.800:

- 5—Excellent
- 4—Good
- 3—Fair
- 2—Poor
- 1—Bad

Speech Quality

Perhaps the most important issue in VoIP is ensuring high speech quality. Anyone who has made calls over the Internet can attest to the fact that the quality is variable at best. In order for VoIP to be a commercial challenge to circuit-switched technology, the voice quality must be at least as good as experienced in today's telephony networks and should not vary. The reason why this goal is such a challenge for VoIP networks is the fact that IP was not originally designed to carry voice or similar real-time interactive media.

From the outset, IP was designed for data. Among the characteristics of most data traffic is that it is asynchronous, which means that it can tolerate delay. Data traffic is also extremely sensitive to packet loss, which requires mechanisms for the recipient of data packets to acknowledge receipt, such that the sender can retransmit packets if an acknowledgment is not returned in a timely manner.

Delay In many ways, real-time communications (such as voice) are almost the exact opposite of data—at least, from a requirement perspective. First, voice is more tolerant of packet loss, provided that lost packets are kept to a small percentage of the total (e.g., fewer than 5 percent). Voice is extremely intolerant of delay, however. Anyone who has ever made an international call over a satellite knows how annoying the delay can be. In a satellite call, it takes about 120 milliseconds (ms) for the signal to travel from Earth to the satellite and another 120 milliseconds to come back down. If we add some additional delays due to processing and terrestrial transmission, we have a one-way delay from point A to point B of 250ms to 300ms. In other words, we have a round-trip delay of 500ms to 600ms.

The round-trip time is very important. Consider, for example, a conversation between A and B. A is speaking, and B decides to interrupt. Because there is a delay, however, what B is hearing is already old by the time he hears it. Furthermore, when B interrupts, it takes time for A to hear the interruption. Therefore, it seems to B that A has ignored the interruption. Therefore, B stops. Now, however, A finally hears the interruption and stops, so there is silence. Then, they might both try to start again and both back off again. The whole experience can be frustrating if the round-trip delay is any more than about 300ms.

ITU-T Recommendation G.114 states that the round-trip delay should be 300ms or less for telephony. Contrast this value with the delay requirements for transmission of e-mail. As stated previously, it hardly matters if an e-mail takes 30 seconds or three minutes to reach the recipient, provided that the content of the e-mail arrives exactly as sent.

Jitter   Just as important as the actual delay is delay variation, also known as jitter. If there is some delay in a given conversation, then it is possible for the parties to adjust to the delay fairly quickly—provided that the delay is not too large. Adjusting to the delay is difficult, however, if the delay keeps varying. Therefore, another requirement is to ensure that jitter is minimized, so that whatever delay exists is at least kept constant. This goal can be achieved through the use of jitter buffers, where speech packets are buffered so that they can be played to the receiver in a steady fashion. The down side of these buffers, however, is the fact that they add to the overall delay.

Jitter is something that is of concern only for real-time communications, such as voice. Jitter is not an issue for the types of traffic that are traditionally carried over IP networks. Jitter is caused in IP networks in two ways. First, packets take different routes from sender to receiver and consequently experience different delays. Second, a given packet in a voice conversation can experience longer queuing times than the previous packet—even if they both traveled the same route. This situation is due to the fact that the network resources, and particularly the queues, that the voice packets use might also be used by other traffic in the network.

In contrast to the situation in IP networks, jitter is not an issue in circuit-switched networks, because there is an open dedicated pipe from sender to receiver for the duration of the conversation. In other words, all of the speech follows the same path and does not have to share resources with any other traffic. What goes in one side comes out the other side after a fixed delay. That delay is generally tiny—at least, in the absence of satellite links.

Packet Loss   Another factor in high-quality voice is that little or none of the speech should be lost on the way from the speaker to the listener. Obviously, we need to ensure that speech is received as transmitted. The possibility exists, however, that queues might overflow in network nodes between the sender and the receiver, resulting in packet loss. While traditional retransmission mechanisms are used in the case of lost data packets, those same mechanisms cannot be applied to real-time communications (such as voice). First, it takes some time for the two ends to determine that a packet is missing. Second, it takes more time to retransmit the missing packet. If packets 1 through 5 and 7 through 9 have been received, then the receiving end has a choice: it can either present what has been received or hold packets 7 through 9 until packet 6 is received, then play packets 6 through 9. The second of these two approaches simply does not work, because the delay would be intolerable. Therefore, if a packet is missing, then the end systems must carry on without the packet. If a packet turns up out of sequence, however, then the risk exists that the packet can be presented to the recipient out of sequence (another no-no).

In a circuit-switched network, all speech in a given conversation follows the same path and is received in the order in which it is transmitted. If something is lost, the cause is a fault rather than an inherent characteristic of the system. If IP networks are required to carry voice with high quality, then they must emulate the performance of circuit-switched networks for voice traffic. Fortunately, engineers have conducted much work in this area and have developed solutions. Chapter 8, "Quality of Service," provides details regarding a number of these solutions.

Speech-Coding Techniques   Yet another factor in high-quality voice is the choice of speech-coding technique. Reducing bandwidth, minimizing delays, and minimizing packet loss are all well and good; however, these measures are useless if the speech still sounds "tinny" or synthetic. Consequently, we need to select a solution that reduces bandwidth while maintaining natural-sounding speech. In general, coding techniques are such that speech quality degrades as bandwidth reduces. The relationship is not linear, however, so it is possible to significantly reduce bandwidth at the cost of only a small degradation in quality. In general, the objective is to offer (or at least approach) toll-quality voice, where toll quality relates to a *Mean Opinion Score* (MOS) of 4.0 or better on the following scale (as specified in ITU-T Recommendation P.800):

- 5—Excellent
- 4—Good
- 3—Fair
- 2—Poor
- 1—Bad

Table 1-1 shows MOS values for several common coders.

While concepts such as "good" and "fair" do not sound scientific, they are not quite as nebulous as they might first appear. The ratings can be considered a summation of various quality parameters, such as delay, jitter, echo, noise, cross-talk, intelligibility, and overall lack of distortion. The manner in which various coders are tested and the scoring methodology are well standardized to the extent that the MOS scale is the most common standard for voice-quality measurement today. Nonetheless, MOS values are still subjective in nature. Therefore, regardless of how well tests are conducted, the subjective character of the evaluation means that one might find slightly different ratings for a given codec. The variations might

Table 1-1

Mean opinion scores for selected codecs

Speech Coder	Bit Rate (kbps)	MOS
G.711	64	4.3
G.726	32	4.0
G.723 (celp)	6.3	3.8
G.728	16	3.9
G.729	8	4.0
GSM Full Rate (RPE_LTP)	13	3.7

depend on the details of the test environment and the individual test participants. Therefore, anyone who intends to deploy VoIP should perform his or her own analysis of the available speech coders in a test environment that closely matches the expected network conditions. We provide details concerning voice-coding technologies in Chapter 3.

Network Reliability and Scalability

Not only should the voice sound clear and crisp and be free from noticeable delay, but the service should also be available at all times. Recall that the phone system almost never fails, and contrast that statement with office computer networks—some of which seem to fail on a regular basis. If VoIP is to be a commercial challenge to established telephony networks, then the network must be rock solid. In other words, the five-nines reliability standard must become the standard for IP telephony networks.

Fortunately, today's VoIP solutions do offer reliability and resilience. Just because the components are often the same as are found in less-reliable networks does not mean that VoIP networks are inherently unreliable. In fact, today's solutions include the type of redundancy and load sharing that is found in traditional telephony networks and that is designed to meet carrier-grade performance.

In addition, it must be possible to increase the capacity of the networks in order to handle millions of simultaneous calls. That certainly was not the case just a few years ago, when VoIP systems were small and aimed only for a corporate LAN environment. Nowadays, the networks are far more scalable. VoIP gateway products exist on the market today that support hundreds of megabits per second with equipment that occupies just a single 19-inch shelf.

Today's VoIP networks use fiber-optic transport, gigabit routers, and/or a high-speed ATM base.

An advantage of VoIP architectures is that it is relatively easy to start small and expand as the traffic demand increases. Standardized protocols enable small network nodes to be controlled by the same call-processing systems that are used for large network nodes. Therefore, an expansion of the network capacity can be conducted in small or large steps as needs dictate.

Managing Access and Prioritizing Traffic

One of the great advantages that VoIP offers is the capability to use a single network to support a wide range of applications, including data, voice, and video. That situation brings its own set of problems, however. Because of the fact that voice and data have different quality requirements, they must not be treated in the same way within the network. For example, making sure that voice packets are not waiting in a queue somewhere while a large file transfer is occupying network bandwidth is a critical task.

This situation translates to a number of requirements. First, a call must not be connected if there are not sufficient network resources available to handle the call. The last thing that we want is to answer the phone and find that conversation is not possible because there is not enough bandwidth available. Therefore, we must ensure up front that sufficient bandwidth exists before any phone rings.

Second, different types of traffic must be handled on the network in different ways. We must be able to give certain types of traffic priority over other types, so that the most critical traffic is affected the least by network congestion. In practical terms, when a network becomes heavily loaded, e-mail traffic should feel the effects before interactive traffic (such as voice).

All of these issues fall into the realm of QOS. QOS is an area of development within the IP community that has undergone huge efforts in recent times. Solutions exist today for ensuring high QOS in IP networks, and further development is proceeding at an incredible pace. We describe these solutions in Chapter 8.

VoIP Implementations

Before describing the technical details of VoIP in detail, the remainder of this chapter is devoted to a brief overview of current technology and some descriptions of how the technology is being used today.

IP-Based PBX Solutions

One of the most successful areas of IP telephony is in the enterprise market, as opposed to the public telecommunications market. Numerous IP-based PBX solutions are in place, and more are being deployed daily. The idea of an IP-based PBX makes a lot of sense. First, this system integrates the corporate telephone system with the corporate computer network, removing the need for two separate networks. Because a new office will be wired for a LAN anyway, why not use that network for voice communication as well and not bother installing separate cabling for a telephone network? The PBX itself becomes just another server or group of servers in the corporate LAN, which helps to facilitate voice/data integration. Such a scenario is depicted in Figure 1-7.

With an IP-based PBX, all of the services provided by a typical PBX are still provided, plus many more. In many cases, the client terminals use software that can be integrated with existing address-book applications so that making a phone call can be as simple as clicking a contact person's name. In addition, the solutions are often integrated with IP-based voice-mail systems. Connections to the outside world are established via gateways that might be separate from or integrated with the PBX itself. In either case, the

Figure 1-7
IP-based PBX

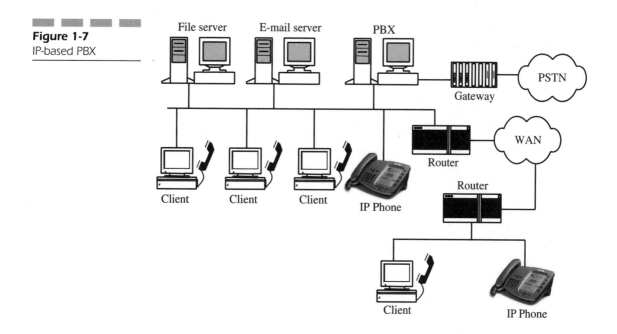

PBX looks similar to any standard PBX as far as the *Public-Switched Telephone Network* (PSTN) is concerned.

IP Voice Mail

Voice mail is perhaps one of the easiest applications to implement by using IP. The reason is because voice mail, as opposed to interactive voice, is a non-real-time application. When playing back stored speech, sufficient buffering can be performed locally to overcome any delay or delay variation in the transport of the stored speech from where it is stored to where it is played, regardless of whether the distance is measured in feet or in thousands of miles. Playing back a voice-mail message actually involves streaming audio, as opposed to real-time audio.

The neat aspect of IP-based voice mail is the fact that like other IP-based applications, voice mail can use a Web interface. A user can be informed when voice mail is deposited and can then access a Web page to see what messages exist. The user might be able to access the page from any location. Moreover, the user receives more than just a flashing light and a stutter dial tone to indicate that voice mail has been deposited and does not have to wade through multiple messages to get to the message that is of most interest to him or her. Instead, a list can be displayed that shows the details of each message, such as when it was deposited and from whom. The user can just click each message in the order of his or her choice. Fast-forward, rewind, skip, and other controls can all be included. Furthermore, the application can be integrated with address-book software so that just a couple of clicks are all that is necessary to forward the message in much the same way as e-mail applications. In fact, some implementations integrate e-mail, voice mail, and fax mail into a single unified messaging service with a single user interface.

Hosted PBX Solutions

For the *small office/home office* (SOHO), the hosted PBX is an attractive alternative to implementing a PBX on the customer's premises. Several companies offer hosted PBX services, and several other companies make hosted PBX systems. Given that an office will contract with an ISP for Internet service and possibly Web hosting, the capability to offer a hosted PBX solution gives the ISP the capability to offer a more complete service package and generate additional revenue. For the customer, the service is likely to be

Figure 1-8
Hosted IP-PBX

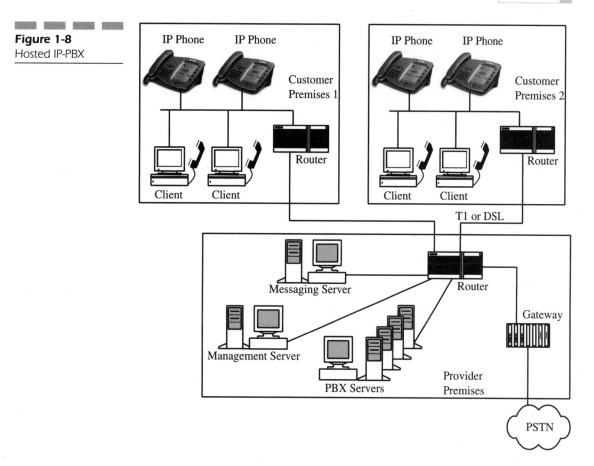

cheaper than a PBX on the premises and might also be cheaper than Centrex service from the PSTN. Figure 1-8 shows an example configuration.

IP Call Centers

Regardless of VoIP, call centers are perhaps one of the best examples today of computer-telephony integration. When someone calls a customer-service center via a regular 800 number, the only information that is initially available to the customer-service representative is the caller's phone number. In some cases, such as at a phone company service center, this information might be sufficient to automatically load the customer's information on the agent's desktop. In many cases, however, this information is not sufficient, and the customer has to provide an account number and/or some other unique information over the phone.

Today, many companies provide the customer with the ability to query account information and to manipulate customer data over the Internet. In the event that the customer has difficulty, however, there is almost always a toll-free number to call. With VoIP, that toll-free number can be replaced with a button on the Web page. When the user clicks the button to initiate a call to a customer-service representative, not only can a free representative be selected, but that agent's desktop can also automatically be loaded with the customer's information. Furthermore, the possibility exists to route the call to the most appropriate personnel at the service center (or perhaps in another service center). This feature can help avoid the hassle of a customer speaking to one representative first, who then has to transfer the customer to someone else.

The results of VoIP in a call-center scenario can include better customer service, reduced queue times, increased productivity, and lower cost.

IP User Devices

VoIP protocols, and the *Session Initiation Protocol* (SIP) in particular, enable a great deal of intelligence to be placed inside the user device, rather than in the network. This feature enables the user to invoke a range of services locally, rather than having to subscribe to equivalent services in the network—as is the case with traditional telephony today. In addition, the user devices can incorporate many of the features already discussed in this chapter, such as the integration of telephony with Web browsing, e-mail, voice mail, address-book applications, etc. Of course, this situation requires more sophisticated devices than the 12-button keypad with which we are all familiar. The overkill of a full-blown workstation is not necessary, however. Instead, a number of devices are already on the market that are more akin to an integrated telephone/Web browser/message system/personal organizer. No longer is it necessary to remember various "star codes" to activate features. Instead, an intuitive *Graphical User Interface* (GUI) is available. Figure 1-9 offers examples of two such IP telephony devices. Contrast these examples with the traditional telephone keypad and ask yourself which user interface you would prefer.

Overview of Following Chapters

This chapter has provided a brief introduction to VoIP and has highlighted some of its attractions and challenges. The remainder of this book is devoted to addressing these issues in further detail.

Figure 1-9
New versus old
phones

pingtel

congruency, Inc.

We begin these detailed discussions in Chapter 2 ("Transporting Voice by Using IP") with a review of IP networking in general. This review will help the reader understand what IP offers, why it is a best-effort protocol, and why there are significant challenges in carrying real-time traffic (such as voice) over IP. We place an emphasis on those protocols that run on top of IP and that help make such real-time transport a possibility. Specifically, we give considerable attention to the *Real-Time Transport Protocol* (RTP), which is the protocol that is used for carrying voice packets over IP.

Chapter 3 focuses on voice-coding techniques. We already mentioned that IP as a transport protocol can take advantage of more sophisticated and efficient voice-coding schemes. Numerous coders exist from which to choose, however, and the choice is not necessarily simple. When implementing a VoIP network, you must understand what coding schemes are available and the advantages and disadvantages of each.

Chapter 4 ("H.323") discusses the H.323 architecture and protocol suite. H.323 is the standard for VoIP networks today. Although VoIP technology is relatively new, it has already undergone significant deployment—both in private VoIP networks and over the Internet. For some time, however, VoIP implementations used proprietary technology—which did not easily enable communication between users of different equipment. The emergence of H.323 has helped change that situation. As a result, H.323 is the most widely deployed VoIP technology at the moment. Whether that will remain the case is a topic of much discussion in the industry.

Chapter 5 ("The Session Initiation Protocol") discusses the rising star of VoIP technology: SIP. Though new and not yet widely deployed, SIP is hailed by many as the standard of the future for VoIP. Simple in design and easy to implement, the simplicity of SIP is one of its greatest advantages. Simplicity does not mean that flexibility is sacrificed, however. Instead, SIP is extremely flexible and supports a range of advanced features. Many people in the VoIP industry are convinced that SIP will rapidly overtake H.323 in the marketplace.

Interoperating with other networks (such as the PSTN) is of major concern in the deployment of a VoIP network. Such internetworking requires the use of gateways. Hand in hand with SIP are two protocols that deal with the control of those gateways: the *Media Gateway Control Protocol* (MGCP) and MEGACO. The two protocols address the same issue. MGCP is today's standard but is to be replaced in the near future with MEGACO. The main advantage of each is that they enable a widely distributed VoIP network architecture. Used in conjunction with SIP, they provide the foundation for the network architecture known as softswitch, which we describe in Chapter 6 ("Media Gateway Control and the Softswitch Architecture").

Of critical importance to a communications system is the signaling system that it uses. In fact, H.323, SIP, MGCP, and MEGACO are all signaling protocols, while the state of the art in PSTN signaling is SS7. SS7 is the signaling system that has made call setup as quick as it is today and that has provided the foundation for numerous services (such as caller-ID, toll-free calling, calling-card services, and mobile telephony). All of these services need to be supported just as well in a VoIP network as in a traditional network. In other words, the VoIP network must interoperate with the SS7 network. In Chapter 7, we explore the technology to support this functionality.

Chapter 8 deals with QOS. We already described the quality challenges that need to be met in order for a VoIP network to meet the stringent performance requirements that define a carrier-grade network. A great deal of work has been conducted in this area, and various solutions are discussed in Chapter 8. These solutions offer the means for VoIP to reach the goal of being the new technology of choice for voice transport.

Chapter 9 ("Accessing the Network") deals with the issue of accessing the network in the first place. If access is to be achieved from a standard telephone via a gateway, then there is no problem—but that situation does not facilitate the integration of voice and data applications. Many people in office environments use a corporate LAN with high-speed connectivity to the network, which is great. But what do they have when they go home—a 28,800bps modem? If the promise of voice and data integration is to be fulfilled, then the last mile to the customer must no longer be a bottleneck. Fortunately, technologies such as cable modems, *Digital Subscriber Lines* (DSLs), and wireless access mean that the access no longer needs to be such a limiting component of the network. These technologies are discussed in Chapter 9.

Finally, recall Figure 1-6 and the revenue potential of IP fax. Given that IP fax is expected to be such a hot item, it is worth having a look at the technology. This technology is addressed in Chapter 10, "Fax over IP."

CHAPTER 2

Transporting Voice by Using IP

Introduction

The Internet is a collection of networks, both public and private. The private networks include the *Local-Area Networks* (LANs) and *Wide-Area Networks* (WANs) of various institutions, corporations, businesses, government agencies, etc. The public networks include those of the numerous *Internet Service Providers* (ISPs). While the private networks might communicate internally by using various communications protocols, the public networks communicate by using the *Internet Protocol* (IP). In order for one private network to communicate with another via a public network, then each private network must have a gateway to the public network—and that gateway must speak the language of that network: IP.

We illustrate this situation in Figure 2-1, where we show a number of private networks that are connected to a number of routers. The various entities within a given private network might communicate internally in any way they want, but when communicating with other networks via the Internet, they do so with IP. The routers provide the connectivity between various networks and form the backbone of the Internet.

We do not mean to imply that private networks such as LANs and WANs do not use IP. In fact, the vast majority of them do. The reasons are numerous, including the fact that all major developers of computer systems embrace IP, and it supports a vast array of applications. Also, numerous

Figure 2-1
Interconnecting
networks

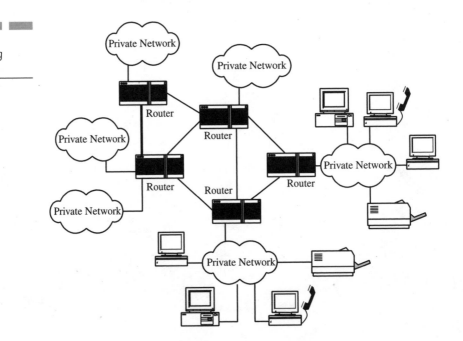

new IP-based applications appear almost daily, and implementing an IP-based network is relatively easy.

IP and some of the related protocols would require a separate book to describe in any detail. In fact, numerous books are available. The main intent in this chapter is only to provide a high-level overview of IP in general and an overview of aspects that are of particular significance to the transport of voice by using IP. This chapter will provide a better understanding of the architectures and protocols that we describe in subsequent chapters. A part of this chapter is dedicated to a brief explanation of the process of developing an Internet standard. Given that at the time of this writing, many new specifications or enhancements for VoIP were being finalized or enhanced, you should have a high-level understanding of the standardization process.

Overview of the IP Protocol Suite

Strictly speaking, IP is a routing protocol for the passing of data packets. Other protocols invoke IP for the purposes of getting those data packets from their origin to their destination. Thus, IP must work in cooperation with higher-layer protocols in order for any application or service to operate. Equally, IP does not incorporate any physical transmission facilities, so it must work in cooperation with lower-layer transmission systems.

When we use the term *layer*, we generally refer to the *Open Systems Interconnection* (OSI) seven-layer model. Figure 2-2 shows the OSI model. Each side of a communications transaction involves a number of layers. The top layer represents the useable information that is to be passed to the other side. In order to pass this information, however, it must be packaged appropriately and routed correctly, and it must traverse some physical medium. Consider, for example, a simple conversation between person A and person B. Person A has a thought that he or she wants to communicate to person B. In order to communicate that thought, person A must articulate the thought in language by using words and phrases. Those are communicated to the mouth by using nerve impulses from the brain. The mouth constructs sounds, which incorporate that language. Those sounds are transmitted through the air through vibrations. At person B's end, the vibrations are collected and the sounds are extracted. These are passed to the brain through nerve impulses. At the brain, the original thought or concept is extracted from the language that is used. Thus, the thought has passed from the mind of person A to the mind of person B but has used other functions in order to do so.

Figure 2-2
OSI seven-layer stack

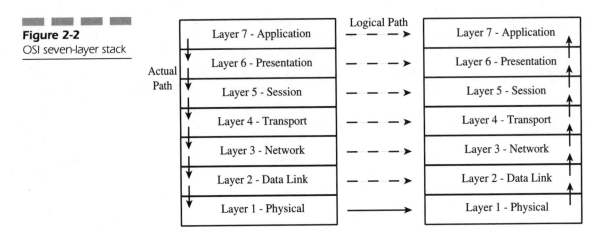

The OSI model includes seven layers, as follows:

- *Physical layer*   Includes the physical media used to carry the information, such as cables and connectors. The physical layer also includes the coding and modulation schemes that are used to represent the actual 1s and 0s.

- *Data link layer*   Enables the transport of information over a single link. This layer includes the packaging of the information into frames and can also include functions for error detection and/or correction and retransmission.

- *Network layer*   Provides functions for routing traffic through a network from source to destination. Because a source of traffic might not be connected directly to the required destination, there needs to be a mechanism for the information to pass through intermediate points, where it can be forwarded in the right direction until it hits the correct destination.

- *Transport layer*   Provides functions that ensure error-free, omission-free, and in-sequence delivery of data. To support multiple streams of information from source to destination, this layer provides mechanisms for identifying each individual stream and ensuring that it is passed upward to the correct application.

- *Session layer*   Deals with the commencement (e.g., login) and completion (e.g., logout) of a session between applications. This layer establishes the type of dialogue (e.g., one way at a time or both ways at the same time). In the case of one way at a time, this layer keeps track of whose turn is next.

■ *Presentation layer*   Specifies the language to be used between applications. This layer includes the encoding to be used for the various pieces of information (e.g., whether certain pieces of information are coded as ASCII or binary).

■ *Application layer*   Provides an interface to the user (or at least, to the user application). For example, file transfer programs and Web browsers belong to the application layer.

Figure 2-3 shows a correlation between the IP protocol suite and the OSI model. We can see that IP itself resides at layer 3. Above IP, we have a choice of protocols: the *Transmission Control Protocol* (TCP) or the *User Datagram Protocol* (UDP).

TCP ensures reliable, error-free delivery of data packets. When multiple packets are being sent, TCP ensures that all of the packets arrive at the destination and that they are all presented in the correct order to the application. UDP is a simpler protocol aimed mainly at one-shot transactions, where a packet is sent and a response is received. UDP does not include sequencing and retransmission functions. Above TCP or UDP, we find a number of different applications and services. The application or service itself determines whether TCP or UDP is used.

Taking the foregoing statement into account, Figure 2-4 shows how an IP packet would appear when being passed from one node in a network to another. The application data itself is passed down to the transport layer, where a TCP or UDP header is applied. The data is then passed to the network layer, where an IP header is applied. Then, the data is passed down to

Figure 2-3
The IP suite and
the OSI stack

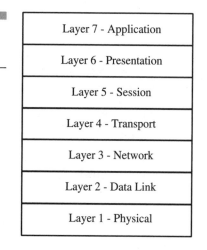

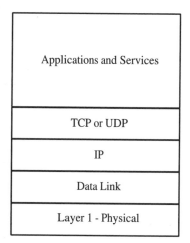

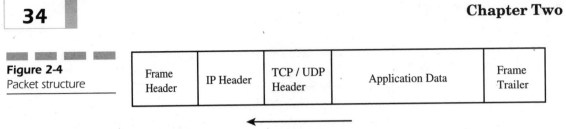

Figure 2-4
Packet structure

the data link layer, where it is packaged into a frame structure for transmission. At the destination, each layer extracts its own header information and passes the remainder to the layer above until the destination application receives only the original data.

Internet Standards and the Standards Process

As we move forward in our discussion, we will make reference to numerous Internet specifications and standards. Before doing so, it is worth discussing Internet standards in general, including naming conventions and the standardization process itself. This information will help put some of the specifications into a better context.

The Internet Society

The Internet Society is a non-profit organization with the overall objective of keeping the Internet alive and growing. Its formal mission statement is "to assure the open development, evolution, and use of the Internet for the benefit of all people throughout the world."

To meet this goal, the Internet Society (among other things) performs the following tasks:

- Supports the development and dissemination of Internet standards
- Supports research and development related to the Internet and internetworking
- Assists developing countries with implementing and evolving Internet infrastructure and use
- Forms a liaison with other organizations, government bodies, and the public for education, collaboration, and coordination that is related to Internet development

A great many of the solutions described in this book have been developed under the auspices of the Internet Society. The Internet Society is the copyright holder for those specifications. Whenever this book makes reference to any specification developed within the Internet Society or any of its subordinate organizations, the following copyright notice is applicable:

Full Copyright Statement
Copyright (c) The Internet Society. All Rights Reserved.
This document and translations of it may be copied and furnished to others, and derivative works that comment on or otherwise explain it or assist in its implementation may be prepared, copied, published and distributed, in whole or in part, without restriction of any kind, provided that the above copyright notice and this paragraph are included on all such copies and derivative works. However, this document itself may not be modified in any way, such as by removing the copyright notice or references to the Internet Society or other Internet organizations, except as needed for the purpose of developing Internet standards in which case the procedures for copyrights defined in the Internet Standards process must be followed, or as required to translate it into languages other than English.

The limited permissions granted above are perpetual and will not be revoked by the Internet Society or its successors or assigns.

This document and the information contained herein is provided on an "AS IS" basis and THE INTERNET SOCIETY AND THE INTERNET ENGINEERING TASK FORCE DISCLAIMS ALL WARRANTIES, EXPRESS OR IMPLIED, INCLUDING BUT NOT LIMITED TO ANY WARRANTY THAT THE USE OF THE INFORMATION HEREIN WILL NOT INFRINGE ANY RIGHTS OR ANY IMPLIED WARRANTIES OF MERCHANTABILITY OR FITNESS FOR A PARTICULAR PURPOSE.

The Internet Architecture Board (IAB)

The IAB is the technical advisory group of the Internet Society. The IAB provides technical guidance to the Internet Society, oversees the Internet standards process, and provides liaisons with other organizations on behalf of the Internet Society in technical matters.

The Internet Engineering Task Force (IETF)

The IETF comprises a huge number of volunteers who cooperate in the development of Internet standards. These individuals come from equipment vendors, network operators, and research institutions and generally have significant technical expertise. They perform the detailed technical work of developing Internet standards.

The work of the IETF is mainly done within working groups that are formed to address specific areas or topics (e.g., security or routing). Most of the work is conducted by individuals in the working groups who share their work with others in the group through the use of mailing lists. The IETF as a whole meets three times a year.

The Internet Engineering Steering Group (IESG)

The IESG is responsible for management of the IETF's activities. In particular, the IESG approves whether a particular specification becomes an official standard or whether the standard can even advance within the standards process.

The Internet Assigned Numbers Authority (IANA)

The IANA is responsible for the administration of unique numbers and parameters that are used in Internet standards. Whenever a new standard proposes to utilize parameters that have specific values and specific meanings associated with those values, those parameters must be registered with the IANA.

The Internet Standards Process

The process for developing an Internet standard is documented in RFC 2026. The following section is meant simply as an overview of what is described in RFC 2026.

An Internet standard begins its life as an Internet Draft, which is just the early version of a specification that can be updated, replaced, or made obsolete by another document at any time. This draft is stored in the IETF's Internet Drafts directory, where it can be viewed by anyone who has Web access. If an Internet Draft is not revised within six months or has not been recommended for publication as a *Request for Comments* (RFC), it is removed from the directory and officially ceases to exist.

Any reference to an Internet Draft must be made with great care and with the emphasis that the draft is a work in progress and that it can change at any time without warning.

Once an Internet Draft is considered sufficiently complete, it can be published as an RFC with an RFC number. This publication does not mean that the draft is a standard, however. In fact, several steps need to be taken along the standards track before the draft becomes a standard. The first step is where the RFC becomes a proposed standard. This standard is a stable, complete, and well-understood specification that has garnered significant interest within the Internet community. No strict requirement exists that the specification should be implemented and demonstrated before becoming a proposed standard, but such implementation and demonstration might be required by the IESG in the case of particularly critical protocols.

The next step is where the RFC becomes a draft standard. To achieve this status, there must have been at least two independently successful implementations of the specification, and interoperability must be demonstrated. If any portion of the specification has not met the requirement for independent and interoperable implementations, then such a portion must be removed from the specification before it can be granted draft-standard status. In this way, there can be a high level of confidence surrounding the details of the specification.

An RFC becomes a standard once the IESG is satisfied that the specification is stable and mature and that it can be readily and successfully implemented on a large scale. This process requires the specification to be mature and requires significant operational experience in using the specification. Once an RFC becomes a standard, it is given a *standard* (STD) number. The RFC retains its original RFC number, however.

Not all RFCs are standards, nor do all RFCs document technical specifications. For example, some document *Best Current Practices* (BCPs), which generally outline processes, policies, or operational considerations that are related to the Internet or to the Internet community itself. Others are known as applicability statements, which describe how a given specification might be utilized to achieve a particular goal or how different specifications can work together. STD 1 lists the various RFCs and indicates whether they are standards, draft standards, proposed standards, applicability statements, etc. The list is updated periodically.

IP

IP itself is specified in RFC 791, with amendments according to RFCs 950, 919, and 920. RFCs 1122 and 1123 give requirements for Internet hosts in their use of IP, and RFC 1812 documents requirements for IP routers (at least, for IP version 4).

IP is a protocol that is used for routing packets through a network, from origin to destination. A packet of data is equipped with an IP header, which contains information about the originator and the destination address. The data packet with the IP header is known as an IP datagram. Routers use the information in the header to route the packet to its destination.

IP itself makes no guarantees that a given packet will be delivered, let alone that multiple packets will be delivered in sequence and without omission. IP is known as a best-effort protocol. In other words, a packet should be delivered under normal circumstances, but a packet could be lost along the way due to transmission errors, congestion in buffers or transmission facilities, link failures, etc.

The IP Header

The IP header appears as shown in Figure 2-5. While all of the fields in the header exist for a reason, a number of the fields are of particular importance. The first of these is the Version. The currently deployed version of IP is 4, and the header shown in Figure 2-5 applies to IP version 4. The next version to be deployed is version 6.

The Identification, Flags, and Fragment Offset fields are used for cases in which a datagram needs to be split into fragments in order to traverse a link, particularly in the case where the size of the datagram is greater than the maximum that a given link can handle. The Identification field is used to identify datagram fragments that belong together. The Flags field indicates whether a datagram can be fragmented or not, and in the case of a fragmented datagram, this field indicates whether a particular fragment is the last fragment. The Fragment Offset is a number describing where the fragment belongs in the overall datagram, to enable the destination to place the different pieces together correctly.

The *Time to Live* (TTL) is an indication of how long a datagram should exist in the network before being discarded. Although this time period was originally meant to represent a number of seconds, it actually represents a number of hops. TTL is set by the originating node and is decremented by one at each subsequent node. If the datagram does not reach its destination by the time that the TTL reaches zero, then the datagram is discarded.

Figure 2-5

IP datagram format

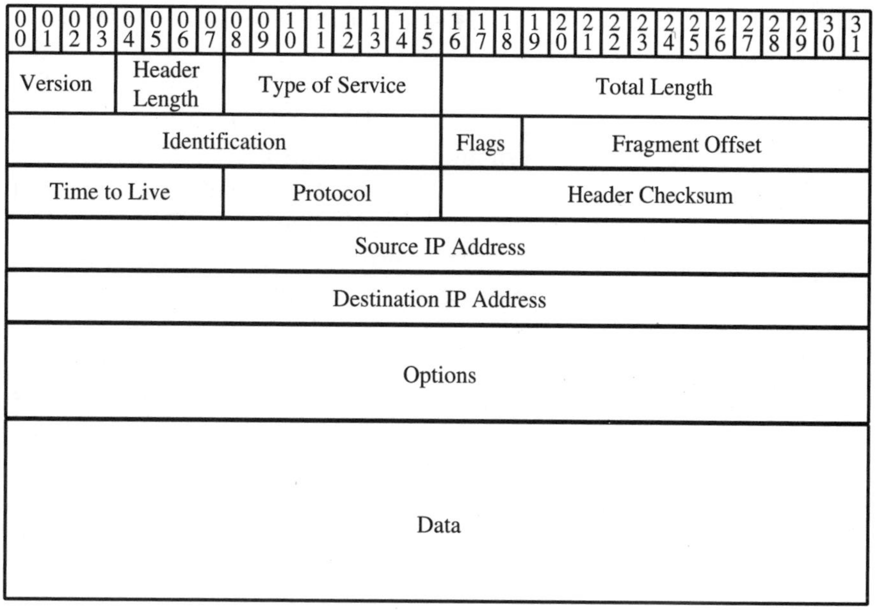

| 0 0 0 0 0 0 0 0 0 0 1 1 1 1 1 1 1 1 1 1 2 2 2 2 2 2 2 2 2 2 3 3 |
| 0 1 2 3 4 5 6 7 8 9 0 1 2 3 4 5 6 7 8 9 0 1 2 3 4 5 6 7 8 9 0 1 |

Version	Header Length	Type of Service	Total Length
Identification		Flags	Fragment Offset
Time to Live	Protocol		Header Checksum
Source IP Address			
Destination IP Address			
Options			
Data			

The Protocol field indicates the higher-layer protocol or application to which this datagram belongs. For example, at the destination node, the value of this field indicates whether this datagram is a TCP or UDP datagram or something else. The protocol value for TCP is 6. The protocol value for UDP is 17.

The source and destination IP addresses in IP version 4 are 32 bits (four octets) long. An IP address is represented in dotted decimal notation as XXX.XXX.XXX.XXX, with the maximum possible value being 255.255.255.255.

IP Routing

The main function of IP is the routing of data from source to destination. This routing is based upon the destination address contained in the IP header and is the principal function of routers.

Routers can contain a range of different interfaces: Ethernet, Frame Relay, ATM, etc. Each interface will provide a connection to another router or perhaps to a LAN hub—or even to a device such as a *Personal Computer* (PC). The function of the router is to determine the best outgoing interface to use for passing on a given IP datagram—i.e., to determine the next hop. The decision is made based on the content of routing tables, which basically match destination IP addresses with interfaces on which the datagrams should be sent.

Of course, routers do not have every possible IP address loaded within them. Instead, a match is made between the first portion of the IP address and a given interface. When we say the first portion of the IP address, how much of the address does that mean? The answer varies. For example, one table entry might specify that any address starting with 182.16.16 should be routed on interface A, whereas another table entry might specify that any address starting with 182.18 should be routed on interface B. In the first situation, there is an IP route mask of 255.255.255.0, while in the second case, there is an IP route mask of 255.255.0.0.

The IP route mask is basically a string of 1s followed by a string of 0s. A logical AND is performed between the mask and the destination address in the packet to determine whether it should be routed over a particular interface. If, after the AND operation, the address of the packet to be routed matches the address in the routing table, then there is a match—and you should use the corresponding interface. For example, if a datagram has a destination address of 182.16.16.7 and a logical AND is performed with mask 255.255.255.0, then the result is 182.16.16.0. This result corresponds to the table entry previously mentioned, and the corresponding route from the router should be used. If two separate table entries begin with the same prefix, then when a destination address is being compared with the table entries, the entry chosen is the one that provides the longest match with the address in question.

Populating Routing Tables   Of course, tables in routers are somewhat more sophisticated than previously suggested, but the basic description still applies and serves the purpose of giving an indication of the large amount of data that might need to be populated in a given router (and, more importantly, kept up to date).

The main issue with routing tables is how to populate them with the correct information in the first place and how to keep the information current in an environment in which new individuals and companies connect to the Internet every day (and where new routers are constantly being added). Another issue arises when several paths exist that could lead to a given destination. Which is the best path to take?

One could certainly determine from an external source what routing information needs to be populated in routing tables and then add the information manually. On the other hand, it would be far better if a given router could somehow automatically keep its own tables up to date with the latest and best routing information. Not only would that take care of the addition of new nodes in the network, but it would also take care of situations in which a link to a router is lost and tables need to be reconfigured to send traffic via another router. Fortunately, such mechanisms exist.

RIP   One such mechanism is the *Routing Information Protocol* (RIP). This protocol is used between routers to pass routing information and to populate routing tables automatically. When a router starts up, it needs to know only about the networks to which it is directly connected. The router proceeds to broadcast what it knows to each of the routers to which it has a direct connection. For example, let's suppose that a new router is directly connected to subnet 182.16.16.0. The router broadcasts this information to its neighboring routers, which now know that they can get to 182.16.16.0 in one hop via the new router. Each of these routers in turn broadcasts to its neighbors that they can get to 182.16.16.0 in one hop, so those neighboring routers now know that they can get to the subnet in two hops, and so on. Strictly speaking, the metric that is provided is not a number of hops; rather, it is a metric that provides an indication of difficulty in getting to the address in question. Thus, a fast link to a destination could be given a metric value of 1, whereas a slow link could be given a higher value metric. In reality, however, most hops are given a value of 1, regardless of the type of hop in question. Once the value 15 has been reached, the destination is considered unreachable (at least, via that path).

Routers send RIP messages to neighbors every 30 seconds. While this time frame might sound pretty quick, it can be slow enough to cause problems. With RIP version 1, this time frame did cause problems because it caused traffic to run in circles at times. Imagine a router (R1) connected to a network (N1) in two ways, as shown in Figure 2-6. The first path uses six hops, and the second path uses just one hop. Routers R2 and R3 are connected to R1, so they know that they can get to N1 in two hops via R1. They broadcast this information to other routers, including R1. Unfortunately, in RIP version 1 messaging, there is not much detail about the routing information being advertised. In particular, R1 does not know that R2 and R3 can get to N1 via R1—which is fine, as long as there is no disruption. What if the direct link from R1 to N1 breaks, however? Assuming that R1 has a packet to send to N1, it sends the packet to R2 because R2 can get to N1 in two hops, and the other alternative is six hops. Of course, R2 sends the packet straight back to R1. Furthermore, R1 might broadcast a RIP message saying that it can now get to N1 in three hops (via R2 or R3). R2 then sends a RIP message saying that it can get to N1 in four hops (via R1), and so on. Meanwhile, R3 is receiving the same RIP messages and is probably sending RIP messages as well, and the whole situation becomes extremely confused.

In order to counteract this problem and other problems, engineers developed RIP version 2 and specified it in RFC 1723. Among other

Figure 2-6
Routing problems
after link failure

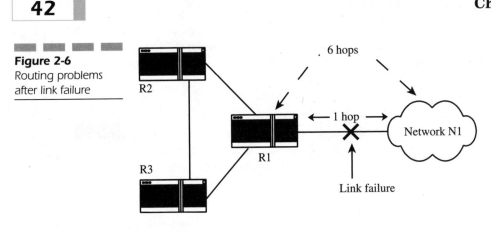

things, version 2 advertises routing information through multicast rather than broadcast, includes next-hop information, and enables the use of a subnet mask—something that was not supported in RIP version 1. Version 2 is not the only choice when it comes to routing protocols, however.

IGRP Another routing protocol is Cisco Systems' *Internet Gateway Routing Protocol* (IGRP), which is more sophisticated than RIP. IGRP has some similarities to RIP in that it involves a router broadcasting its routing table to its neighbors at regular intervals. In addition to hop counts, however, IGRP also provides information related to delays in getting to the address (in the absence of load), the bandwidth available to the address, the current load along the path, the reliability observed along the path, and the *Maximum Transmission Unit* (MTU) that can be carried on the path. These metrics enable routers to make more intelligent decisions about taking one particular route to a destination versus another. Both RIP and IGRP are known as distance-vector protocols.

OSPF Yet another routing protocol is *Open Shortest Path First* (OSPF). OSPF is known as a link-state protocol. With routing protocols such as OSPF, we discuss the *Autonomous System* (AS) concept, where an AS is a group of routers that exchange routing information using one particular protocol, such as OSPF. In OSPF, an AS is divided into areas, where an area is defined as a set of networks and where a network is defined by an IP address and mask. For example, the address 182.16.0.0 with mask 255.255.0.0 could represent a network. A router will belong to an area if it has direct connectivity to one of the networks in the area.

Area 0 is a special area and is known as the backbone area. This area comprises routers that connect the different areas. As shown in Figure 2-7,

■■■ ■■■ ■■■ ■■

Figure 2-7
OSPF areas

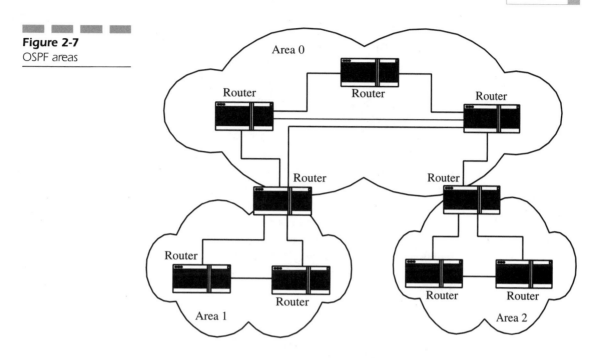

certain border routers will exist in a given area and also in the backbone area to enable routing between different areas.

Within an area, every router will have an identical database describing the topology of the area. The topology describes what is connected to what (i.e., the best way to get from one point to another). Whenever the topology changes, such as an interface (link) failure, this information is transmitted from the router that is attached to the failed interface to all other routers in the same area. Thus, each router within an area will always know the shortest path to a given destination in the area. If the router needs to send packets out of the area, then it only needs to know the best border router to use. The router gathers this information from the border router itself, which passes back information regarding the backbone to the routers in its own area and also passes information about its own area to other routers in the backbone.

BGP   Yet another routing protocol is the *Border Gateway Protocol* (BGP). Recall that an AS is a collection of networks whose routers communicate routing information by using a single protocol, such as OSPF. The obvious question is how routing information is passed from one AS to another so that a router in one AS knows how to get to an IP address that is outside the AS. BGP is designed to take care of this situation. Within an AS, a BGP system has visibility of all of the networks within the AS. The BGP system, which could be a host or a router, is connected to the BGP system of a neighboring AS and tells the neighboring BGP system about the networks that are contained in its AS. That system relays the information to the BGP systems of its neighboring ASs, and so on. The smart part is that each BGP update that is sent from BGP peer to BGP peer for a given network lists all of the ASs that need to be crossed to reach the network in question. Thus, if AS number 6 receives information about the path to network N1, then the information could list AS numbers 5, 4, and 3 as the route that needs to be taken. If AS number 6 is to pass the information to AS number 7, then the route will specify AS numbers 6, 5, 4, and 3. If a given BGP system receives an update from a neighbor and finds that its own AS number is already in the information that it receives, then a loop has occurred—in which case the information is discarded.

The current version of BGP is BGP-4. This version is specified in RFC 1771.

The Transmission Control Protocol (TCP)

TCP resides in the transport layer, above the IP layer. Its primary function is to ensure that all packets are delivered to the destination application in sequence, without omissions and without errors. Recall that IP is a best-effort protocol and that packets might be lost in transit from source to destination. The primary function of TCP is to overcome the lack of reliability that is inherent in IP through end-to-end confirmation that packets have been received. Packets must be retransmitted when necessary. TCP also includes flow control, so that an application at one end does not overwhelm a slower application at the other end. Applications that use the services of TCP include Web browsing, e-mail, and file transfer sessions. TCP is specified in RFC 793.

At a high level, TCP works by breaking up a data stream into chunks called segments. Attached to each segment is a TCP header, which includes a sequence number for the segment in question. The segment and header are then sent down the stack to IP, where an IP header is attached and the datagram is dispatched to the destination IP address. At the destination, the segment is passed from IP to TCP. TCP checks the header for errors, and if all is well, TCP sends back an acknowledgement to the originating TCP. Back at the originating end, TCP knows that the data has reached the far end. On the other hand, if the originating end does not receive an acknowledgement within a given timeout period, the segment is transmitted again.

The TCP Header

The TCP header has the format shown in Figure 2-8. The first fields we notice are the source and destination port numbers.

TCP Port Numbers   Contrary to what the name implies, a port number does not represent any type of physical channel or hardware. Instead, a port number is a means of identifying a specific instance of a given application.

For example, imagine a computer in which three things are happening at the same time: browsing the Web, checking e-mail, and conducting a file transfer. Because each of these functions uses TCP, how does the computer distinguish between data received for one application versus data received for another? That is where port numbers come into play. Whenever a given application wishes to begin a TCP session, a unique port number will be assigned for that particular session. TCP will pass that port number to the far end in any TCP messages relating to that session that are sent.

The near end knows its own port number, but how does it know the port number being used by the far end in advance of any communication with the far end? The near end knows this information in two ways. The easiest is through the use of well-known port numbers. These are port numbers from 0 to 1023 that are reserved by the IANA for certain applications. For example, port number 23 is used for Telnet, and port number 25 is used for the *Simple Mail Transfer Protocol* (SMTP, used for e-mail).

Figure 2-8
TCP header format

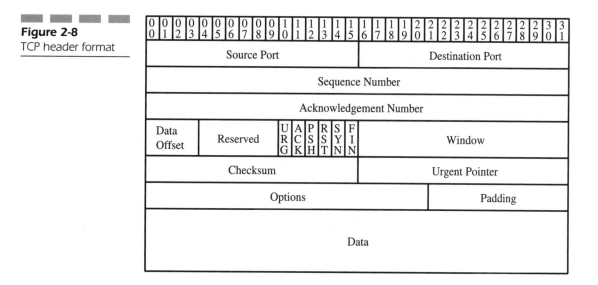

If a client wishes to access a service at a server, and that service is allocated a well-known port number, then that port number will be used as the destination port number. Of course, for a given server, there might be many clients trying to access the same type of service. The server can distinguish between the various clients because TCP and IP together include both the destination address and port number and the source address and port number. The combination of all four will be unique. The combination of an IP address and a port number is called a socket address.

Sequence and Acknowledgement Numbers   The next field we encounter is the sequence number. Although, as the name suggests, a sequence number is used to identify individual segments in a data stream, the sequence number does not actually number the segments. In other words, if a given segment has sequence number 100, then the next sequence will not necessarily have sequence number 101. The reason is because the sequence number is actually used to count data octets. The sequence number is a number that is applicable to the first data octet in the segment. Thus, if a given segment has a sequence number of 100 and contains 150 octets of data, then the next segment will have a sequence number of 250.

The acknowledgement number indicates the next segment number that the receiver expects. If, for example, a segment has been received with a sequence number of 100 and if the segment contains 150 octets of data, then the acknowledgement returned will contain an acknowledgement number of 250—indicating that the next segment expected should have a sequence number of 250.

Other Header Fields   The data offset, also known as the header length, indicates the length of the TCP header in 32-bit words. This data offset also indicates where the actual user data begins; hence the name given to the field.

A number of flags exist as follows:

- *URG*   Set to 1 if urgent data is included, in which case the urgent pointer field is of significance.

- *ACK*   Set to 1 to indicate an acknowledgement. During communication, this flag will be set in all segments that are sent, except for the first segment that is sent.

- *PSH*   Indicates a Push function. This flag will be set in cases where the data should be delivered promptly to the recipient.

- *RST*   Reset; used to indicate an error and to abort a session.

■ *SYN*   Synchronize; used in the initial messages when setting up a data transfer.

■ *FIN*   Finish; used to close a session gracefully.

The next field is the window field, which is used to indicate the amount of buffer space that is available for receiving data. Data that is received is stored in a buffer until the application retrieves the information. When TCP receives data and sends an acknowledgement, it uses the window field to indicate the amount of empty buffer space that is available at that instant. This feature enables the sending side to ensure that it does not send more data in the next segment(s) than the receiving end can handle.

The checksum field contains the 1s complement of the 1s complement sum of all 16-bit words in the TCP header, the data, and a pseudo header, as shown in Figure 2-9. Of course, the pseudo header is not part of the TCP header but actually comprises components of the IP header. The checksum field is used to verify that the header and data have been received without error.

The urgent pointer contains an offset from the current segment number to the first segment after the urgent data and indicates the length of the urgent data. This function is useful to enable critical information to be sent to the user application as soon as possible.

The only real option defined in RFC 793 is the maximum segment size, which indicates the maximum size chunk of data that the receiving system is capable of handling.

TCP Connections

TCP involves the exchange of data streams between a client and a server. This exchange is accomplished through the establishment of a connection. The application at the client's end decides that it needs to exchange some data with an application at a server. The application sends data to TCP,

Figure 2-9
TCP pseudo header

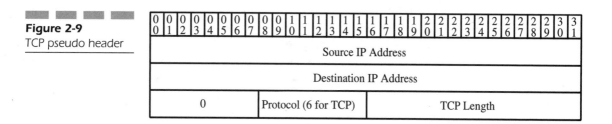

| 0 0 0 0 0 0 0 0 0 0 1 1 1 1 1 1 1 1 1 1 2 2 2 2 2 2 2 2 2 2 3 3 |
0 1 2 3 4 5 6 7 8 9 0 1 2 3 4 5 6 7 8 9 0 1 2 3 4 5 6 7 8 9 0 1
Source IP Address
Destination IP Address
0   Protocol (6 for TCP)   TCP Length

Figure 2-10
Establishing a TCP
connection

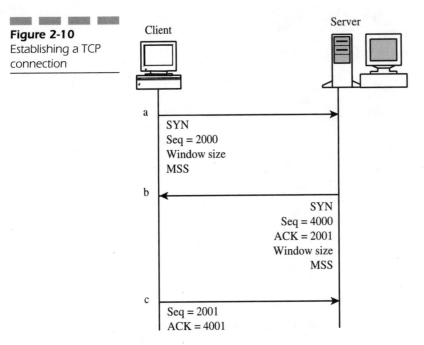

MSS = Maximum Segment Size

where the data is buffered. TCP constructs a TCP segment including source and destination port numbers, an initial sequence number (which is a random number, such as 2,000), the window size, and optionally the maximum segment size. TCP sets the SYN flag to 1 and sends the segment to the far end, as shown in Figure 2-10.

The server receives the segment and responds with a segment of its own. This segment includes the window size, the maximum segment size, a random sequence number of its own (such as 4,000), and an acknowledgement number that is one greater than the received sequence number (2,001 in our example). The segment has the SYN and ACK flags set.

The client issues another segment containing a sequence number that is equal to the original sequence number plus one (i.e., the same as the acknowledgement number received, which would be 2,001 in our case). The acknowledgement number in the segment is one plus the sequence number received from the server (i.e., 4,001). The ACK flag is now set.

At this point, the TCP connection is open, and both ends are ready to exchange media. If the client were to now send a segment, the segment would contain a sequence number of 2,001 and an acknowledgement number of 4,001.

During a connection, a server or client will acknowledge segments that are received. Every received segment does not have to be acknowledged individually, however—nor is it necessary for a sender to wait for a given segment to be acknowledged before sending another segment. Rather, a given segment must be acknowledged before a timeout occurs at the sender's end. For example, let's assume that a client sends three segments with segment numbers 100, 200, and 300. Let's also assume that the third of these segments contains 100 octets of data. The server can send a single acknowledgement for all three, in which case the acknowledgement number would be 400—indicating the next sequence number expected. This acknowledgement also indicates that all three segments have been safely received.

The client will have a timer running for each segment that has not been acknowledged. In our example, let's suppose that the first of the three segments was lost. In that case, the server would *not* return an ACK with an acknowledgement number of 400. In fact, let's assume that the server returns nothing. Upon the expiration of the timer at the client end, the segment will be retransmitted. The server might then indicate that it has received all three segments by sending an ACK, with an acknowledgement number of 400.

Closing a connection is initiated by sending a segment with the FIN flag set. Upon receipt of such a segment at the receiver, an ACK will be returned, followed by a FIN. This second FIN (in the reverse direction to the first) will be acknowledged, and the connection closure is complete.

The User Datagram Protocol (UDP)

UDP performs a simple function. At the sending end, UDP simply passes individual pieces of data from an application to IP to be routed to the far end. At the receiving end, it simply passes incoming pieces of data from IP to the appropriate application.

Unlike TCP, UDP provides no acknowledgement functionality. Therefore, there is no guarantee that anything sent from an application that uses UDP will actually get to the desired destination. UDP is inherently unreliable. Why, then, should there be any interest in using UDP instead of using a more reliable protocol, such as TCP? The answer is that there are many applications that require a quick, one-shot transmission of a piece of data or a simple request/response. An example is the Domain Name System, where a client sends a domain name to a domain name server, which responds with an IP address. A simple transaction requiring the establishment and tear-down of a TCP connection just for the sake of two packets of real information would be wasteful and cumbersome.

What happens if a UDP packet is lost? If the application expects a reply to a particular request and does not receive an answer within a reasonable period of time, then the application should retransmit the request. Of course, this process could then lead to the far end receiving two requests. In such situations, it would be smart for the application itself to include a request identifier within the data to enable duplicates to be recognized.

Figure 2-11 shows the format of the UDP header. As is the case for TCP, the header contains source and destination port numbers. The source port number is optional and should be used in the case where the sending application expects a response. If not, then a value of zero is inserted. You should note that a TCP port number of a given value and a UDP port number of the same value represent different ports. Thus, if two applications are running at the same time on a given host (one using UDP and one using TCP), both applications can use the same port number without the risk of data being sent to the wrong application.

In the UDP header, we also notice a length field that gives the total length of the datagram, including the header and the user data. The checksum gives the 16-bit 1s complement of the 16-bit 1s complement sum of the header, the data, and a pseudo header. The pseudo header is not actually part of the UDP header or data; rather, it includes the IP header fields source IP address, destination IP address, UDP length, and protocol, as shown in Figure 2-12. The checksum is simply a means of checking that the received data is error-free.

Figure 2-11
UDP header format

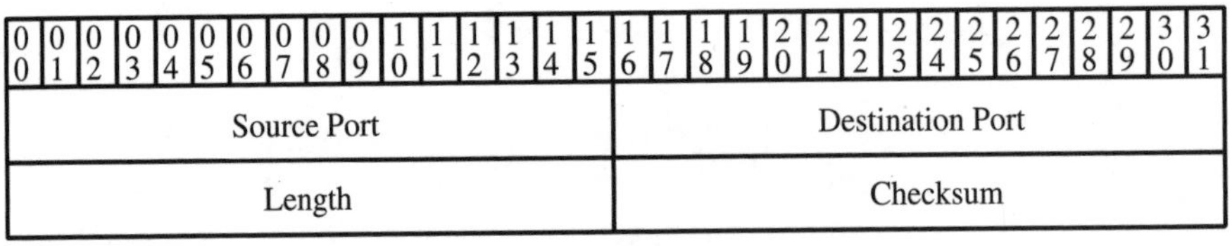

Figure 2-12
UDP pseudo header

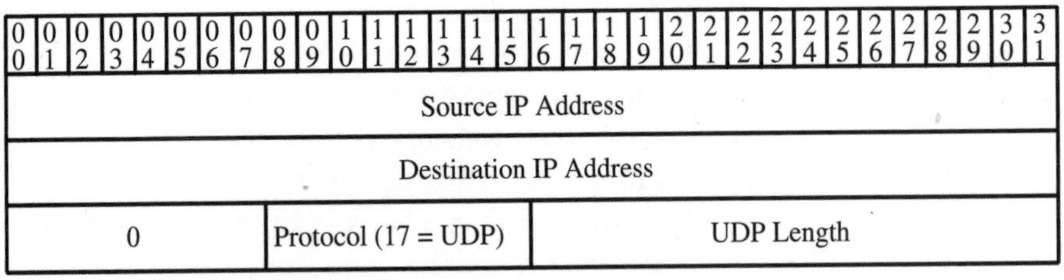

Voice over UDP, not TCP

When voice is to be carried on IP, UDP is used rather than TCP. Why should UDP, an inherently unreliable protocol, be chosen over TCP, a protocol that ensures ordered delivery without the loss of data? The answer is related to the characteristics of speech itself. In a conversation, the occasional loss of one or two packets of voice, while not desirable, is certainly not a catastrophe, since voice traffic will generally use small packets (10ms to 40ms in duration). Obviously, modern voice-coding techniques operate better if there are no lost packets, but the coding and decoding algorithms do have the capability to recover from losses such that the occasional lost packet does not cause major degradation.

On the other hand, voice is extremely delay-sensitive. Anyone who has made a phone call over a satellite connection, where the round-trip time takes hundreds of milliseconds, knows how annoying delay can be. Unfortunately, the connection-setup routine in TCP and TCP's acknowledgement routines introduce delays, which must be avoided. Even worse, in the case of lost packets, TCP will cause retransmission and thereby introduce even more delays. Tolerating some packet loss is far better than introducing delays in voice transmission.

So how much packet loss should you tolerate? Packet loss of about 5 percent is generally acceptable, provided that the lost packets are fairly evenly spaced. A serious problem can arise if groups of successive packets are lost, however. This problem can be alleviated if you use resource management and reservation techniques. These techniques generally involve the allocation of a suitable amount of resources (bandwidth and buffer space) before a call is established. We discuss these issues in some detail in Chapter 8, "Quality of Service." Remember that we focus here on VoIP as an alternative to traditional circuit-switched voice. We are not necessarily addressing Voice over the Internet. While it might be unreasonable to expect controlled access and resource reservation on a grand scale on the Internet, such an expectation is far more reasonable on a managed IP network (where a limited number of providers can manage the traffic).

What if packets arrive at a destination in the wrong order? If you are using UDP, packets will be passed to the application above, regardless of the order of arrival. In fact, UDP has no concept of packet order. Most packets should arrive in sequence, however. The reason is because all packets in a given session usually take the same route from source to destination. Of course, it is possible that different packets will take different routes, but this situation will usually occur only if a link fails if there is congestion or if some topology change occurs on the network. While these situations do happen, the chances of such an event occurring during a particular session are relatively small. Furthermore, QOS management techniques can involve establishing a set path through the network for a given session, further reducing the probability of out-of-sequence arrival of packets. Again, this situation applies to managed networks and not necessarily to the Internet as a whole (at least, not for the time being).

The foregoing arguments are more related to the problems of carrying voice by using TCP than they are to the suitability of UDP for voice. In fact, UDP was not designed with voice traffic in mind, and voice traffic over UDP is not exactly a marriage made in heaven. This match just happens to be better than voice traffic over TCP. Nevertheless, it would be nice if some of the shortcomings of UDP could be overcome without having to resort to TCP. Fortunately, technologies are available to help mitigate some of the inherent weaknesses in UDP when it is used for real-time applications, such as voice or video. We discuss these technologies in the following sections of this chapter.

The Real-Time Transport Protocol (RTP)

RTP is specified in RFC 1889. The official title for this RFC is "RTP: A Transport Protocol for Real-Time Applications." The reason for the somewhat lengthy title is that the RFC actually describes two protocols: the *Real-Time Transport Protocol* (RTP) and the *RTP Control Protocol* (RTCP). Between the two, they provide the network-transport services that are required for the support of real-time applications (such as voice and video).

As previously discussed, UDP does nothing in terms of avoiding packet loss or even ensuring ordered delivery. RTP, which operates on top of UDP, helps address some of these functions. For example, RTP packets include a sequence number, so that the application using RTP can at least detect the occurrence of lost packets and can ensure that received packets are presented to the user in the correct order. RTP packets also include a timestamp that corresponds to the time at which the packet was sampled from its source media stream. The destination application can use this timestamp to ensure synchronized play-out to the destination user and to calculate delay and jitter.

NOTE: *RTP itself does not do anything about these issues; rather, it simply provides additional information to a higher-layer application so that the application can make reasonable decisions about how the packet of data or voice should be best handled.*

As mentioned, RTP has a companion protocol, RTCP. This protocol provides a number of messages that are exchanged between session users and that provide feedback regarding the quality of the session. The type of information includes details such as the numbers of lost RTP packets, delays, and interarrival jitter. While the actual voice packets are carried within RTP packets, RTCP packets are used for the transfer of this quality feedback.

Whenever an RTP session is opened, an RTCP session is also implicitly opened. In other words, when a UDP port number is assigned to an RTP session for the transfer of media packets, a separate port number is assigned for RTCP messages. An RTP port number will always be even, and the corresponding RTCP port number will be the next-highest number (odd). RTP and RTCP can use any UDP port pair between 1025 and 65535; however, the ports of 5004 and 5005 have been allocated as default ports in the case where port numbers are not explicitly allocated.

RTP Payload Formats

RTP carries the actual digitally encoded voice by taking one or more digitally encoded voice samples and attaching an RTP header so that we have RTP packets, comprising an RTP header and a payload of the voice samples. These RTP packets are sent to UDP, where a UDP header is attached. The combination is then sent to IP, where an IP header is attached and the resulting IP datagram is routed to the destination. At the destination, the various headers are used to pass the packet up the stack to the appropriate application.

Given that there are many different voice- and video-coding standards, RTP must include a mechanism for the receiving end to know which coding standard is being used so that that payload data can be correctly interpreted. RTP performs this task by including a payload type identifier in the RTP header. The interpretation of the payload type number is specified in RFC 1890 ("RTP Profile for Audio and Video Conferences with Minimal Control"), which is a companion specification to the RTP specification itself.

This specification allocates a payload type number to various coding schemes and provides high-level descriptions of the coding techniques themselves. This specification also provides references to the formal specifications for the coding schemes.

Since the publication of RFC 1890, a number of new coding schemes have become available. For example, while RFC 1890 describes the GSM full rate coder, it makes no reference to the GSM *Enhanced Full Rate* (EFR) coder. Therefore, recognizing that a number of new coding schemes have become available, a new Internet draft has been prepared that provides an updated RTP profile. This draft is effectively a revised version of RFC 1890 and includes a number of new coding schemes. At the time of this writing, this Internet draft had the same title as RFC 1890 and was planned to supersede RFC 1890 at some point in the future. As with all Internet drafts, it should be read with the understanding that it is a work in progress and is subject to change at any time.

Tables 2-1 and 2-2 list the various RTP audio and video payload types that are respectively defined in the new profile draft. An asterisk indicates payload types that are defined in the new document but are not defined in RFC 1890.

As we see in Tables 2-1 and 2-2, some coding schemes are given a specific payload type number. These schemes are known as static payload types. Other payload types are marked "Dyn," meaning dynamic, and are not allocated any specific payload type number. Several reasons exist, but the main reason is historical. RTP has existed for several years. When it was originally developed, separate signaling standards for real-time applications did not exist. Therefore, it was not possible for a sending application to enable a receiving application to know in advance (and out-of-band) what type of coding was to be used in a given session. Therefore, specific payload type numbers were allocated to a number of popular coding schemes. When an RTP packet arrived at a receiver, the payload type number would unambiguously indicate the coding scheme that the sender chose so that the receiver would know which actions to take in order to decode the data. That was fine, provided that the chosen coding scheme was supported at the receiving end. Unfortunately, there was no way for a sender to find out in advance what coding schemes a particular receiver could handle.

In more recent times, separate signaling systems have been developed to enable the negotiation of media streams as part of the call setup process and in advance of actually sending any voice or video data. For example, the *Session Initiation Protocol* (SIP) and the *Session Description Protocol* (SDP), both described in Chapter 5 ("The Session Initiation Protocol"), support such capabilities. Therefore, the two ends of a conversation can now

Table 2-1

Payload types for
Audio Encodings

Payload Type	Encoding Name	Media Type	Clock Rate (Hz)	Channels
0	PCMU	Audio	8,000	1
1	1016	Audio	8,000	1
2	G726-32	Audio	8,000	1
3	GSM	Audio	8,000	1
*4	G723	Audio	8,000	1
5	DVI4	Audio	8,000	1
6	DVI4	Audio	16,000	1
7	LPC	Audio	8,000	1
8	PCMA	Audio	8,000	1
9	G722	Audio	8,000	1
10	L16	Audio	44,100	2
11	L16	Audio	44,100	1
*12	QCELP	Audio	8,000	1
13	Reserved			1
14	MPA	Audio	90,000	1
15	G728	Audio	8,000	1
*16	DVI4	Audio	11,025	1
*17	DVI4	Audio	22,050	1
*18	G729	Audio	8,000	1
19	Reserved			1
20	Unassigned			1
21	Unassigned			1
22	Unassigned			1
23	Unassigned			1
*dyn	GSM-HR	Audio	8,000	1
*dyn	GSM-EFR	Audio	8,000	1
*dyn	L8	Audio	Variable	1
*dyn	RED	Audio	Conditional	1
*dyn	VDVI	Audio	Variable	1

Table 2-2

Payload types for
Video and
Combined
Encodings

Payload Type	Encoding Name	Media Type	Clock Rate (Hz)
24	Unassigned	Video	
25	CelB	Video	90,000
26	JPEG	Video	90,000
27	Unassigned	Video	
28	nv	Video	90,000
29	Unassigned	Video	
30	Unassigned	Video	
31	H261	Video	90,000
32	MPV	Video	90,000
33	MP2T	Audio/Video	90,000
*34	H263	Video	90,000
35-71	Unassigned	?	
72-76	Reserved	N/A	N/A
77-95	Unassigned	?	
96-127	Dynamic	?	90,000
*Dyn	BT656	Video	90,000
*Dyn	H263-1998	Video	90,000
*Dyn	MP1S	Video	90,000
*Dyn	MP2P	Video	90,000
*Dyn	BMPEG	Video	90,000

agree in advance on the type of coding scheme to be used. As part of that agreement, they can also decide on the payload type number to be applied to the coding scheme for that call. For example, let's assume that party A wishes to send voice data that is coded according to some coding scheme XXX, and party B is able to handle that coding. They can agree at the start of the call that RTP packets carrying voice that is coded according to XXX will have a payload type of 120. When packets arrive with payload type 120, party B knows exactly how to interpret the data. After the call is over, payload type 120 could be used to indicate some other coding scheme in a subsequent call. This example shows the usage of a dynamic payload type.

The RTP audio/video profile allocates the payload type numbers 96 to 127 to be used for dynamic payload types. Furthermore, some of the newer coding schemes referenced by the profile specification are not given a fixed (static) payload type; rather, they are considered dynamic. Dynamic payload types are useful. They enable many new coding schemes to be developed in the future without running the risk of exhausting the available payload type number space (maximum of 127).

NOTE: *The encoding name is also significant. Whenever a new coding scheme is to be used with RTP, particularly if it is to be a dynamic payload type, then the encoding name should be specified in a separate payload specification document, and the encoding name should be registered with the IANA. The reason is because external signaling systems, such as SDP, make use of the encoding name in order to unambiguously refer to a particular payload specification.*

Of the dynamic payload types, one is a little different from the rest and deserves special mention. This type is the Redundant payload type (RED) and is specified in RFC 2198. Normally, a single RTP packet contains one or more samples of coded voice or video, with each sample in the packet coded in the same way. If the packet is lost, the receiving application must cope with the loss of data as best it can, such as by replaying the content of a previous packet to fill the void in time. The idea with the redundant payload type is for a single packet to contain one or more voice samples as usual, plus one or more copies of previous samples that were already sent to the receiver. In other words, each packet contains the current sampled speech plus a backup copy of the previously sampled speech. If a given packet is lost, then a copy of the lost speech sample arrives in the next packet. Note that the backup copy might use a different (e.g., more bandwidth-efficient) encoding scheme than is used for the primary copy. If the backup copy uses a lower-bandwidth coding scheme, then it provides for redundancy to cope with packet loss without the need for excessive bandwidth.

The RTP Header

As mentioned, RTP carries the actual encoded voice. In other words, one or more digitally coded samples make up the RTP payload. An RTP header is attached to this payload, and the packet is sent to UDP.

The RTP header includes information that is necessary for the destination application to reconstruct the original voice sample (at least to as close a match as is enabled by the coding scheme being used).

The RTP header is shown in Figure 2-13. The meanings of the various header fields are as follows:

- Version (*V*)   This header should be set to the value 2, which is the current version of RTP.

- Padding (*P*)   This header is a one-bit field indicating whether the packet contains one or more padding octets at the end of the payload. Because the payload needs to align with a 32-bit boundary, it is possible that the end of the payload contains padding to align with the boundary. If such padding exists, then the P bit is set, and the last octet of the payload contains a count of how many padding octets are included. Of course, such padding octets should be ignored.

- Extension (*X*)   This header is a one-bit field indicating whether the fixed header, as shown in Figure 2-13, contains a header extension. If this bit is set, then the header is followed by exactly one header extension (described later) and with the format shown in Figure 2-14.

Figure 2-13
RTP header format

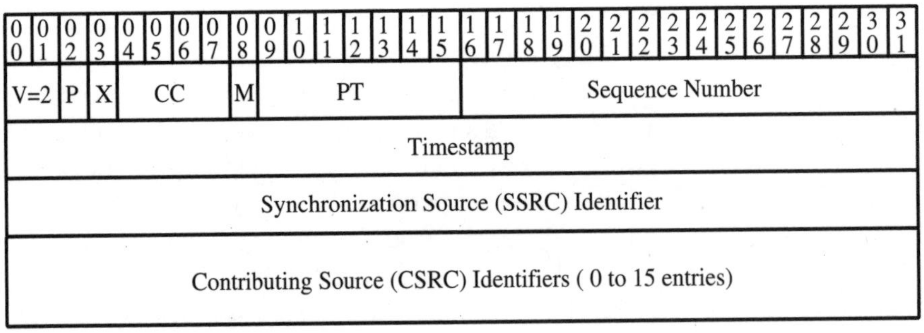

Figure 2-14
RTP header extension

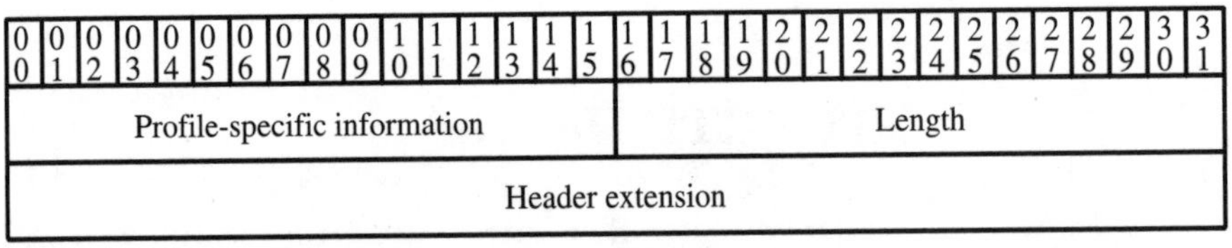

- CSRC Count (*CC*)   This header is a four-bit field indicating the number of contributing source identifiers included in the header and takes a value of 0 to 15. Refer to Contributing Source Identifiers, described later.

- Marker (*M*)   This header is a one-bit field, the interpretation of which is dependent upon the payload being carried. RFC 1889 does not mandate any specific use of this bit, but RFC 1890 (audio/video profile) does. RFC 1890 states that for an application that does not send packets during periods of silence, this bit should be set in the first packet after a period of silence (i.e., the first packet of a talkspurt). Applications that do not support silence suppression should not set this bit.

- Payload Type (*PT*)   This header is a seven-bit field that indicates the format of the RTP payload. In general, a single RTP packet will contain media that is coded according to only one payload format. The exception to this rule is the RED profile, where several payload formats might be included. In the case of RED, however, the different payload segments are managed by the addition of headers within the payload and not by the PT field. In the case of a RED payload, the PT value would simply indicate that it is a RED payload. Refer to RFC 2198.

- *Sequence Number*   This header is a 16-bit field and is set to a random number by the sender at the beginning of a session, then incremented by one for each successive RTP packet that is sent. This header enables the receiver to detect packet loss and/or packets that are arriving in the wrong order.

- *Timestamp*   This header is a 32-bit field that indicates the instant at which the first sample in payload was generated. The sampling instant must be derived from a clock that increases monotonically and linearly in time, so that far-end applications can play out the packets in a synchronized manner and so that jitter calculations can be performed. The resolution of the clock needs to be adequate in order to support synchronized play-out. The clock frequency is dependent on the format of the payload data. For static payload formats, the applicable clock frequency is defined in the RTP profile. For example, the frequency for typical voice-coding schemes is 8,000Hz. The increase in the timestamp value from packet to packet will depend on the number of samples contained in a packet. The timestamp is tied to the number of sampling instants that occurred from one packet to the next. For example, if one packet contains 10 voice samples and a timestamp value of 1, then the next packet should have a timestamp value of 11. Given that the sampling occurs at a rate of 8,000Hz (every 0.125ms), then the difference of 10 in the timestamp indicates a difference in time of 1.25ms. If no packets are sent during periods of silence, then the next RTP packet might have a timestamp that is significantly greater than the previous RTP packet. The difference will reflect the number of samples contained in the first packet and the duration of silence. The initial value of the timestamp is a random number chosen by the sending application.

■ Synchronization Source (*SSRC*) This header is a 32-bit field indicating the synchronization source, which is the entity that is responsible for setting the sequence number and timestamp values and is normally the sender of the RTP packet. The identifier is chosen randomly by the sender so as not to be dependent upon the network address. This header field is meant to be globally unique within an RTP session. In the case where the RTP stream is coming from a mixer, then the SSRC identifies the mixer, not the original source of media. Refer to the discussion of mixers and translators later in this chapter.

■ Contributing Source (*CSRC*) This header is a 32-bit field containing an SSRC value for a contributor to a session. The field is used when the RTP stream comes from a mixer and is used to identify the original sources of media behind the mixer. Zero to 15 CSRC entries might exist in a single RTP header. Refer to the discussion of mixers and translators later in this chapter.

RTP Header Extensions The RTP header is designed to accommodate the common requirements of most, if not all, media streams. Certain payload formats might require additional information, however. This information can be contained within the payload itself; e.g., by specifying in the payload profile that the first *n* octets of the payload have some specific meaning. Alternatively, you can add an extension to the RTP header.

The existence of a header extension is indicated by setting the X bit in the RTP header to 1. If the X bit is set to 1, then a header extension of the form shown in Figure 2-14 will exist between the CSRC fields and the actual data payload. Note that RTP only mandates that the header extension should contain a length indicator in a specific location. Beyond that requirement, RTP does not specify what type of data is to be included in the header or even how long the extension can be. Provided that a length indication exists, it is possible for the boundary between the header extension and the payload to be clearly identified.

Mixers and Translators

Mixers are applications that enable multiple media streams from different sources to be combined into one overall RTP stream. A good example is an audio conference, as depicted in Figure 2-15. In the example, there are four participants, and each is sending and receiving audio at 64kbps. If the bandwidth that is available to a given participant is limited to 64kbps in each direction, then that participant does not have sufficient bandwidth to send/receive individual RTP streams to/from each of the other participants. In the example, the mixer takes care of this problem by combining the individual streams into a single stream that runs at 64kbps. A given participant does not hear the other three participants individually but hears the combined audio of everyone. Another example would be where one of the participants is connected to the conference via a slow-speed connection, while the other participants have a high-speed connection. In such a case, the mixer could also change the data format on the slower link by perhaps applying a lower-bandwidth coding scheme. In terms of RTP header values, the RTP packets received by a given participant would contain an SSRC value specific to the mixer and three CSRC values corresponding to the three other participants. The mixer sets the timestamp value.

A translator is used to manage communications between entities that do not support the same media formats or bit rates. An example is shown in Figure 2-16, where participant A can support only 64kbps G.711 mu-law encoded voice (payload type 0), while participant B can support only 32kbps G.726 ADPCM (payload type 2). In terms of RTP headers, the RTP packets received by participant A would have an SSRC value that identified

Figure 2-15
An RTP mixer

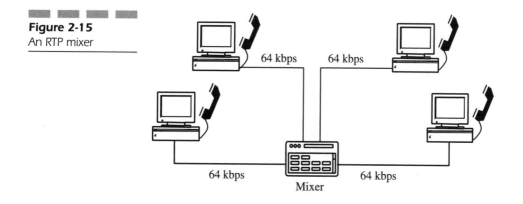

64 kbps   64 kbps

64 kbps   64 kbps

Mixer

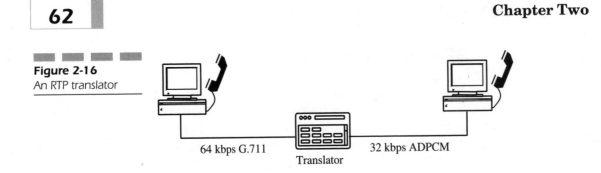

Figure 2-16
An RTP translator

64 kbps G.711   Translator   32 kbps ADPCM

participant B, not a value identifying the translator. The difference between a mixer and a translator is that a mixer combines several streams into a single stream, perhaps with some translation of the payload format, whereas a translator performs some translation of the media format but does not combine media streams.

The advantage of a mixer over a translator is that in the case of several streams, extra bandwidth does not need to be allocated for each stream. The disadvantage is that the receiver of the RTP stream cannot mute the stream from one source while continuing to listen to the stream from another source.

The RTP Control Protocol (RTCP)

As mentioned previously, RTP comes with a companion control protocol, RTCP. This protocol enables periodic exchange of control information between session participants, with the main goal of providing quality-related feedback. This feedback can be used to detect and potentially correct distribution problems. By using RTCP and IP multicast, a third party (such as a network operator who is not a session participant per-se) can also monitor session quality and detect network problems.

RTCP defines five different types of RTCP packets:

- Sender Report (*SR*) Used by active session participants to relay transmission and reception statistics
- Receiver Report (*RR*) Used to send reception statistics from those participants who receive media but do not send media
- Source Description (*SDES*) Contains one or more descriptions related to a particular session participant. In particular, the SDES must contain a *canonical name* (CNAME), which is used to identify session

participants. This name is separate from the SSRC, because the SSRC might change in the case of a host reset. Furthermore, within a given session, a given source can generate multiple RTP streams, as would be the case if both audio and video were being transmitted. In such a case, the two streams would have different SSRC values but would have the same CNAME value. At a receiver, the CNAME would be used to link the two streams in order to provide a synchronized play-out. Within a given session, the CNAME must be unique. Other information might also exist within the SDES packet, such as a regular name, e-mail address, or phone number for the participant in question.

- *BYE* Indicates the end of participation in a session
- *APP* Application-specific functions. APP enables RTCP to send packets that convey information that is specific to a particular media type or application. RTCP does not specify the detailed contents of an APP packet.

Although these RTCP packet types are individually defined, they are never sent individually from one session participant to another. In reality, two or more RTCP packets will be combined into a compound packet. Reasons for this action include the fact that SRs and/or RRs should be sent as often as possible to enable better statistical resolution and that new receivers in a session must receive the identity (CNAME) of various media sources quickly in order to enable a correlation between media sources and the received media. In fact, the specification states that all RTCP packets will be sent as part of a compound packet and that every packet must begin with a report packet (SR/RR) and must contain an SDES packet. When a SDES packet needs to be sent, a report packet (SR/RR) is sent even if there is no data to report. In such a case, an empty RR packet is sent as the first packet in the compound RTCP packet.

An RTCP compound packet would appear as shown in Figure 2-17. In this particular example, the sender of the compound packet happens to be leaving the session, as indicated by the inclusion of the BYE packet.

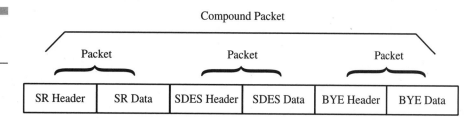

Figure 2-17
Example of an RTCP compound packet

RTCP Sender Report (SR)

Session participants who actively send and receive RTP packets use the SR. The format of the sender report is as shown in Figure 2-18. The SR has three distinct sections: header information, sender information, and a number of receiver report blocks. Optionally, the SR can also have a profile-specific extension.

We can see that the RTCP packet format bears quite a resemblance to the format of an RTP packet. As, is the case for an RTP packet, we first encounter the version field (V), which has the value 2 for the current version of RTP and RTCP. We also encounter a padding bit (P), which is used to indicate whether there are padding octets at the end of the packet. If the P bit is set, then the final octet of the packet will contain a count of the number of padding octets that exist and the number that should be ignored.

Figure 2-18

RTCP sender report

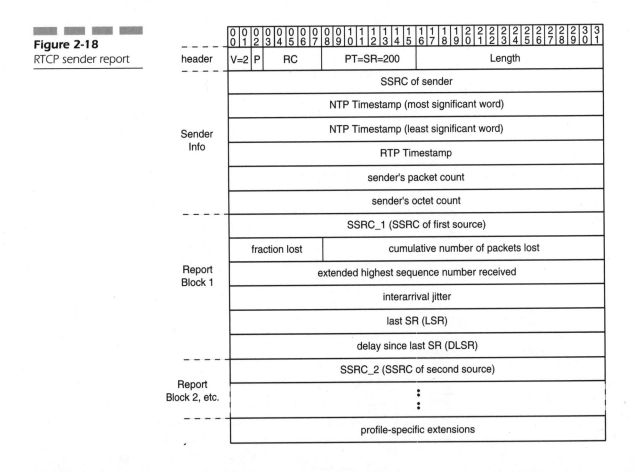

The first difference between an RTP packet and an RTCP packet is noticeable in the next field (the RC field), which indicates how many reception report blocks are contained in this report. This field is a 5-bit field, which means that a maximum of 31 RR blocks can be included in a single RTCP SR packet. What happens in a large conference where a given participant needs to provide RR data for more than 31 other session participants? This situation is one in which the compounding of RTCP packets is useful, because it is perfectly valid to have a SR followed by a RR in the same RTCP compound packet.

The next field we encounter is the payload type field, which has the value 200 for a sender report. A length indicator follows this field.

The sender information contains the SSRC of the sender, plus some timing information and statistics regarding the number of packets and octets sent by the receiver. Specifically, the sender information includes the following fields:

- SSRC of sender

- *NTP Timestamp*   Network Time Protocol Timestamp. This field is a 64-bit field that indicates the time elapsed in seconds since 00:00 on January 1, 1900 (GMT). The most-significant 32 bits represent the number of seconds, while the 32 least-significant bits represent a fraction of a second, enabling a precision of about 200 picoseconds. Using 32 bits to represent the number of seconds since 1900 means that we will run into difficulty in the year 2036, when the 32-bit space will no longer be adequate. Before that time, we will need to create some other version of NTP.

  The timestamp information itself is obtained in the network from a primary time server, which propagates timing information by using NTP. The time servers in turn receive clocking information from one or more master sources. In the United States, these sources include Station WWV, operated in Fort Collins, Colorado by the *National Institute of Standards and Technology* (NIST). Another such source is Station WWVB in Boulder, Colorado, also operated by NIST. The NTP protocol version 3 is specified in RFC 1305.

- *RTP Timestamp*   A timestamp corresponding to the NTP timestamp but that uses the same units and has the same offset as used for RTP timestamps in RTP packets. Recall that RTP timestamps in RTP packets begin at a random value and increase with each packet sent in units that are dependent on the payload format. Including both the RTP timestamp and the NTP timestamp in the same report enables receivers of the report to better synchronize with the sender of the report, particularly if they share the same timing source.

- *Sender's packet count*   The total number of RTP packets transmitted by the sender from the start of the session to the time this report was issued. Thus, the packet count is cumulative and is only reset within a given session if the sender's SSRC value is being changed.

- *Sender's octet count*   The total number of RTP payload octets transmitted by the sender since the beginning of the session. As for the packet count, it is only reset within a given session if the sender's SSRC value changes.

One or more RR blocks follow the sender information. These provide other session participants with information as to how successfully the RTP packets that they have previously sent have been received. For each RR block, the following fields are included:

- *SSRC_n*   The source identifier of the session participant to which the data in this particular receiver report block pertains

- *Fraction lost*   An eight-bit field indicating what fraction of packets have been lost since the last report issued by this participant. The fraction lost is the number of packets lost divided by the number of packets expected. The number of lost packets can be determined by examination of the sequence numbers in RTP packet headers.

- *Cumulative number of packets lost*   The total number of packets from the source in question that have been lost since the beginning of the RTP session

- *Extended highest sequence number received*   The sequence number of the last RTP packet received from the source. The low 16 bits contain the last sequence number received. The high 16 bits indicate the number of sequence number cycles, used in the case where the sequence numbers from a particular source cycle through 0 one or more times.

- *Interarrival jitter*   An estimate of the variance in RTP packet arrival. Refer to the section on calculating jitter later in this chapter.

- Last SR Timestamp (*LSR*)   The middle 32 bits of the 64-bit NTP timestamp used in the last SR received from the source in question. This timestamp is a means of indicating to the source whether sender reports issued by that source are being received.

- Delay Since Last SR (*DLSR*)   The duration, expressed in units of 1/65,536 seconds, between the reception of the last sender report from the source and issuance of this receiver report block.

RTCP does not define the profile-specific extensions. If they are to be used with a given type of media (i.e., RTP payload type), then the profile definition of the payload in question should include a specification of the format and usage of any extensions.

RTCP Receiver Report (RR)

The RR is issued by a session participant who receives RTP packets but does not send, or has not yet sent, RTP packets of its own to other session participants. The RR has the format shown in Figure 2-19. The report packet is almost identical to the sender report, with the exception that the payload type has the value 201 and the sender-specific information is not included.

RTCP Source Description Packet (SDES)

The SDES packet provides identification and other information regarding session participants. The SDES packet must exist in every RTCP compound packet. The general format of the SDES packet is shown in Figure 2-20.

Figure 2-19
RTCP receiver report

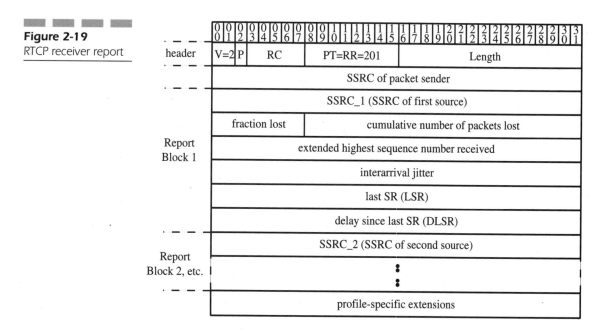

Figure 2-20
SDES packet

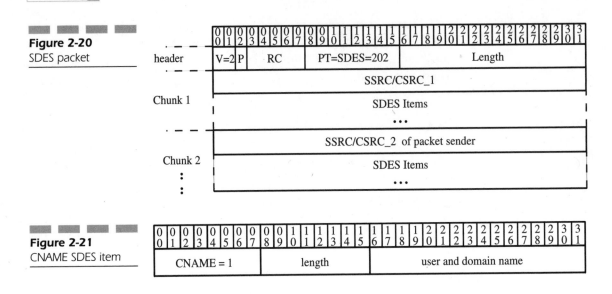

Figure 2-21
CNAME SDES item

The SDES packet comprises two main parts: a header and zero or more chunks of information. The header contains a length field, a payload type value field with value 202, and a *source-count* (SC) field. The SC field occupies five bits and indicates the number of chunks in the packet.

Each chunk contains an SSRC or CSRC value, followed by one or more identifiers and pieces of information that are applicable to that SSRC or CSRC. These values are known as SDES items and include such items as e-mail address, phone number, and name (e.g., Jane Doe), among others. All of these SDES items and their structures are defined in RFC 1889. Of the possible SDES items, only one is mandatory: the canonical name (CNAME).

The CNAME is meant to be a unique identifier for a session participant and does not change within a given session. Normally, the SSRC value will not change either. The value of the SSRC might change during a session, however, if it found that two participants have chosen the same value of SSRC or if a host resets during a session. In order to ensure that the CNAME is unique, it should have the form `user@host`, where "host" is a fully qualified domain name. The format of the CNAME parameter is shown in Figure 2-21.

RTCP BYE Packet

The BYE packet is used to indicate that one or more media sources are no longer active. This packet can also include a text string that indicates a reason for leaving the session.

The BYE packet has the format shown in Figure 2-22. The SC field indicates the number of source identifiers included in the packet; i.e., the number of SSRC or CRSC entries. After the list of SSRC/CSRC values, an optional text string might be provided, indicating the reason for leaving. In such a case, the text string is preceded by a length field indicating the number of octets in the text string.

Application-Defined RTCP Packet

Analysis of the various RTP and RTCP packet types reveals that RTP and RTCP are written so that packets can be extended with application-specific data. In keeping with this flexible approach, RTCP includes an application-defined packet with the format shown in Figure 2-23.

Obviously, this type of packet is used for some non-standardized application; hence, the format of the application data is not specified. Some identifier is necessary, however, to enable a packet recipient to understand the data; the payload type value of 204 is not sufficient. If, for example, an experimental application were to require two new types of RTCP packet,

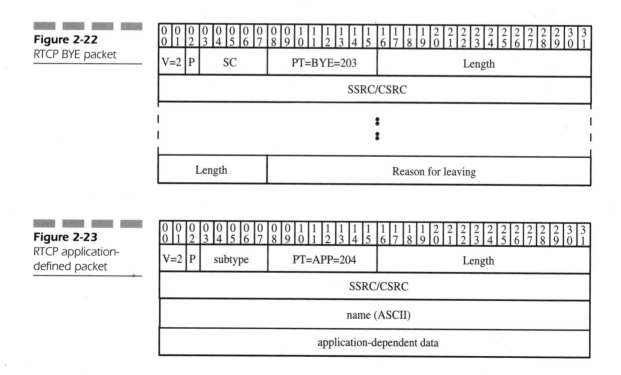

Figure 2-22
RTCP BYE packet

Figure 2-23
RTCP application-defined packet

each carrying different information, then there must be a means to differentiate between the two types. At this point, the Name field comes into play. This field contains an ASCII string specified by the creator (i.e., designer) of the new application and is used to imply a particular coding and significance for the application data. The subtype field is also available for the application creator to indicate the use of any options within the particular application.

Calculating Round-Trip Time

One of the functions enabled by the SRs and RRs of RTCP is the calculation of round-trip time. This time refers to the time that would be necessary for a packet to be sent from a particular source to a particular destination and back again. This metric is useful when measuring voice quality and is calculated from the data in RTCP reports as follows.

Let's assume that participant A issues report A at time T1. Participant B receives this report at time T2. Subsequently, participant B issues report B at time T3, and participant A receives this report at time T4. Participant A can calculate the round-trip time as simply

$$T4 - T3 + T2 - T1$$

or

$$T4 - (T3 - T2) - T1$$

The data within the RTCP reports either explicitly provides the values for T1, T2, T3, and T4 or provides participant A with sufficient information to determine the values. For example, report B in our example contains an NTP timestamp for when the report is issued (T3). Report B also contains the LSR timestamp, which points participant A to the timestamp included in report A and hence to the value T1. The DLSR timestamp in report B contains the delay between reception at B of report A and the sending from B of report B (T3 − T2). Finally, participant A takes note of the instant that report B is received (T4).

Calculating Jitter

As mentioned previously, delay is significant to the perceived voice quality and should be minimized. Some delay will always exist, however. Given that some delay will always exist, the delay should be constant or close to constant. In other words, if a given packet takes 50ms to get from its source to its destination, then the next packet should take about the same time. Unfortunately, this delay is not always constant, and the variation in delay is known as jitter.

Jitter is defined as the mean deviation of the difference in packet spacing at the receiver, compared to the packet spacing at the sender, for a pair of packets. This value is equivalent to the deviation in transit time for a pair of packets.

If Si is the RTP timestamp from packet i, and Ri is the time of arrival in RTP timestamp units for packet i, then for the two packets i and j, the deviation in transit time, D, is given by

$$D(i,j) = (Rj-Ri) - (Sj-Si) = (Rj-Sj) - (Ri-Si)$$

The interarrival jitter is calculated continuously as each data packet, i, is received from a given source SSRC_n, using the difference D for that packet and the previous packet, $i - 1$, in order of arrival according to the formula.

$$J(i) = J(i-1) + (|D(i-1, i)| - J(i-1))/16$$

The current value of J is included in sender and receiver reports.

Timing of RTCP Packets

RTCP provides useful feedback regarding the quality of an RTP session. RTCP enables participants and/or network operators to obtain information about delay, jitter, and packet loss and to take corrective action where possible to improve quality. Of course, the better the information, the better the decisions that can be made—and, of course, the quality of the information is linked to how often RTCP reports are issued. This statement suggests that RTCP reports should be sent as often as possible. RTCP packets consume bandwidth, however, and if a large number of RTCP reports are being sent, then these reports themselves contribute to delay, jitter, and RTP packet loss. In other words, too many RTCP packets can make the quality worse. Therefore, you need to strike a balance between the need to provide sufficient feedback and the need to minimize the bandwidth consumed by RTCP packets.

Determining how often to send RTCP packets involves keeping the following considerations in mind, as specified in RFC 1889. The control traffic should be limited to a small and known fraction of the session bandwidth—

small so that the primary function of the transport protocol to carry data is not impaired, and known so that the control traffic can be included in the bandwidth specification that is given to a resource reservation protocol. Also, each participant should be able to independently calculate its share. The fraction of the session bandwidth allocated to RTCP should be fixed at 5 percent.

RFC 1889 provides an algorithm for calculating the interval between RTCP packets. The algorithm is designed to meet the foregoing criteria, and it is also designed to ensure that RTCP can scale from small to large sessions. The algorithm includes the following main characteristics:

- Senders are collectively allowed at least 25 percent of the control-traffic bandwidth so that in sessions with a small number of participants, newly arriving participants can quickly receive the CNAME for sending participants.

- The calculated interval between RTCP packets is to be greater than a minimum of five seconds, which is the mean interval.

- The interval between RTCP packets varies between 0.5 and 1.5 times the calculated interval. This interval helps avoid unintended synchronization where all participants send RTCP packets at the same instant (hence, clogging the network).

- A dynamic estimate of the average compound RTCP packet size is calculated, including all those packets received and sent, to automatically adapt to changes in the amount of control information that is carried. In other words, if RTCP compound packets start becoming bigger, perhaps because of more participants, then the rate at which the RTCP packets are sent needs to be adapted in order to ensure that excessive bandwidth is not consumed.

IP Multicast

Many situations exist where a given IP datagram needs to be sent to multiple hosts. Perhaps the example that is most applicable to VoIP is a conference, where voice streams from one or more participants need to be distributed to many listeners. Sending packets to each destination individually is possible, but a far better solution is to send the same packet to a single address that is associated with all listeners. This type of solution is known as multicasting. The following section provides a brief overview of the concept of multicasting.

Multicasting involves the definition of multicast groups, and a given group is assigned a multicast address. All members of the group retain their own IP addresses but also receive datagrams that are sent to the multicast address.

In order to support multicasting, a number of requirements must be met. There must be an address space allocated for multicast addresses so that multicast addresses are not treated as ordinary addresses. Hosts that wish to join a particular group must be able to do so and must inform local routers of their membership in a group. Routers must have the capacity to recognize when one or more attached hosts belong to a particular group, as identified by a particular multicast address. Routers must support the capability to forward multicast datagrams correctly. Finally, routing protocols such as OSPF must support propagation of routing information for multicast addresses. Furthermore, routing tables should be set up so that the minimum number of datagrams is sent.

For example, let's assume that a particular host joins a group and that the host is connected to router A. Let's also assume that other members of the group are connected to router B. Finally, let's assume that some host attached to router C wishes to send a datagram to the group. In that case, we need a means for router C to be aware that datagrams for the group's multicast address must be sent to both router A and router B.

As a second example, let's assume that multicast router D exists in the path between router C and routers A and B. In that case, router C should not send two datagrams to routers A and B via router D. Instead, the router should forward the datagram to router D, and router D should branch out the datagram to routers A and B. In other words, we want the number of copies of a particular multicast datagram to be kept to a minimum.

The IP version 4 address space 224.0.0.0 to 239.255.255.255 is allocated for multicast addresses. Therefore, if a router receives a datagram for such an address, it must have the capability to route it appropriately. Of these addresses, a number of addresses are allocated for special purposes. Examples include the following:

224.0.0.1 All hosts on a local subnet

224.0.0.2 All routers on a local subnet

224.0.0.5 All routers supporting OSPF

224.0.0.9 All routers supporting RIP version 2

The *Internet Group Message Protocol* (IGMP) is used by hosts in a particular group to advertise their membership in a group to routers, so that multicast datagrams can be forwarded to the appropriate hosts. IGMP version 1 is specified in RFC 1112. IGMP version 2 is specified in RFC 2236.

Support for multicasting within routing protocols has also been developed. For example, RFC 1584 specifies extensions to OSPF in order to accommodate the propagation of multicast routing information.

IP Version 6

The foregoing discussion regarding IP and higher-layer protocols has been based upon IP version 4. This version of the protocol is currently deployed throughout the world. The next generation of IP, however, IP version 6, is starting to be deployed. Accordingly, a brief analysis of version 6 is appropriate here.

Among the biggest drivers for the creation of IP version 6 has been the explosive growth of the Internet and IP-based networks in general. Clearly, such growth cannot be sustained indefinitely with an address space of only 32 bits, particularly because the allocation of addresses has not been hierarchical. Furthermore, the IP networks are being used more and more for real-time and interactive applications—applications that were not strongly considered in the initial IP design. IP version 6 is designed to address these issues and to provide greater flexibility for the support of future applications that are as yet unknown.

Compared to version 4, IP version 6 offers the following enhancements:

- *Expanded address space* Each address is allocated 128 bits instead of 32 bits.
- *Simplified header format* Enables easier processing of IP datagrams
- *Improved support for headers and extensions* Enables greater flexibility for the introduction of new options
- *Flow-labeling capability* Enables the identification of traffic flows (and therefore, better support at the IP level) for real-time applications
- *Authentication and privacy* Support for authentication, data integrity, and data confidentiality are supported at the IP level, rather than through separate protocols or mechanisms above IP.

IP Version 6 Header

The IP version 6 header is shown in Figure 2-24, which clearly shows that the header—at least, in the basic form shown in Figure 2-24—is less complex than the version 4 header. The fields are as follows:

Version The version number is 6.

Traffic Class This field is an eight-bit field, which enables certain types of traffic to be differentiated from others—and potentially, to be given a high priority. For example, real-time applications might be given a higher priority than other applications, such as e-mail. In fact, we will see in Chapter 8 ("Quality of Service") how this field is used to ensure better quality of service for real-time applications.

Flow Label This field is a 20-bit field that can be used by a source to label sequences of packets that belong to a single flow, where a flow is identified by a unique combination of source address, destination address, and flow label. Packets belonging to a given flow might all be subject to special handling by intervening routers. The flow label should be set to zero for packets that do not belong to a particular sequence that needs special handling.

Payload Length This field is a 16-bit unsigned integer indicating the length of the payload in octets. Note that IPv6 enables header extensions to be included after the fixed header. As far as the length field is concerned, such extensions are part of the payload.

Next Header This field is an eight-bit field that serves the same purpose and uses the same values as the protocol field in the IP version 4 header. Thus, this field is used to indicate the next higher-layer protocol to which the payload data should be passed at the destination node. This field is also used to indicate the existence of IP version 6 header extensions, as explained in the section of this chapter called "IP Version 6 Header Extensions."

Hop Limit This field is an eight-bit unsigned integer that indicates the maximum number of hops that an IP packet should experience before being discarded. This number is decremented by one at each node that forwards the packet, and the packet is discarded if the value reaches zero. This field is analogous to the TTL field in the IP version 4 header.

Source and Destination Addresses Each of these is a 128-bit IP version 6 address. These addresses are described in the following section.

IP Version 6 Addresses

As previously described, IP version 4 addresses are generally represented by a dotted decimal notation with a maximum value of 255.255.255.255. IP

Figure 2-24

IP version 6 header

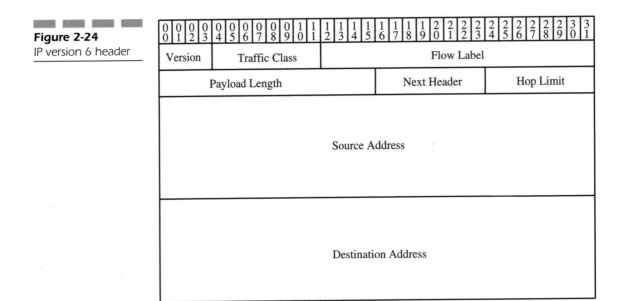

version 6 addresses are significantly different. They have the format XXXX:XXXX:XXXX:XXXX:XXXX:XXXX:XXXX:XXXX, where each *x* is a hexadecimal character. Thus, the address consists of eight fields, with each field consisting of four hexadecimal characters. The maximum possible value for an IP version 6 address is therefore FFFF:FFFF:FFFF:FFFF:FFFF:FFFF:FFFF:FFFF. This value is clearly a greater order of magnitude than the corresponding IP version 4 maximum.

Leading zeros do not have to be included in a given field, but there must be an entry for every field. Therefore, for example, 1511:1:0:0:0:FA22:45:11 is a valid address. The one exception to this rule regarding leading zeros is the case where several contiguous fields have the value zero. In such a case, the symbol "::" can be used to represent a number of contiguous fields with zero values. For example, the address 1511:1:0:0:0:FA22:45:11 could be represented as 1511:1::FA22:45:11. This notation can also be applied in the case where the first several fields contain zero values. For example, the address 0:0:0:0:AA11:50:22:F77 could be represented as ::AA11:50:22:F77. If the "::" notation appears once in an address, then it is possible to determine the correct content of each of the eight fields of the address. That is no longer possible, however, if the "::" notation appears more than once. Therefore, the notation is not allowed to appear more than once in an address. The formatting and usage of IP version 6 addresses is specified in RFC 2373.

One important point to note is that the use of IP version 6 should not generally require significant changes to upper-layer protocols. An important exception, however, is the use of pseudo headers when calculating TCP and UDP checksums. Recall that the checksum calculation in UDP and TCP includes a pseudo header that incorporates certain fields of the IP header, including the source and destination addresses. Because these are significantly different in IP version 6 compared to IP version 4, TCP and UDP checksum calculations are suitably adapted when IP version 6 is used as the routing protocol instead of IP version 4.

IP Version 6 Special Addresses IP version 6 identifies a number of special addresses, including the following:

- *The all-zeros address* An unspecified address; sometimes used when a node does not yet know its address. This address can be represented as either 0:0:0:0:0:0:0:0 or simply :: when using the short notation. This address cannot be used as a destination address in an IP version 6 packet.

- *The loopback address* Used by a node to send an IP version 6 packet to itself. This address's last field is set to 1; i.e., 0:0:0:0:0:0:0:1 or simply ::1 when using the short notation. This address can be used to route a packet on a virtual internal interface but should never be used as a destination address in a packet that is sent outside a physical node.

- *IPv6 address with embedded IPv4 address (type 1)* An IP version 6 address in which the first 96 bits are 0 and the last 32 bits contain an IP version 4 address. This address is termed an IP version 4-compatible IP version 6 address and is used by IP version 6 hosts and routers that tunnel IP version 6 packets through an IP version 4 infrastructure. Such an address can be represented in a manner that indicates the embedded IP version 4 address in the familiar dotted-decimal format. For example, the address 0:0:0:0:0:0:145.180.15.15, or simply ::145.180.15.15, would be used to represent an IP version 6 address with the embedded IP version 4 address 145.180.15.15.

- *IPv6 address with embedded IPv4 address (type 2)* An IP version 6 address in which the first 80 bits are 0, the next 16 bits are set to FFFF, and the last 32 bits are set to an IP version 4 address. This address is used to give an IP version 6 address representation to a pure IP version 4 address, which would apply to nodes that do not support IP version 6 at all. Using the same IP version 4 address as in the previous example, the IP version 6 representation would be 0:0:0:0:0:FFFF:145.180.15.15 or simply ::FFFF:145.180.15.15.

IP Version 6 Header Extensions

IP version 6 enables extension header fields to be placed between the fixed header and the actual data payload. With one exception, these header fields are not acted upon by intermediate nodes and are only considered at the destination node. Recall that the IP version 6 header contains a field called Next Header, which is used to indicate the type of payload carried in the IP datagram (and therefore to what upper-layer protocol the packet should be passed at the destination node). For example, if the value in the header is 6, the protocol identifier for TCP, then the payload is a TCP packet and the overall datagram would appear as in Figure 2-25.

The field can also be used to indicate the existence of a header extension, however, which can be used to carry certain optional information. If there is a header extension, then the Next Header value will indicate the type of extension. Furthermore, each extension will have its own Next Header field, which enables several contiguous header extensions to be included between the IP header and the upper-layer data. The value of Next Header in the IP header will point to the first extension; the value of Next Header in the first extension will point to the second extension; and so on. Finally, the value of Next Header in the last extension will indicate the upper-layer protocol being used. Figure 2-26 shows an example in which the three extensions "Hop-by-Hop," "Destination Options," and "Routing" are used with a TCP packet.

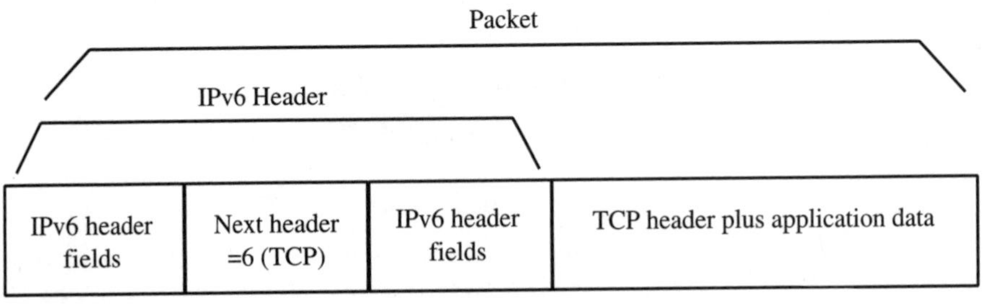

Figure 2-25
Use of Next
Header field

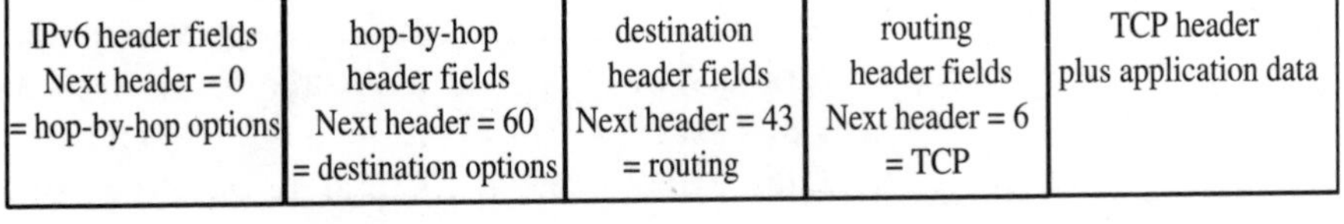

IPv6 header fields	Next header =6 (TCP)	IPv6 header fields	TCP header plus application data

Figure 2-26
Concatenated
header extensions

IPv6 header fields Next header = 0 = hop-by-hop options	hop-by-hop header fields Next header = 60 = destination options	destination header fields Next header = 43 = routing	routing header fields Next header = 6 = TCP	TCP header plus application data

Note that with a few exceptions, IP version 6 itself does not usually specify the details of the various extension headers—at least, not the specifics of what optional information and what corresponding functions are to be invoked. Rather, the header extension mechanisms specified by IP version 6 offer a means to support applications and requirements that might be as yet unknown.

Hop-by-Hop Extension   In general, header extensions are not analyzed by nodes along the delivery path and are meant to be acted upon only by the destination node. The one exception to this rule is the hop-by-hop extension, which must be examined and processed by every node along the way. Because of the fact that it must be examined at each node, the hop-by-hop extension, if present, must immediately follow the IP header.

The hop-by-hop extension is of variable length, with the format shown in Figure 2-27. The Next Header value is used to indicate the existence of a subsequent header extension or the upper-layer protocol to which the packet payload applies. The length field gives the length of this header extension in units of eight octets, not including the first eight octets. The individual options themselves, whatever they might signify, are of the format *Type-Length-Value* (TLV), where the type field is eight bits long, the length field is eight bits long, and the data is a variable number of octets according to the length field. This scenario is shown in Figure 2-28.

Within the Option Type field, the first three bits are of special significance. The first two bits indicate how the node should react if it does not understand the option in question, as follows:

Figure 2-27
IP version 6 hop-by-hop header extension

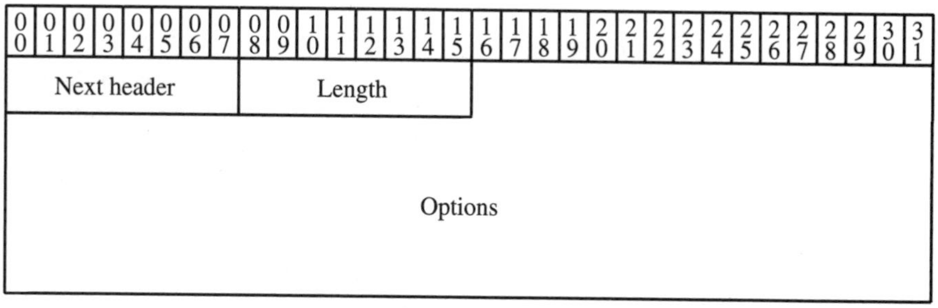

Figure 2-28
Format of options

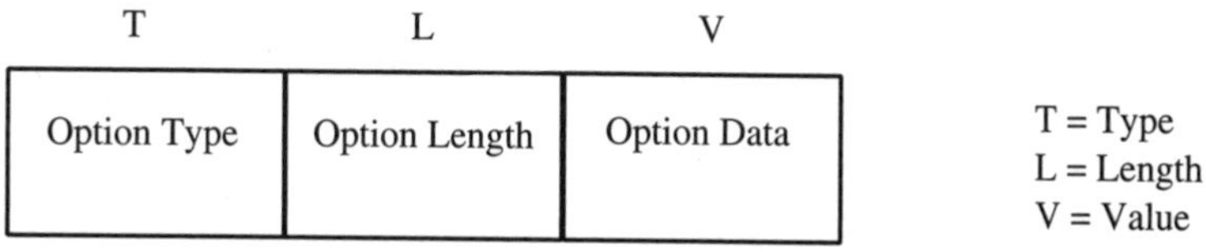

T = Type
L = Length
V = Value

- *00* Skip this option and continue processing the header
- *01* Discard the packet
- *10* Discard the packet and send an ICMP Parameter Problem, Code 2 message to the originator of the packet
- *11* Discard the packet and send an ICMP Parameter Problem, Code 2 message to the originator of the packet only if the destination address in the IP header is not a multicast address

The third bit indicates whether the option data can be changed en route. If set to 0, then it cannot change. If set to 1, then it can change.

The remainder of the options field is used to carry whatever options the sender of the datagram requires. The IP version 6 specification does not specify what those options might be.

Destination Options Extension   The Destination Options extension has the same format as the hop-by-hop extension in that it is of variable length and contains a number of options in the TLV format. The destination node only will examine these options. The Destination Options header extension is identified by the use of the value 60 in the Next Header field in the header immediately preceding.

Routing Extension   The Routing header extension includes a Routing type field to enable various routing options to take place. For the moment, however, IP version 6 only defines the routing type 0. The function of this routing type is to specify the nodes that should be visited on the path from the source of a packet to the ultimate destination. When used with routing type 0, the Routing header extension has the format shown in Figure 2-29.

As with all header extensions, the first two fields are the Next Header field and the length field. The next field is the Routing Type field, which is an eight-bit field indicating a particular variant of the routing header (and which is zero in the figure). The Segments Left field indicates the number of nodes that still need to be visited before the packet reaches its ultimate destination. Finally, a list of IP addresses needs to be visited along the way.

At first glance, one might think that this type of header would need to be analyzed by intermediate nodes between the source IP address and the destination IP address. After all, the purpose of the header is to specify a particular route—i.e., what nodes to visit en route. Hence, one would think that this header is another exception to the rule that only destination nodes examine header extensions.

In fact, the way this header is used is quite clever. Let's assume that a packet is to be sent from address A to address Z and must visit addresses

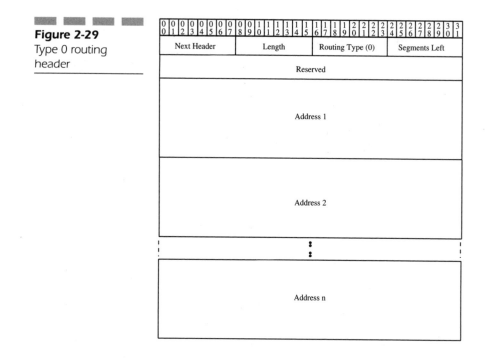

Figure 2-29
Type 0 routing
header

B, C, and D (in that order) along the way. In such a case, the source address in the IP header is address A, and the destination address in the IP header is address B. Addresses C, D, and Z are listed in the Route header. Hence, address B is the destination of the IP version 6 packet and as such is required to examine the route header. Upon examination of the header, it swaps address C into the destination address of the IP header, decrements the segments left field, and forwards the packet. At address C, the same process is repeated, but this time address D is placed into the destination address of the IP header. The process continues until the packet reaches address Z, at which point the value in the Segments Left field is zero—indicating that the packet is at its final destination. Furthermore, the Route header now provides a list of nodes visited en route.

Interoperating IP Version 4 and IP Version 6

IP version 6 is the new version of IP. IP version 4 is currently widely used, however, and it is not reasonable to expect that all IP networks will be changed from IP version 4 to IP version 6 at the same time. In fact, it is certain that IP version 4 and IP version 6 will coexist for a long time, and it is

Figure 2-30
Interoperating
between IP version 6
and IP version 4

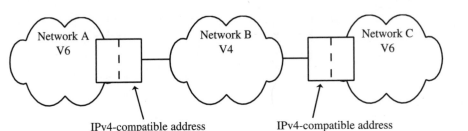

likely that IP version 6 networks will initially appear as islands in a sea of IP version 4. Therefore, a clear requirement exists to support seamless interoperation between the two versions. In other words, IP version 6 nodes need to have the capability to send datagrams to IP version 4 nodes. Also, IP version 4 nodes need to have the capability to send datagrams to IP version 6 nodes, and IP version 6 nodes need the capability to send datagrams to each other via IP version 4 networks. Finally, IP version 4 nodes should have the capability to send datagrams to each other via IP version 6 networks.

The use of IP version 4-compatible addresses makes this situation possible. Through the use of IP version 4-compatible nodes with IP version 4-compatible addresses at the boundaries of IP version 6 networks, you can accomplish internetworking. Figure 2-30 provides an example of how this goal is achieved. By allocating an IP version 4-compatible address to the interfaces on the version 6 networks towards the version 4 network, the version 6 networks can appear as version 4 networks. Consider a scenario in which a packet has to go from one IP version 6 network to another IP version 6 network via an IP version 4 network. At the edge of the source network, the packet can be wrapped in an IP version 4 header. That header would contain an address for the remote IP version 6 network, which would cause the packet to tunnel through the IP version 4 network. At the destination network, the IP version 4 header is removed, and the remaining IP version 6 packet is routed locally within the destination network.

Other Supporting Functions and Protocols

The purpose of this chapter has been to provide a brief overview of IP networking in general, with some extra focus on those protocols (such as RTP) that help to enable real-time applications (such as voice). Many different functions and protocols are involved with making IP communications work, however, and these must be supported in a VoIP network as well as in any other IP network. The following section provides a brief summary of just a few such protocols and functions (many others exist). Implementers should refer to the appropriate RFCs for details.

IP Security

For some time, IP-based networks were significantly lacking in security, which made IP-based transactions vulnerable to attack from less-than-honorable individuals. Such attacks have included eavesdropping and capturing data such as credit-card details and other sensitive information, plus impersonation, whereby an unauthorized individual gains access to services by pretending to be someone else. In fact, IP has historically been so vulnerable that many people still refuse to disclose sensitive information over the Internet for fear that it will be exposed to prying eyes. To some degree, such caution is warranted even with the security mechanisms that have been deployed recently.

Security mechanisms have been developed that support the three main aspects of security:

Authentication Making sure that a user is who he or she claims to be or that a system (e.g., host) is what it claims to be

Integrity Ensuring that data is not inappropriately altered on its way from sender to receiver

Confidentiality Ensuring that data is not subject to unwanted disclosure

The overall approach to security in IP networks is described in RFC 2401 (Security Architecture for the Internet Protocol). This specification is supported by other specifications such as RFC 2402 (IP Authentication Header), which addresses authentication, and RFC 2406 (IP Encapsulating Security Payload), which addresses confidentiality.

Obviously, support for security is extremely important in a VoIP network. We have seen in the past that IP networks are vulnerable to attack. We have also seen the rampant abuse that can occur if a network does not provide security. A great example is cloning and other fraud activities that were widespread in first-generation cellular networks. If a VoIP network is to be carrier grade, then customers should have a high level of assurance that conversations or other transactions cannot be overheard and that no unauthorized persons can steal a service for which the customer pays. We certainly do not want the threat of fraud to be a reason why VoIP should fail to be the huge success that we expect.

Of course, technical security mechanisms are only a part of the solution, and they are only as good as the processes and practices surrounding their implementation. A secure network must apply security at every level, ensuring that security is an integral part of all systems, all processes, and of personnel training.

Domain Name Service (DNS)

We discussed IP addresses in some detail in this chapter. A given source of IP datagrams, however, often does not initially know the IP address of the destination to which those packets should be sent. In most cases, the application will only know a name for the remote application. This name might take the form of a domain name, such as `host.somenet.net`, or it might take the form of `name@domain`, as is common for e-mail addresses. Therefore, a need exists for a translation service, whereby the name of the application to which a packet should be sent can be translated to an IP address for inclusion in the IP header. This translation service is known as the *Domain Name Service* (DNS).

DNS servers are scattered all over the Internet. They provide a translation from domain name to IP address. Of course, not every DNS server has a copy of every domain name with a corresponding address. Instead, DNS supports the concept of delegating, whereby one DNS server might query another for information. Typically, a local DNS server will be queried to provide an address for a given domain name. If it does not have an entry for the name in question, then it queries a root DNS server in its own country. That server might have an address for the domain name, or if not, it will provide an address for another server that should be able to help. For example, assume that a host in the United States needs to send a packet to a destination in the United Kingdom. The local DNS server is queried, and that server in turn queries a root server in the United States. The root server does not have an entry for the domain name but returns an address for a root server in the United Kingdom, which should be queried instead. The root server in the United Kingdom is queried. This server might provide an address or might provide the address of a local DNS server in the United Kingdom that should be queried. Ultimately, the correct address is returned and can be used in the IP header.

Path MTU Discovery

Recall that it is possible to fragment large datagrams into smaller pieces in the case that a given datagram is too big to be handled all at once by a given node or link. Fragmentation can have undesirable effects for real-time traffic applications, however, and should be avoided. Therefore, you must determine in advance the maximum packet size that can be transmitted between sender and receiver. This maximum is known as the Path *Maximum Transmission Unit* (MTU).

One simple way to find out the MTU is for a source to send an initial packet with the "don't fragment" flag set to 1. The size of this probing packet will be set to the MTU of the outgoing interface from the source. If the packet reaches a point in the network where it is too big to be forwarded, then the router that discovers the problem will return the ICMP Destination Unreachable message with code 4, indicating that the message is too big. Furthermore, according to the MTU discovery process described in RFC 1191, the message should contain the MTU of the outgoing interface that the router wished to use. The source can then use that MTU as the size of the next packet. Therefore, the next packet will have the capacity to get past the router where there originally was a problem—and hopefully make it to its destination. If the packet hits a point in the network that has an even lower MTU, then the process is repeated until the packet arrives at its destination. Note that a packet of size 68 octets will always get through, because every router is required to support packets of 68 octets or less.

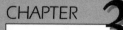

Speech-Coding Techniques

Introduction

Chapter 1 mentioned that *Voice over IP* (VoIP) can take advantage of a variety of efficient speech-coding techniques, and Chapter 2 delved into some detail regarding how digitally encoded speech is carried over an IP network by using the *Real-Time Transport Protocol* (RTP). Chapter 2 made reference to various RTP payload types, which in turn referred to various speech-coding standards. If it were not already obvious, the discussions in these previous chapters should indicate that speech-coding techniques are of particular significance in VoIP networks. In fact, speech coding is an extremely important topic.

Although some analog systems still remain, it has been quite a long time since analog transmission was the norm in voice communications. These days, voice is digitally encoded and shipped around the network (any network) as a stream of ones and zeros. Speech/voice coding is simply the process by which a digital stream of ones and zeros is made to represent an analog voice waveform. This process involves converting the incoming analog voice pattern into a digital stream and converting that digital stream back to an analog voice pattern at its ultimate destination. One of the reasons for implementing VoIP is the opportunity to take advantage of efficient voice coding—where fewer bits are used to represent the voice being transmitted, thereby reducing bandwidth requirements and reducing cost.

The initial conclusion could be that one should implement the most bandwidth-efficient coding scheme possible, thereby saving the most money possible. Unfortunately, however, there is no such thing as the proverbial "free lunch." As a rough guide, the lower the bandwidth, the lower the quality. So, if we reduce the bandwidth significantly, we run the risk of providing substandard voice quality—not a noble or smart goal, and certainly not something that is consistent with carrier-grade operation.

One should not assume, however, that anything similar to a linear relationship exists between bandwidth and quality. Generally, the lower the bandwidth, the lower the quality—but that is a rough guide. Another component to the relationship exists: the speech-coding scheme used, which relates to processing power. For example, voice transmission at 64kbps sounds better than any voice transmission at 16kbps. Several different coding techniques might exist, however, that result in a bit rate of 16kbps—and one of those techniques might provide better quality than another. You can be certain that the technique that provides the better quality uses a more complex algorithm and hence needs more processing power. Consequently, this technique might be somewhat more expensive to implement, emphasizing again that we receive nothing for nothing. In the end, the choice of coding scheme is a balance between quality and cost.

In this chapter, we examine some of the issues surrounding speech coding in general and consider some of the coding standards that exist today. You should note that this analysis is not by any means exhaustive as far as coding standards. This analysis is meant only as an overview of some of the most common standards, so that the reader can gain an understanding of some of the technologies involved. Given that this area of telecommunications has undergone great activity and great advances in recent years, further advances will likely be made in the near future in this area.

Voice Quality

We would like to minimize bandwidth while maintaining sufficiently good voice quality. Bandwidth is easily quantified, but how do we quantify voice quality? Surely, this measurement is subjective rather than objective. Standardized methods exist for measuring voice quality, however, including a standardized ranking system called the *Mean Opinion Score* (MOS). This system is described in ITU-T Recommendation P.800. Basically, MOS is a five-point scale as follows:

Excellent—5

Good—4

Fair—3

Poor—2

Bad—1

The objective with any coding technique is to achieve as high a ranking on this scale as possible while keeping the bandwidth requirement relatively low. To determine how well a particular coding technique ranks on this scale, a number of people (a minimum of about 30 people and possibly many more) listen to a selection of voice samples or participate in conversations, with the speech being coded by using the technique to be evaluated. They rank each of the samples or conversations according the five-point scale, and a mean score is calculated to give the MOS.

On the surface, this system does not seem too scientific. While the basis of the MOS test is a matter of people listening to voice samples, there is more to the system. ITU-T P.800 makes a number of recommendations regarding the selection of participants, the test environment, explanations to listeners, analysis of results, etc. Consequently, different MOS tests performed on the same coding algorithm tend to give roughly similar results. Because the tests are subjective, however, variations exist. One test might yield a score of 4.1 for a particular coding algorithm, whereas another test might yield a score of 3.9 for the same algorithm. Therefore, one should treat MOS values with care—and at the risk of sounding cynical, they should perhaps be treated with some skepticism, depending on who is claming a particular MOS for a particular codec.

In general, and within design restrictions such as bandwidth, most coding schemes aim to achieve or approach toll quality. Although the definition of this term varies, toll quality generally refers to a MOS of 4.0 or higher.

A Little About Speech

Speech is generated when air is pushed from the lungs past the vocal cords and along the vocal tract. The basic vibrations occur at the vocal cords, but the sound is altered by the disposition of the vocal tract—e.g., by the position of the tongue or by the shape of the mouth. The vocal tract can be considered a filter, and many codec technologies attempt to model the vocal tract as a filter. As the shape of the vocal tract changes relatively slowly, the transfer function of the filter needs to be changed relatively infrequently (e.g., every 20ms or so). If the vocal tract can be modeled as a filter, then the vibrations at the vocal cords correspond to the excitation signal applied to the filter.

Speech sounds can be classified into three main types[1]:

- *Voiced sounds* are produced when the vocal cords vibrate open and closed, thus interrupting the flow of air from the lungs to the vocal tract and producing quasi-periodic pulses of air as excitation occurs. The rate of the opening and closing gives the pitch of the sound. This pitch can be adjusted by varying the shape of (and the tension in) the vocal cords and the pressure of the air behind them. Voiced sounds show a high degree of periodicity at the pitch period, which is typically between 2ms and 20ms. This long-term periodicity can be seen in Figure 3-1, which shows a segment of voiced speech sampled at 8kHz. Here, the pitch period is about 8ms or 64 samples. The power spectral density for this segment is shown in Figure 3-2.

- *Unvoiced sounds* result when the excitation is a noise-like turbulence produced by forcing air at high velocities through a constriction in the vocal tract while the glottis is held open. Such sounds show little long-term periodicity, as shown in Figures 3-3 and 3-4, although short-term correlations due to the vocal tract are still present.

---

[1] From www_mobile.ecs.soton.ac.uk/speech_codecs

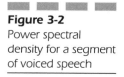

Figure 3-1
Typical segment of
voiced speech

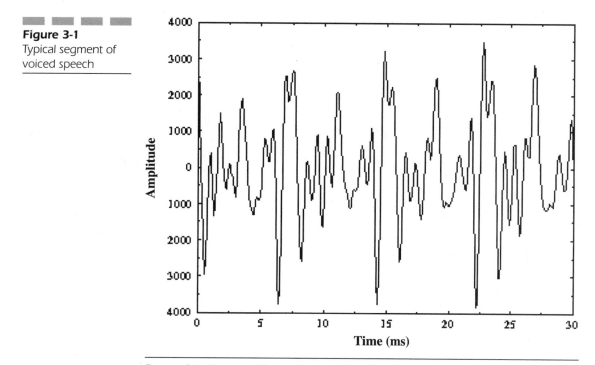

Source: http://www-mobile.ecs.soton.ac.uk/speech_codecs/speech_props_figs.html[2]

Figure 3-2
Power spectral
density for a segment
of voiced speech

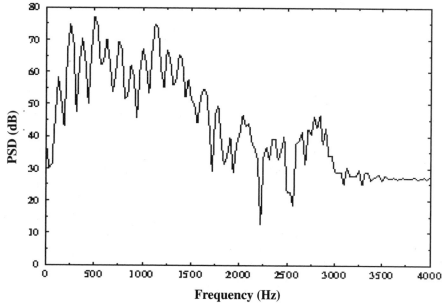

Source: http://www-mobile.ecs.soton.ac.uk/speech_codecs/speech_props_figs.html[2]

Figure 3-3
Typical segment of
unvoiced speech

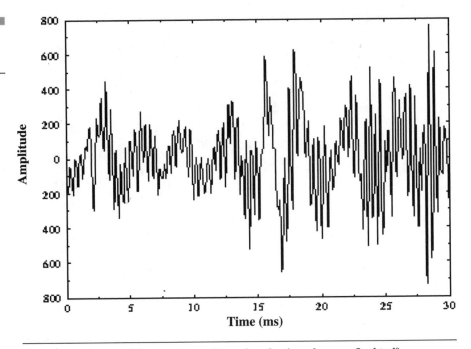

Source: http://www-mobile.ecs.soton.ac.uk/speech_codecs/speech_props_figs.html[2]

- *Plosive sounds* result when a complete closure is made in the vocal tract and air pressure is built up behind this closure and released suddenly.

Although the combined efforts of the vocal cords and the vocal tract can be made to create a vast array of sounds, the type of vibration at the vocal cords and the shape of the vocal tract change fairly slowly. We can see from the figures that the speech signal is relatively predictable over time. This predictability is something that can be used in the design of voice-coding systems to minimize the amount of data to be transferred (and hence, to reduce bandwidth). For example, if the systems at each end of a conversation can both make reasonably accurate predictions of how the speech signal will change over time, then we do not have to transfer the same amount of data as would be required if speech were totally unpredictable. With good voice-coding schemes, the reduction in bandwidth can be significant—without a huge degradation in perceived quality.

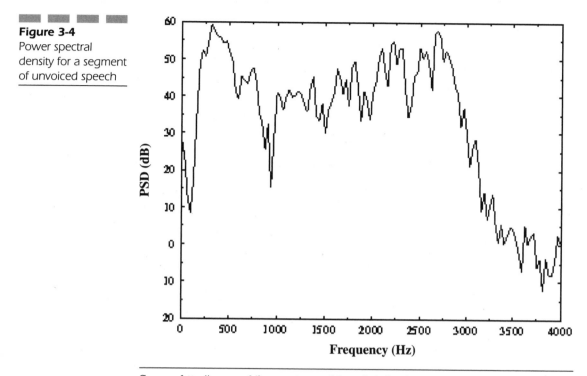

Figure 3-4
Power spectral
density for a segment
of unvoiced speech

Source: http://www-mobile.ecs.soton.ac.uk/speech_codecs/speech_props_figs.html[2]

Voice Sampling

In order to create a digital representation of an analog waveform (such as voice), we must first take a number of discreet samples of the waveform and then represent each sample by some number of bits.

Figure 3-5 gives an example of the sampling of a simple sine wave and what the samples might look like. Obviously, it would take an infinite number of samples in order for those samples to completely recreate the original signal in all respects. Of course, that is not possible, nor is it desirable. Instead, we would prefer to take a sufficient number of samples so that we can use the sample values and suitable mathematics to the recreate the

---

[2] These world wide web pages constitute a supplement of L. Hanzo, FCA, J. P. Woodard: *Voice Compression and Communication,* to be published by IEEE Papers, 2000.

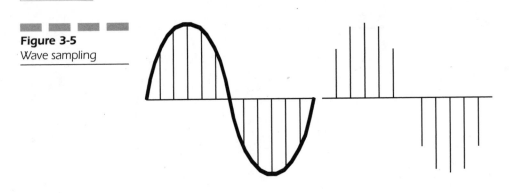

Figure 3-5
Wave sampling

original signal (or at least, come very close). The Nyquist Sampling Theorem can help here. This theorem basically states that a signal can be reconstructed if it is sampled at a minimum of twice the maximum frequency of the signal. Thus, if the maximum frequency of a given signal is 4,000Hz, then we need to take at least 8,000 samples per second.

In fact, we conduct this process with almost all speech codecs. Human speech is generally in the frequency range between about 300Hz and 3,800Hz. We assume a maximum of less than 4,000Hz, apply a low-pass filter to take out any frequency components higher than 4,000Hz, then take 8,000 samples per second. Note that this procedure will not properly capture Luciano Pavarotti in full voice or even a less-talented individual singing in the shower—just typical conversational speech. For this reason, someone singing on the telephone never sounds quite as good as they should, regardless of whether the singer has talent.

Quantization

We sample at 8,000Hz. The next concern is how to represent each sample. In other words, how many bits are used to represent each sample? The answer is not as simple as one might think, largely because of quantization noise.

When we take a sample, we have a limited number of bits to represent the value of the sample. By using a limited number of bits to represent each sample, however, we *quantize* the signal. Figure 3-6 shows an example of quantization, where there are three bits to represent various signal levels. Because we only have three bits, we can represent a maximum of eight different signal levels. If at a given sampling instant in our example the analog signal has a value of 5.3, then with our limited number of bits, we can only apply the value 5 to the digitized representation. When we transmit the digitized representation to the far end, all that can be recovered is the

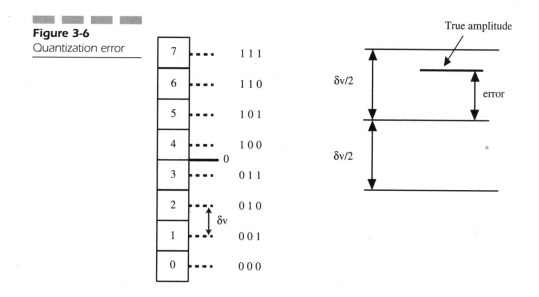

Figure 3-6
Quantization error

value 5. The real signal level can never be recovered. The difference between the actual level of the input analog signal and the digitized representation is known as *quantization noise*.

The easiest way to minimize this effect is to use more bits, thereby providing better granularity. So, if we were to use six bits instead of three, we could at least tell that the sampled value is closer to 5.5 than to 5.0, and we could send the bit representation for 5.5. This process, however, can approach the stage of diminishing returns. The more accurate we wish the samples to be, the more bits we need to represent each sample—and the more bandwidth we need. Alternatively, we can use relatively few bits, but we can use them in a more sophisticated manner.

If we look at the possible range of signal levels in speech, and if we wish to apply uniform quantization levels, then we have two effects. First, we would need many bits to represent each sample; and second, loud talkers would sound better than quiet talkers. Loud talkers would sound better because the effect of quantization noise is less at higher levels than at lower levels. For example, the detrimental effect of digitizing a sample of 11.2 as the value 11 (about 1.8 percent too low) is a lot less than the effect of digitizing a sample of 2.2 as the value 2 (about 9 percent too low). More formally, the signal-to-noise ratio is better for loud talkers than for quiet talkers.

Therefore, we use non-uniform quantization. This process involves the usage of smaller quantization steps at smaller signal levels and larger quantization steps for larger signal levels. This procedure gives greater granularity at low signal levels and less granularity at high signal levels.

The effect is to spread the signal-to-noise ratio more evenly across the range of different signals and to enable fewer bits to be used, compared to uniform quantization. In a coder in which the quantization levels are simply sent from one end to the other, this process translates to a lower bandwidth requirement than for uniform quantization.

Types of Speech Coders

Before we get into descriptions of the various types of codecs (coder/decoders) available, we should first examine the various categories of codecs. Three types exist: waveform codecs, source codecs (also known as vocoders), and hybrid codecs.

Waveform codecs basically sample and code the incoming analog signal without any thought as to how the signal was generated in the first place. They then transmit quantized values of the samples to the destination end, where the original signal is reconstructed at least to a good approximation. Generally, waveform codecs produce a high-quality output and are not too complex. The big disadvantage is that they consume large amounts of bandwidth compared to other codecs. When waveform codecs are used at lower bandwidths, the speech quality degrades significantly.

Source codecs attempt to match the incoming signal to a mathematical model of how speech is produced. They usually use a linear-predictive filter model of the vocal tract, with a voiced/unvoiced flag to represent the excitation that is applied to the filter. In other words, the filter represents the vocal tract, and the voiced/unvoiced flag represents whether a voiced or unvoiced input is received from the vocal cords. The information that is sent to the far end is a set of model parameters, rather than a representation of the signal itself. The far end, using the same modeling technique in reverse, takes the values received and reconstructs an analog signal.

Vocoders operate at low bit rates but tend to produce speech that sounds synthetic. Using higher bit rates does not offer much improvement, due to the limitations in the underlying model. While vocoders are used in private communications systems and particularly in military applications, they are generally not used in public networks.

Hybrid codecs attempt to provide the best of both worlds. While they attempt to perform a degree of waveform matching, they also utilize the knowledge of how people produce sounds in the first place. They tend to provide quite good quality at lower bit rates than waveform codecs. Figure 3-7 shows a comparison of the three types of codecs with respect to quality and bandwidth.

Figure 3-7
Speech quality versus
bit rate for common
classes of codecs

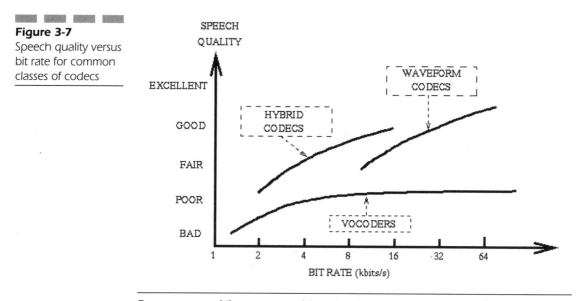

Source: www-mobile.ecs.soton.ac.uk/speech_codecs/common_props_classes.html

G.711

G.711 is the most commonplace coding technique used today. G.711 is a waveform codec and is the coding technique that is used in circuit-switched telephone networks all over the world. G.711 has a sampling rate of 8,000Hz. If uniform quantization were to be used, then the signal levels commonly found in speech would be such that at least 12 bits per sample would be needed, leading to a bit rate of 96kbps. Non-uniform quantization is used, however, with eight bits used to represent each sample. This quantization leads to the well-known 64kbps DS0 rate that we have come to know and love. These eight-bit quantized samples are transmitted to the far end. G.711 is often called *Pulse-Code Modulation* (PCM).

G.711 has two variants: A-law and mu-law. mu-law is used primarily in North America, and A-law is used in most other countries. The difference between the two is the manner in which non-uniform quantization is performed. Both are symmetrical around zero. A-law is skewed to be a little friendlier to lower signal levels than mu-law in that it provides small quantization intervals for a longer range of low-level signals at the expense of larger quantization levels for a longer range of higher-level signals.

Both A-law and mu-law provide good quality and have an MOS of about 4.3. The main drawback with G.711 however, is the 64kbps bandwidth requirement. Receiving comparable speech quality at a lower bandwidth would be nice.

Adaptive DPCM (ADPCM)

PCM codecs such as G.711 send individual samples to the far end, where they are reconstructed to form something close to the original waveform. Because voice changes relatively slowly, however, it is possible to predict the value of a given sample based on the values of previous samples. In such a case, we only have to transmit the difference between the predicted value and the actual value of the sample. Because the far end is performing the same predictions, it can determine the original sample value if told the difference between the prediction and the actual sample value. This technique is known as *Differential PCM* (DPCM) and can significantly reduce the transmission bandwidth requirement without drastic degradation in quality. In its simplest form, DPCM does not even make a prediction of the next sample; rather, it simply transmits the difference between sample N and sample N+1, providing all of the information necessary for the far end to recreate sample N+1.

A slightly more advanced version of DPCM is *Adaptive DPCM* (ADPCM). ADPCM typically predicts sample values based on past samples while factoring in some knowledge of how speech varies over time. The error between the actual sample and the prediction is quantized, and this quantized error is transmitted to the far end. Assuming that predictions are fairly accurate, fewer bits are required to describe the quantized error; hence, the bandwidth requirement is lower. The ITU-T Recommendation G.721, which offers ADPCM-coded speech at 32kbps, is a good example of ADPCM.

G.721 has now been superseded by ITU-T recommendation G.726, which is a more advanced ADPCM codec. G.726 takes A-law or mu-law coded speech and converts it to or from a 16kbps, 24kbps, 32kbps, or 40kbps channel. Running at 32kbps, G.726 has an MOS of about 4.0, which is pretty good.

Both PCM and ADPCM are waveform codecs. As such, they have effectively no algorithmic delay. In other words, the functioning of the algorithm itself does not require quantized speech samples or quantized error values to be held up for a brief time before being sent to the far end. Given that voice is a delay-sensitive form of communication, this offers a significant advantage. The disadvantage, of course, is the bandwidth that these codecs consume—especially at the higher end.

Analysis-by-Synthesis (AbS) Codecs

Hybrid codecs attempt to fill the gap between waveform and source codecs. As described previously, waveform coders are capable of providing good-quality speech at higher bit rates. They can even provide relatively acceptable quality at rates down to about 16kbps. They are of limited use at rates below these, however. Vocoders, on the other hand, can provide intelligible speech at 2.4kbps and lower but cannot provide natural-sounding speech at any bit rate. The objective with hybrid codecs is to provide good-quality speech at a lower bit rate than is possible with waveform codecs.

Although other forms of hybrid codecs exist, the most successful and most commonly used are time-domain *Analysis-by-Synthesis* (AbS) codecs. Such codecs use the same linear-prediction filter model of the vocal tract as is found in LPC vocoders. Instead of applying a simple two-state, voiced/unvoiced model to find the necessary input to this filter, however, the excitation signal is chosen by attempting to match the reconstructed speech waveform as closely as possible to the original speech waveform. In other words, different excitation signals are attempted, and the one that gives a result that is closest to the original waveform is selected (hence the name AbS). AbS codecs were first introduced in 1982 with what was to become known as the *Multi-Pulse Excited* (MPE) codec. Later, the *Regular-Pulse Excited* (RPE) and *Code-Excited Linear Predictive* (CELP) codecs were introduced. Variations of the latter are discussed briefly here.

G.728 LD-CELP

CELP codecs implement a filter (the characteristics of which change over time), and they contain a codebook of acoustic vectors. Each vector contains a set of elements in which the elements represent various characteristics of the excitation signal. This setup is obviously far more sophisticated than a simple voiced/unvoiced flag.

With CELP coders, all that is transmitted to the far end is a set of information indicating filter coefficients, gain, and a pointer to the excitation vector that was chosen. Because the far end contains the same codebook and filter capabilities, it can reconstruct the original signal to a good degree of accuracy.

ITU-T Recommendation G.728 specifies a *Low Delay-Code-Excited Linear Predictor* (LD-CELP), which is a backward-adaptive coder because it uses previous speech samples to determine the applicable filter coefficients. G.728 operates on five samples at a time. In other words, five samples (sampled at 8,000Hz) are needed to determine a codebook vector and filter coefficients to best match the five samples. The choice of filter coefficients is based upon previous and current samples. Because the coder operates on five samples at a time, we have a delay of less than one millisecond.

Because the receiver also has access to the previous samples and can perform the same determination, all that is necessary to transmit is an indication of the excitation parameters. This procedure simply involves transmitting an index to the excitation vector that is chosen. Because there are 1,024 vectors in the code book (and these are available at both the sender and receiver ends), all that needs to be transmitted is a 10-bit index value. G.728 uses five samples at a time, and these samples are taken at a rate of 8,000 per second. For each of the five samples, G.728 needs to transmit only 10 bits. A little arithmetic shows that G.728 results in a transmitted bit rate of 16kbps.

The following text is taken directly from the G.728 specification and provides some further information about the coding mechanism. The text and the content of Figures 3-8 and 3-9 are reproduced here with the prior authorization of the ITU as copyright holder.

Figure 3-8
Simplified block diagram of an LD-CELP code

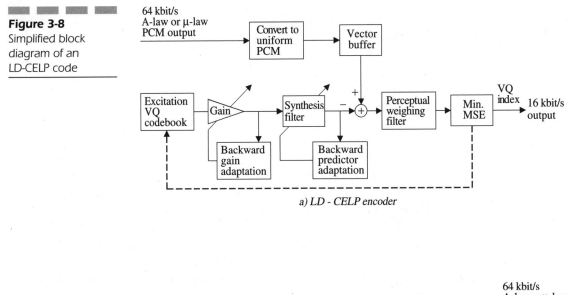

a) LD - CELP encoder

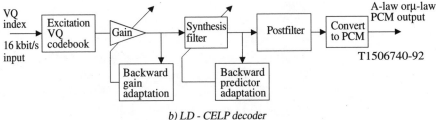

b) LD - CELP decoder

After the conversion from A-law or μ-law PCM to uniform PCM, the input signal is partitioned into blocks of five-consecutive input signal samples. For each input block, the encoder passes each of 1024 candidate codebook vectors (stored in an excitation codebook) through a gain scaling unit and a synthesis filter. From the resulting 1024 candidate quantized signal vectors, the encoder identifies the one that minimizes a frequency-weighted mean-squared error measure with respect to the input signal vector. The 10-bit codebook index of the corresponding best codebook vector (or "codevector"), which gives rise to that best candidate quantized signal vector, is transmitted to the decoder.

The decoding operation is also performed on a block-by-block basis. Upon receiving each 10-bit index, the decoder performs a table look-up to extract the corresponding codevector from the excitation codebook. The extracted code-vector is then passed through a gain scaling unit and a synthesis filter to produce the current decoded signal vector.

The decoded signal vector is then passed through an adaptive postfilter to enhance the perceptual quality. The postfilter coefficients are updated periodically using the information available at the decoder. The five samples of the postfilter signal vector are next converted to five A-law or μ-law PCM output samples.

The reason why the G.728 coder is termed LD-CELP is because of the small algorithmic delay that it introduces. Due to the fact that it operates on five PCM samples at a time, and given that these samples are taken at a rate of 8,000Hz, the coder introduces a delay of 0.625ms. This delay is so small that it would not be noticeable even to the keenest ear.

The fact that only a 10-bit pointer value is transmitted for every five incoming PCM samples means that the actual bandwidth consumed between encoder and decoder is one-quarter of that used by regular PCM (i.e., 16kbps). Given that the G.728 codec gives an MOS score of about 3.9, the quality reduction is not enormous compared to the bandwidth that we save. Figure 3-8 provides a simplified block diagram of the LD-CELP encoder and decoder.

G.723.1 ACELP

ITU-T Recommendation G.723.1 specifies a speech codec that can operate at either 6.3kbps or 5.3kbps, with the higher bit rate providing higher speech quality. Both rates are mandatory parts of the coder and decoder, and we can change from one mode to another during a conversation.

The coder takes a band-limited input speech signal that is sampled at 8,000Hz and that undergoes uniform PCM quantization, resulting in a 16-bit PCM signal. The encoder then operates on blocks or frames of 240 samples at a time. Thus, each frame corresponds to 30ms of speech, which means that the coder automatically causes a delay of 30ms. A look-ahead of 7.5ms also exists, resulting in a total algorithmic delay of 37.5ms. Of course, there will be other small delays within the coder itself as a result of the processing effort involved.

Each frame is passed through a high-pass filter to remove any DC component and then is divided into four subframes of 60 samples each. Various operations are performed on these subframes in order to determine the appropriate filter coefficients. *Algebraic Code-Excited Linear Prediction* (ACELP) is used in the case of the lower bit rate of 5.3kbps, and *Multi-Pulse Maximum Likelihood Quantization* (MP-MLQ) is used in the case of the higher rate of 6.3kbps.

The information transmitted to the far end includes linear prediction coefficients, gain parameters, and excitation codebook index values. The information transmitted comprises 24-octet frames in the case of transmission at 6.3kbps and 20-octet frames in the case of transmission at 5.3kbps.

Normal conversation involves significant periods of silence (or at least silence from one of the parties). During such periods of silence, it is desirable not to consume significant bandwidth by transmitting the silence at the same rate that speech is transmitted. For this reason, G.723.1 Annex A specifies a mechanism for silence suppression, whereby *Silence Insertion Description* (SID) frames can also be used. SIDs are only four octets in length, which means that transmission of silence occupies about 1kbps. This rate is significantly better than G.711, where silence is still transmitted at 64kbps.

Therefore, three different types of frames can be transmitted by using G.723.1: one for 6.3kbps, one for 5.3kbps, and a SID frame. Within each frame, the two least-significant bits of the first octet indicate the frame size, and the codec version in use (as shown in Table 3-1).

G.723.1 has an MOS of about 3.8, which is good considering the vastly reduced bandwidth that it uses. G.723.1 does have the disadvantage of a minimum 37.5ms delay at the encoder, however. While this delay is well within the bounds of what is acceptable for good-quality speech, we must remember that round-trip delay is what is important, not just one-way delay. Moreover, there will be various other delays in the network, including processing delays and queuing delays—particularly at routers in a VoIP network.

	Bits	Meaning	Octets/Frame
Table 3-1			
G.723.1 frame types	00	High-rate speech (6.3kbps)	24
	01	Low-rate speech (5.3kbps)	20
	10	SID frame	4
	11		N/A

G.729

ITU-T Recommendation G.729 specifies a speech coder that operates at 8kbps. This coder uses input frames of 10ms, corresponding to 80 samples at a sampling rate of 8,000Hz. This coder also includes a 5ms look-ahead, resulting in an algorithmic delay of 15ms (significantly better than G.723.1). From each input frame, we can determine linear prediction coefficients, excitation code book indices, and gain parameters, and these are transmitted to the far end. They are included in an 80-bit frame. Given that the input signal corresponds to 10ms of speech and results in a transmission of 80 bits, the transmitted bit rate is 8kbps. Figure 3-9 shows a high-level block diagram of the G.729 encoder.

G.729 is a complex codec. In order to reduce the complexity in the algorithm, a number of simplifications were introduced in Annex A to G.729. These include simplified code book search routines and a simplification to the postfilter at the decoder, among other items. G.729A uses exactly the same transmitted frame structure as G.729 and therefore uses the same bandwidth. In other words, the encoder might be operating according to G.729, while the decoder might operate by using G.729A or vice-versa. Note that G.729A might result in slightly lower quality than G.729.

An Annex B to G.729 also exists, which is a recommendation for the *Voice Activity Detection* (VAD), *Discontinuous Transmission* (DTX), and *Comfort Noise Generation* (CNG). VAD is simply the decision as to whether voice or noise is present at the input. The decision is based on analysis of several parameters of the input signal. Note that this determination is not conducted simply on the basis of one frame; rather, this decision is made on the basis of the current frame plus the two preceding frames. This process ensures that transmission occurs for at least two frames after a person has stopped speaking.

The next decision involves whether to send nothing at all or to send a SID frame. The SID frame contains some information to enable the decoder

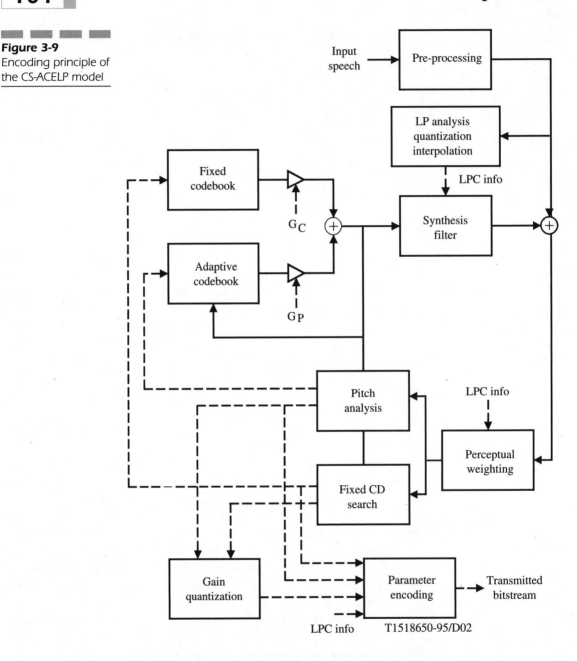

Figure 3-9
Encoding principle of
the CS-ACELP model

to generate comfort noise that corresponds to the background noise at the transmission end. The G.729B SID frame is a mere 15 bits long, which is significantly shorter than the 80-bit speech frame.

Assuming that the silence continues for some time, the encoder keeps watch over the background noise. If there is no significant change, then nothing is sent and the decoder continues to generate the same comfort noise. If the encoder notices a significant change in the background-noise energy, however, then an updated SID frame is sent to update the decoder based on the characteristics of the background noise. This procedure avoids a comfort noise that is constant and that, if it persists for some time, might no longer be comforting to the listener.

G.729 offers an MOS of about 4.0, whereas G.729A has a lower value of about 3.7.

Selecting Codecs

The codecs that we just discussed constitute a small selection of the coding techniques that are available. Many others add to the menu of choices when choosing a coding technology for a given network. Examples of other choices include the following:

- CDMA QCELP, as defined in IS-733. This coder is a variable-rate coder and is currently used in IS-95-based CDMA wireless systems. CDMA QCELP can operate at several rates, but the two most common are the high rate of 13,300bps and a lower rate of 6,200bps. The codec supports silence suppression, however, such that the net bit rate transmitted is less in each case. The packetization of the QCELP codec for use with RTP is defined in RFC 2658.

- GSM *Enhanced Fullrate* (EFR), as defined in GSM specification 06.60. This coder is an enhanced version of the original GSM *Full-Rate* (FR) coder. GSM EFR is enhanced in that it uses an ACELP-based coding scheme instead of the RPE-LTP scheme of the full-rate coder but operates at the same bit rate and uses the same overall packing structure as used for full rate in order to ensure compatibility. This coder operates at a rate of 12,200bps. Like the QCELP codec, however, it supports discontinuous transmission, such that the actual consumed bit rate is lower. Packetization of GSM EFR for use with RTP is accomplished in a similar manner to that which is used for GSM FR, as described in RFC 1890.

- GSM *Adaptive Multi-Rate* (AMR) codec, as defined in GSM specification 06.90. GSM AMR is a multi-mode codec that can operate in any of eight different modes with bit rates from 4.75kbps to 12.2kbps. When operating at 12.2kbps, GSM AMR is effectively the GSM EFR coder. When operating at 7.4kbps, it is effectively the IS-641 coder that is used in TDMA cellular systems. The coder can change to any of its modes at any time in response to channel conditions and offers discontinuous transmission support, minimizing the bandwidth that is consumed during silent periods and thereby reducing the net bandwidth consumption. The AMR codec is the coding choice for many third-generation wireless systems. The packetization rules for use of the codec with RTP are in the form of an Internet draft as of this writing.

Clearly, many options exist from which to choose. Furthermore, new coding schemes are being developed at a rapid pace. The task of selecting the best codec or set of codecs to use for a given network implementation is a matter of balancing quality with bandwidth consumption. Given the range of codecs available, however, the choice might not be that easy.

When looking at MOS values, one should remember that the MOS value presented for a particular codec is normally the value that is achieved under laboratory conditions. The likelihood is great that a given codec will not give the same quality in a particular network implementation. For example, G.711, while offering the highest quality under ideal circumstances, does not incorporate any logic to deal with lost packets. In contrast, G.729 does have the capability to accommodate a lost frame by interpolating from previous frames. On the other hand, this situation causes subsequent speech frames at the decoder to generate errors because of the fact that the decoder bases its output on the parameters of previous frames, and it can take some time to recover from a lost or errored frame.

An important consideration is the delay that a codec will introduce. While waveform codecs do not introduce delay, predictive codecs introduce a delay at the coder and a somewhat smaller delay at the decoder.

Another consideration is the *Digital Signal Processor* (DSP) power that is required, which is measured in *million instructions per second* (MIPS). Whereas G.728 or G.729 might need to run on a 40MIPS processor, G.726 could run on a 10MIPS processor. In general, the more MIPS, the greater the cost of the DSP. Fortunately, the price/performance of DSPs continues to improve such that this problem is becoming less of a factor.

Cascaded Codecs

In general, we should minimize the number of times that a given speech segment is coded and decoded. In the ideal case, the original analog signal should be coded only once and extracted only at the ultimate destination. Coding a particular analog stream by using one codec at, say, 16kbps, then later converting that to, say, G.711, and later encoding the stream again by using, say,

G.729 is not a good idea. In such cases, the voice quality might not even come close to the MOS of the lowest-quality codec in the chain. The simple reason is that each coder can only generate an approximation of the incoming signal that it receives. At each step along the chain, the approximation bears fewer and fewer resemblances to the original analog waveform.

We understand that implementing VoIP in certain situations might require such cascaded codecs. If that is the case, then a thorough trial of different configurations and codecs is probably well worth the effort in order to ensure that the ultimate quality does not degrade to an extent that the loss of quality would negate any benefits that VoIP might give.

Tones, Signals, and DTMF Digits

The more sophisticated codecs available today achieve bandwidth efficiency without losing significant quality, due to smart algorithms and powerful DSPs. Those smart algorithms, however, are based upon a detailed understanding of how voice is produced in the first place. In other words, the codecs are optimized for human speech. Unfortunately, that human speech is not always the only data that needs to be transmitted.

Within the telephone network, in addition to voice, all sorts of tones and beeps are transmitted—not the kind of sounds produced naturally as part of human conversation. These tones include fax tones, various tones such as a busy tone, and the DTMF digits used in two-stage dialing, voice-mail retrieval, and other applications. G.711 can generally handle these tones and beeps quite well, because that coder samples the incoming signal without considering the origin of the signal. Because G.711 is the coder used in traditional telephony, such tones present no problem for circuit-switched networks. The same tones cannot be faithfully reproduced by lower-bandwidth codecs such as G.723.1 or G.729, however. If such tones are used as input, the output from those codecs can be unintelligible.

Many VoIP systems use gateways to provide an interface between the IP network and the circuit-switched network. Incoming speech, either in analog form or coded according to G.711, is converted to packetized speech at a lower bandwidth by using an efficient coder. The speech then passes though the IP network to its destination. The destination might be a native VoIP device or another gateway, in which case the packetized speech needs to be converted back to G.711. In either case, if DTMF digits or other tones arrive at the ingress gateway, they need to be intercepted by the gateway and communicated to the far end in a different manner. They should not be encoded in the same way as speech.

Assuming that a gateway has the capability to detect tones of different types, there are two ways to convey those tones to the destination. The first is to use an external signaling system that is separate from the media stream itself. While various signaling protocols exist (as discussed in Chapters 4, 5, and 6), they are not optimized for transmitting tones under any and all circumstances. For example, collecting DTMF digits at the start of a call and conveying them in a signaling system as a called-party number is only one part of the process. That procedure is relatively straightforward if the VoIP network is expecting to receive digits at the start of a call and if it understands that such digits represent a dialed number. DTMF digits in the middle of a call, however, such as those that are produced when someone is trying to retrieve voice mail, are more difficult to handle. The VoIP network itself is not involved in the interpretation of the real meaning of such digits.

The other alternative is to encode the tones differently from the speech but send them along the same media path. We can perform this action in two ways. First, if a gateway can recognize a given tone, then it can send one or more RTP packets that simply provide a name or other identifier for the tone, plus an indication of how long the tone is applied. This action involves using an RTP packet to signal the occurrence of a tone, rather than carrying the tone itself. Assuming that the far end understands the identifier, it can reproduce the actual tone. The second approach is to use a dynamic RTP profile, where RTP packets carry information regarding the frequency, volume, and duration of the tone. That approach means that the ingress gateway does not need to recognize a particular tone; rather, it only needs to detect and measure the characteristics of a tone and then send a packet describing what was measured.

RFC 2198 describes an RTP payload format for redundant audio data in which speech quality can be maintained in the presence of packet loss by sending redundant codings for the same speech signal. Having two methods of communicating the same information regarding a tone means that we can use one of the methods as a backup for the other, in accordance with RFC 2198. Thus, the entity (e.g., gateway) that detects a tone can send both types of RTP payload in a given RTP packet. The receiver can then select from the two representations of the tone.

An Internet draft titled "RTP Payload Format for DTMF Digits, Telephony Tones, and Telephony Signals" was prepared to address the issue of carrying tones within the RTP stream. This draft addresses both of the two methods just described. To support the sending of RTP packets with identifiers for the individual tones, the draft specifies a large number of tones and events (such as DTMF digits, a busy tone, a congestion tone, a ringing tone, etc.) and provides a number for each.

The payload format for these named events appears in Figure 3-10. The event field contains a number indicating the event in question. The R bit is reserved. The E bit indicates the end of the tone. A sender should delay setting this bit until the tone has stopped. The volume field indicates the power level of the tone from 0dBm0 to –63dBm0. The duration indicates (in timestamp units) how long the tone has lasted so far. Therefore, the tone event starts at the time indicated in the RTP timestamp, and the duration field indicates how long the tone lasted. The size of this field can accommodate a duration of up to eight seconds, assuming a sampling rate of 8,000Hz.

NOTE: *A gateway that detects a tone should send an event packet as soon as the tone is recognized and every 50ms thereafter while the tone lasts. The gateway should send at least three event packets—even if the tone is extremely brief—just to make sure that the far end has been informed of the event occurrence, even in the case of packet loss. When several packets are sent for the same tone, the RTP timestamp of each packet should be the same—indicating the instant that the tone was detected, with the duration field increased in each packet.*

Figure 3-11 shows the payload format to be used for the second method of representing a tone—i.e., when sending a description of the tone characteristics. The modulation field indicates the modulation in Hz, up to a

Figure 3-10 RTP payload format for named events

00	01	02	03	04	05	06	07	08	09	10	11	12	13	14	15	16	17	18	19	20	21	22	23	24	25	26	27	28	29	30	31
event								E	R	volume						duration															

Figure 3-11 RTP payload format for tones

00	01	02	03	04	05	06	07	08	09	10	11	12	13	14	15	16	17	18	19	20	21	22	23	24	25	26	27	28	29	30	31
modulation									T	volume						duration															
R	R	R	R	frequency												R	R	R	R	frequency											
R	R	R	R	frequency												R	R	R	R	frequency											
⋮																															
R	R	R	R	frequency												R	R	R	R	frequency											

maximum of 511Hz. If no modulation exists, this field is zero. The T bit indicates whether the modulation frequency is to be divided by three. Some tones are in use where the modulation frequency is not a whole number of Hz. For example, there are cases where we use modulation at 16 2/3 Hz. The volume measures the volume of the tone from 0dBm0 to –63dBm0. The duration is the duration in RTP timestamp units. The frequency indicates the frequency, in Hz, of one component of the tone. This field support values up to 4095Hz, which exceeds the frequency range of telephone systems. Note that a given tone can have multiple frequency components. For example, MF (as used in R1-MF signaling) stands for multi-frequency.

4

H.323

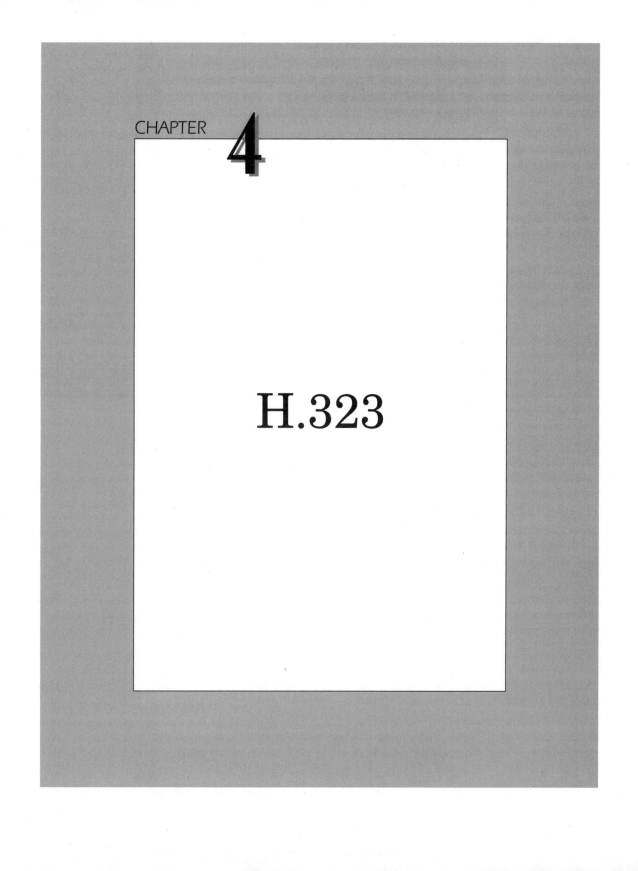

Introduction

The chapters thus far addressed the issues related to *Internet Protocol* (IP) networks in general and the mechanisms for transporting digitally encoded voice across those networks. Chapter 2 in particular, "Transporting Voice by Using IP," described how voice is carried in RTP packets between session participants. What we did not address, however, is the setup and tear-down of those voice sessions. We assume that session participants know of each other's existence and that media sessions are somehow created such that they can exchange voice by using RTP packets. So how are those sessions created and ended? How does one party indicate to another party a desire to set up a call, and how does the second party indicate a willingness to accept the call? The answer is signaling.

In traditional telephony networks, specific signaling protocols exist that are invoked before and during a call, in order to communicate a desire to set up a call, to monitor call progress, and to gracefully bring a call to an end. Perhaps the best example is the *Integrated Services Digital Network User Part* (ISUP), a component of the *Signaling System 7* (SS7) signaling suite. In *Voice over IP* (VoIP) systems, signaling protocols are also necessary for exactly the same reasons.

The first VoIP systems used proprietary signaling protocols. The immediate drawback was that two users could communicate only if they both used systems from the same vendor. This lack of interoperability between systems from different vendors was a major inconvenience and impeded the early adoption of VoIP. In response to this problem, the *International Telecommunications Union* (ITU)-T recommendation H.323 originated, which served as a standardized signaling protocol for VoIP. H.323 is the most widely deployed standard in VoIP networks today and is the focus of this chapter.

The first version of H.323 appeared in 1996 and bore the title, "Visual Telephone Systems And Equipment For Local Area Networks Which Provide A Non-Guaranteed Quality Of Service" (quite a mouthful). Its scope was multimedia communications over *Local-Area Networks* (LANs). As such, many found H.323 to be lacking the functions needed for supporting VoIP in a broader environment. Consequently, engineers revised H.323, and in 1998 they released H.323 version 2. This recommendation had a much friendlier title: "Packet-based Multimedia Communications Systems." This version of H.323 has received more support than its predecessor, particularly among those network operators and equipment vendors who have a background in more traditional telephony. H.323 version 2 is widely implemented in VoIP solutions, and in many ways, this version is the standard for VoIP systems today. H.323 version 2 is the focus of this chapter. Therefore, when we make any references to the H.323 recommendation, we are implying version 2 (unless otherwise stated).

The H.323 Architecture

H.323 is one of those ITU recommendations that specifies an overall architecture and methodology and that incorporates several other recommendations. You should read this recommendation in conjunction with other recommendations, and equally, if you read other recommendations, you should also read H.323. Among the other most important recommendations are H.225.0 and H.245, although many others exist.

We illustrate the scope of H.323 in Figure 4-1. The architecture involves H.323 terminals, gateways, gatekeepers, and *Multipoint Controller Units* (MCUs). The overall objective of H.323 is to enable the exchange of media streams between H.323 endpoints, where an H.323 endpoint is an H.323 terminal, a gateway, or an MCU.

Figure 4-1

Scope of H.323 and interoperability of H.323 terminals

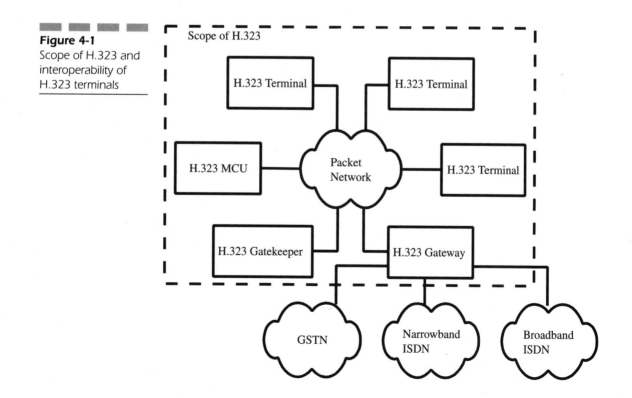

An H.323 terminal is an endpoint that offers real-time communication with other H.323 endpoints. Typically, this terminal is an end-user communications device that supports at least one audio codec and might optionally support other audio codecs and/or video codecs.

A gateway is an H.323 endpoint that provides translation services between the H.323 network and another type of network, such as an ISDN network or the regular phone network—which in ITU parlance is known as the *General-Switched Telephone Network* (GSTN). One side of the gateway supports H.323 signaling and terminates packet media according to the requirements of H.323. The other side of the gateway interfaces with a circuit-switched network and supports the transmission characteristics and signaling protocols of the circuit-switched network. On the H.323 side, the gateway has the characteristics of an H.323 terminal. On the circuit-switched side, the gateway has the characteristics of a node in the circuit-switched network. Translation between the signaling protocols and media formats of one side and those of the other side is performed internally within the gateway. The translation is totally transparent to other nodes in the circuit-switched network and in the H.323 network. Gateways can also serve as conduits for communications between H.323 terminals that are not on the same network, where the communication between the terminals needs to pass via an external network (such as the *Public-Switched Telephone Network* (PSTN)).

A gatekeeper is an optional entity within an H.323 network. When present, the gatekeeper controls a number of H.323 terminals, gateways, and *Multipoint Controllers* (MCs). By control, we mean that the gatekeeper authorizes network access from one or more endpoints and can permit or deny any given call from an endpoint within its control. The gatekeeper might offer bandwidth-control services, which can help to ensure service quality if used in conjunction with bandwidth and/or resource-management techniques. A gatekeeper also offers address-translation services, enabling the use of aliases within the network.

The set of terminals, gateways, and MCs that a single gatekeeper controls is known as a zone. Figure 4-2 shows a representation of such a zone. A zone can span multiple networks or subnetworks, and all entities within a zone do not have to be contiguous.

An MC is an H.323 endpoint that manages multipoint conferences between three or more terminals and/or gateways. For such conferences, the MC denotes the media that can be shared between entities by transmitting a capability set to the various participants. Also, the MC can change the capability set in the event that other endpoints join the conference or if existing endpoints leave the conference. An MC can reside within a separate

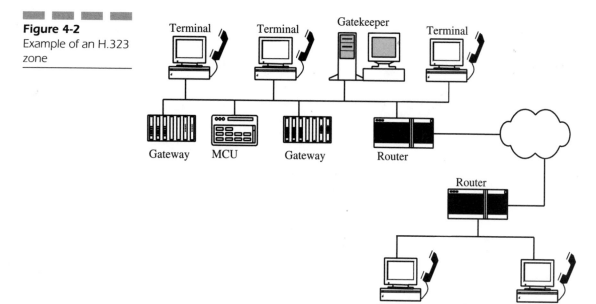

Figure 4-2

Example of an H.323 zone

MCU or might be incorporated within the same platform as a gateway, gatekeeper, or H.323 terminal.

For every MC, at least one *Multipoint Processor* (MP) exists that operates under the control of the MC. The MP processes the actual media streams, creating N output media streams from M input streams. The MP performs this action by switching, mixing, or a combination of the two. The control protocol between the MC and MP is not standardized.

MCs can support two types of multipoint conferences: centralized and decentralized. These two arrangements appear in Figure 4-3. In a centralized configuration, every endpoint in the conference communicates with the MC in a sort of hub-and-spoke arrangement. In a decentralized configuration, each endpoint in the conference exchanges control signaling with the MC in a point-to-point manner but might share media with the other conference participants through multicast.

Overview of H.323 Signaling

Figure 4-4 shows the H.323 protocol stack. Upon examination, we find a number of protocols already discussed, such as RTP and RTCP. We also see

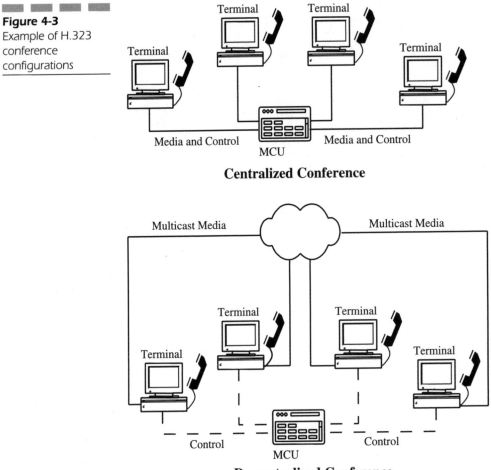

Figure 4-3
Example of H.323
conference
configurations

reliable and unreliable transport protocols. In an IP network, these terms refer to the *Transport Control Protocol* (TCP) and *User Datagram Protocol* (UDP), respectively. Clearly from the figure, the exchange of media is performed by using RTP over UDP, and of course, wherever RTP exists, RTCP is also present. In Figure 4-4, we also find two protocols that we have not yet discussed: H.225.0 and H.245. These two protocols define the actual messages that are exchanged between H.323 endpoints. They are generic protocols in that they could be used in any of a number of network architectures. When it comes to the H.323 network architecture, the manner in

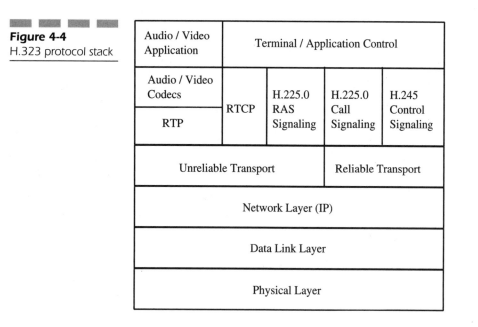

Figure 4-4
H.323 protocol stack

which the H.225.0 and H.245 protocols are applied is specified by recommendation H.323.

Overview of H.323 Protocols

As mentioned previously, the actual signaling messages exchanged between H.323 entities are specified in ITU recommendations H.225.0 and H.245.

H.225.0 is a two-part protocol. One part is effectively a variant of ITU-T recommendation Q.931, the ISDN layer 3 specification, and should be quite familiar to those who have a knowledge of ISDN. This part is used for the establishment and tear-down of connections between H.323 endpoints. This type of signaling is known as call signaling or Q.931 signaling. The other part of H.225.0 is known as *Registration, Admission, and Status* (RAS) signaling. This signaling is used between endpoints and gatekeepers and enables a gatekeeper to manage the endpoints within its zone. For example, an endpoint uses RAS signaling to register with a gatekeeper, and a gatekeeper uses this signaling to permit or deny an endpoint access to network resources.

H.245 is a control protocol used between two or more endpoints. The main purpose of H.245 is to manage the media streams between H.323

session participants. To that end, H.245 includes functions such as ensuring that the media to be sent by one entity is limited to the set of media that can be received and understood by another. H.245 operates through the establishment of one or more logical channels between endpoints. These logical channels carry the media streams between the participants and have a number of properties, such as media type, bit rate, etc.

All three signaling protocols—RAS, Q.931, and H.245—can be used to establish a call, perform maintenance on a call, and tear down a call. The various messages can be interleaved. For example, consider an endpoint that wishes to establish a call to another endpoint. First, the endpoint might use RAS signaling to obtain permission from a gatekeeper. The endpoint might then use Q.931 signaling to establish communication with the other endpoint and to set up the call. Finally, the endpoint might use H.245 control signaling to negotiate media parameters with the other endpoint and to set up the media transfer.

H.323 messages are sent over various different types of channels, depending on the message (and, in some cases, the context). For example, RAS messages are sent over the RAS channel; call-signaling messages are sent over the call-signaling channel; H.245 control messages are sent over the H.245 control channel; and the actual media streams are sent over one or more logical channels. While this process might seem to involve a lot of different channels, you should note that these channels are not necessarily related to particular physical interfaces or hardware. Instead, a channel—when used in an IP environment—is simply a reference to a socket address (i.e., IP address and port number). For example, if a given endpoint uses a particular IP address and port number to receive RAS messages, then any messages that arrive at that IP address and port are said to arrive on the endpoint's RAS channel.

Protocol Syntax As we move forward in this chapter, we will describe the main points of each of the protocols. We will provide a number of illustrations and examples in order to explain how the different protocols fit together. We will not, however, describe the detailed syntax of the protocols in any detail—and for good reason.

As is the case for many ITU-T recommendations, H.323 signaling is specified by using *Abstract Syntax Notation 1* (ASN.1). While this syntax is eminently suitable for interpretation by software tools and is in fact a formalized language, this syntax is certainly not overly friendly to the human reader. Explaining the syntax of the protocols in detail would require a great deal of ASN.1 syntax in this book and a tutorial on the interpretation of the syntax. That material would most likely clutter the text and detract from the main objective, which is to explain the functionality offered and how the protocols support that functionality. Consequently, little information about ASN.1 syntax is included here; instead, we give a textual description.

H.323 Addressing

Every entity in the H.323 network has a network address that uniquely identifies that particular entity. In an IP environment, the network address is an IP address. If *Domain Name Service* (DNS) is available, this IP address might be specified in the form of a *Uniform Resource Locator* (URL), according to RFC 822. For example, the URL `ras://GK1@somedomain` would be a valid URL for a gatekeeper, because RAS is the protocol supported by a gatekeeper. As with any URL, this URL might have a port number appended. In the case of RAS, a default port number exists (1719) that should be used if a port number is not specified.

For convenience of identification, entities such as terminals, gateways, and MCUs should have a domain name that is common with their controlling gatekeeper.

NOTE: *While applying URLs to endpoints and gatekeepers is a convenient means of identification, the actual IP address is passed in messages between H.323 entities.*

For each network address, an H.323 entity might have one or more *Transport Service Access Point* (TSAP) identifiers. In generic terms, a TSAP identifier is an identifier for a particular logical channel at a given entity. In IP terms, the TSAP identifier is the same as a socket address (i.e., an IP address and port number).

In general, the port numbers to be used for signaling transactions or media exchanges are assigned dynamically. Important exceptions are the Gatekeeper UDP Discovery Port with value 1718, the Gatekeeper UDP Registration and Status Port with value 1719, and the Call-Signaling TCP port with value 1720. These port numbers are registered with IANA. The first port is used when an endpoint wishes to determine the gatekeeper with which it should be registered. The second port is used for RAS signaling to a gatekeeper. The third port is used when sending call signaling messages (i.e., Q.931).

In addition to network addresses and TSAP identifiers, H.323 also enables terminals and gateways to have one or more aliases. Given that messages must be addressed with real IP addresses, the capability for nodes to have aliases also means that there must be a means to translate between an alias and an IP address. This translation is one of the functions of a gatekeeper and is supported by RAS signaling.

H.323 is flexible when it comes to the allocation of alias addresses. They can take any number of forms, and a given endpoint might have multiple aliases. The only real restriction is that a given alias must be unique within a zone—i.e., within the set of nodes controlled by a given gatekeeper (which makes sense). One useful application of aliases is the use of E.164 numbers as aliases. For example, a given gateway that is connected to a regular *Private-Branch Exchange* (PBX) over a PRI interface could be allocated a set of E.164 numbers as aliases. These numbers could correspond to the telephone numbers that are reachable at the PBX. If another H.323 endpoint wished to call a number on the PBX, the use of the alias makes this procedure much easier. Only the gatekeeper and the gateway need to know the relationship between the alias and the gateway address. The calling endpoint and other entities avail of the gatekeeper to take care of the necessary translation.

Codecs

In Figure 4-4, we saw a reference to audio and video codecs. Support of video is optional. Where video is supported, then an H.323 endpoint must (at a minimum) support video according to H.261 *Quarter Common Intermediate Format* (QCIF).

Support of audio is mandatory, and H.323 mandates that the G.711 codec must be supported (both A-law and mu-law). Given its relatively high bandwidth requirement, G.711 is certainly not considered the first choice for audio codecs. G.711 is specified as mandatory, however, so that there is at least one audio codec that all H.323 endpoints support. In reality, most endpoints will support a range of more efficient codecs. After all, if a system implements only G.711, then the potential bandwidth efficiencies of VoIP are negated—and some of the reasons for implementing VoIP in the first place are lost.

RAS Signaling

RAS signaling is used between a gatekeeper and the endpoints that it controls. This protocol is the signaling protocol through which a gatekeeper controls the endpoints within its zone. Note that a gatekeeper is an optional entity in H.323. Consequently, RAS signaling is also optional. If an endpoint wishes to use the services of a gatekeeper, then it must implement RAS. If the endpoint chooses not to use the services of a gatekeeper, then the functions that a gatekeeper normally handles need to be performed within the endpoint itself.

RAS signaling is defined in H.225.0 and supports the following functions:

- *Gatekeeper Discovery* Enables an endpoint to determine which Gatekeeper is available to control it.

- *Registration* Enables an endpoint to register with a particular gatekeeper and hence join the zone of that gatekeeper.

- *Unregistration* Enables an endpoint to leave the control of a gatekeeper or enables a gatekeeper to cancel the existing registration of an endpoint, thereby forcibly removing the endpoint from the zone.

- *Admission* Used by an endpoint to request access to the network for the purpose of participating in a session. A request for admission specifies the bandwidth to be used by the endpoint, and the gatekeeper can choose to accept or deny the request based on the bandwidth requested.

- *Bandwidth Change* Used by an endpoint to request the gatekeeper to allocate extra bandwidth to the endpoint.

- *Endpoint Location* A function in which the gatekeeper translates an alias to a network address. An endpoint will invoke this function when it wishes to communicate with a particular endpoint for which it only has an alias identifier. The gatekeeper will respond with a network address to be used in order to contact the endpoint in question.

- *Disengage* Used by an endpoint to inform a gatekeeper that it is disconnecting from a particular call. Disengage can also be used by a gatekeeper to endpoint to force the endpoint to disconnect from a call.

- *Status* Used between the gatekeeper and endpoint to inform the gatekeeper about certain call-related data, such as current bandwidth usage.

- *Resource Availability* Used from a gateway to a gatekeeper in order to inform the gatekeeper of the gateway's currently available capacity, such as the amount of bandwidth available. A gateway can also use Resource Availability to inform a gatekeeper that the gateway has run out of (or is about to run out of) capacity.

■ *Non-Standard*   A mechanism by which proprietary information can be passed between an endpoint and a gatekeeper. Of course, the message contents and functions to be invoked are not defined in H.225.0.

Table 4-1 provides a list of the various RAS messages that are used to support these functions and gives a brief description of the purpose of each message. In subsequent descriptions, we provide further details regarding some of the more common messages.

The next sections of this chapter describe the RAS messages and the associated functions in a little more detail.

Gatekeeper Discovery

In a network that has one or more gatekeepers, an endpoint should register with one of the gatekeepers. In order to perform this registration, the endpoint first needs to find a suitably accommodating gatekeeper—one that is willing to take control of the endpoint. Of course, the endpoint might be configured in advance with the address of the gatekeeper that should be used. In that case, there is no discovery process per se. Rather, the endpoint simply registers with the gatekeeper in question. Although such an approach might lead to a quicker registration, this technique lacks flexibility. In a given network, several gatekeepers might exist in a load-sharing mode, or a backup gatekeeper might need to assume the role of a failed gatekeeper. A static endpoint-gatekeeper relationship is not best suited for such scenarios. Therefore, an automatic gatekeeper discovery process is available.

The gatekeeper discovery process is available for an endpoint to determine its gatekeeper in the event that it does not already have that information. In order to discover which gatekeeper is to exercise control over a given endpoint, the endpoint sends a *Gatekeeper-Request* (GRQ) message. This message can be sent to a number of known addresses, or it can be sent to the gatekeeper's discovery multicast address and port 224.0.1.41:1718.

The GRQ message contains a number of parameters, one of which is a gatekeeper identifier. If this parameter is empty, the GRQ indicates to those gatekeepers that receive the message that the endpoint is soliciting gatekeepers and is willing to accept any available gatekeeper. In other words, the endpoint is asking, "Will someone be my gatekeeper?" In the case that the parameter contains an identifier, the endpoint is asking a specific gatekeeper to be the endpoint's gatekeeper.

One or more of the gatekeepers might respond with a *GatekeeperConfirm* (GCF) message, indicating that the gatekeeper is willing to control

	Message	Function
Table 4-1	*GatekeeperRequest* (GRQ)	Used by an endpoint when trying to discover its gatekeeper
RAS messages	*GatekeeperConfirm* (GCF)	Used by a gatekeeper to indicate that it will be the gatekeeper for a given endpoint
	GatekeeperReject (GRJ)	Used by a gatekeeper to indicate that it will not be the gatekeeper for a given endpoint
	RegistrationRequest (RRQ)	Used by an endpoint to register with a gatekeeper
	RegistrationConfirm (RCF)	A positive response from gatekeeper to endpoint, indicating a successful registration
	RegistrationReject (RRJ)	A negative response to a RegistrationRequest
	UnregistrationRequest (URQ)	Used either by a gatekeeper or an endpoint to cancel an existing registration
	UnregistrationConfirm (UCF)	Used by a gatekeeper or an endpoint to confirm cancellation of a registration
	UnregistrationReject (URJ)	A negative response to an UnregistrationRequest
	AdmissionRequest (ARQ)	Sent from an endpoint to a gatekeeper to request permission to participate in a call
	AdmissionConfirm (ACF)	Used by a gatekeeper to grant permission to a endpoint in order to participate in a call
	AdmissionReject (ARJ)	Used by a gatekeeper to deny an endpoint permission to participate in a call
	BandwidthRequest (BRQ)	Sent from an endpoint or a gatekeeper to request a change in allocated bandwidth
	BandwidthConfirm (BCF)	A positive response to a BandwidthRequest
	BandwidthReject (BRJ)	Used by a gatekeeper or endpoint to deny a change in bandwidth; used only by an endpoint if the new bandwidth cannot be supported
	InfoRequest (IRQ)	Sent by a gatekeeper to an endpoint in order to request status information
	InfoRequestResponse (IRR)	Sent from an endpoint to a gatekeeper in order to provide status information; can be sent on demand or autonomously
	InfoRequestAck (IACK)	Sent by a gatekeeper as confirmation of the receipt of an IRR
	InfoRequestNak (INAK)	Sent by a gatekeeper upon receiving an IRR in an error situation, such as from an unregistered endpoint

(continues)

Table 4-1

Continued

Message	Function
DisengageRequest (DRQ)	Sent by an endpoint or gatekeeper to request disconnection of a call at an endpoint
DisengageConfirm (DCF)	A positive response to a DRQ
DisengageReject (DRJ)	A negative response to a DRQ; e.g., from an unregistered endpoint
LocationRequest (LRQ)	Sent to a gatekeeper to request translation from an alias to a network address
LocationConfirm (LCF)	A response to an LRQ, including the required address
LocationReject (LRJ)	A response to an LRQ when translation was not successful
Non-Standard Message (NSM)	Vendor-specific
UnknownMessageResponse (XRS)	A response to a message that is not recognized
RequestInProgress (RIP)	Sent by an endpoint or gatekeeper as an interim response if a request is taking a long time to process
ResourceAvailableIndicate (RAI)	Sent by a gateway to a gatekeeper in order to inform the gatekeeper of the gateway's current capacity
ResourceAvailableConfirm (RAC)	Sent by a gatekeeper as an acknowledgement of an RAI

the endpoint. Alternatively, a gatekeeper might respond with a *Gatekeeper-Reject* (GRJ) message, indicating that the gatekeeper is not willing to assume control of that endpoint. In the case that a gatekeeper sends a GRJ message, it will include a reason for the rejection (such as a lack of resources at the gatekeeper's end).

This process is illustrated in Figure 4-5. In this scenario, the terminal that wishes to determine its GK sends just one GRQ message but sends it to the gatekeeper's discovery multicast address. Therefore, more than one gatekeeper can receive the message. In Figure 4-5, gatekeeper 1 sends a GRJ message, whereas gatekeeper 2 sends a GCF message. The terminal will now register with gatekeeper 2.

Optionally, a given gatekeeper might send a GCF message indicating one or more other gatekeepers to try. This process is indicated within the GCF message by the presence of the optional parameter `AlternateGatekeeper`.

Figure 4-5
Gatekeeper discovery

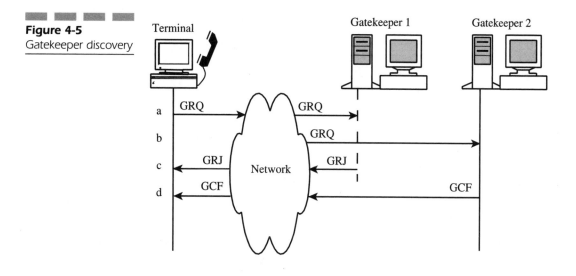

In such an event, the GCF message does *not* mean that the gatekeeper issuing the GCF message is willing to take control of the endpoint. Rather, the GCF message is somewhere between a positive response and a complete rejection and is equivalent to saying something like, "I can't help you, but try the gatekeeper next door." This response is useful for load sharing or redundancy schemes.

Only one gatekeeper at a time can control an endpoint. If an endpoint receives multiple GCF messages from gatekeepers that are willing to accept control, then the choice of gatekeeper is left to the endpoint. If on the other extreme an endpoint receives no positive response to the GRQ message with a timeout period, then the endpoint can retry the GRQ but should not issue a GRQ within five seconds of the previous request.

Endpoint Registration and Registration Cancellation

Gatekeeper discovery is simply a means for an endpoint to determine which gatekeeper is available to assume control of the endpoint. The fact that a given gatekeeper has responded to a GRQ with a GCF does not mean that the endpoint is now under control of that gatekeeper. The process by which an endpoint becomes controlled by a gatekeeper and joins the zone of that gatekeeper is known as registration.

Registration begins with the issuance of a *RegistrationRequest* (RRQ) message from an endpoint to a gatekeeper. The message is directed to a gatekeeper at an address that is configured within the endpoint or that is determined by the endpoint through the gatekeeper discovery procedure. The port number to be used for the message is 1719—the well-known RAS signaling port of a gatekeeper.

In the RRQ, the endpoint includes an address to be used by the endpoint for RAS messages and an address to be used for call-signaling messages. The RRQ might include an alias—a name by which the endpoint wishes to be called. The RRQ might also include one or more alternative addresses that are available for use as backup or redundant addresses. In such cases, these addresses might each have their own aliases.

The gatekeeper can choose to reject the registration for any number of reasons and does so by responding with a *RegistrationReject* (RRJ) message. One reason, for example, would be if the endpoint wishes to use an alias that is already in use within the zone. If all is well, however, and the gatekeeper is willing to accept the registration, then it responds with a *RegistrationConfirm* (RCF) message. If the original RRQ did not specify an alias for the endpoint, then the gatekeeper can assign one—in which case the assigned alias is returned as a parameter in the RCF message.

Registrations can optionally have a limited lifetime. In an RRQ, an endpoint can include a `timeToLive` parameter, indicating the requested duration for the registration in seconds to a maximum value of 4,294,967,295 (hexadecimal FFFFFFFF), which translates to about 136 years. In the RCF, the gatekeeper might choose a lower `timeToLive` value than received in the RRQ. If after some time the registration period is about to expire, then the endpoint might send another RRQ that includes the optional parameter `keepAlive`—a request to reset the time-to-live timer and to extend the registration.

Once registered with a gatekeeper, the endpoint can choose to cancel its registration at a later time. The endpoint does so by sending an *UnregistrationRequest* (URQ) message. Normally, a gatekeeper will respond with a positive confirmation by sending an *UnregistrationConfirm* (UCF) message. The gatekeeper might, however, send a rejection if the endpoint was not registered with that gatekeeper in the first place. The rejection takes the form of an *UnregistrationReject* (URJ) message.

The gatekeeper might also choose to cancel the registration of a particular endpoint, in which case the gatekeeper will send a URQ message to the endpoint. One reason might be because the registration time to live has expired. Upon receipt of a URQ from a gatekeeper, the endpoint should respond with a UCF. Figure 4-6 depicts an example of registration and registration cancellation.

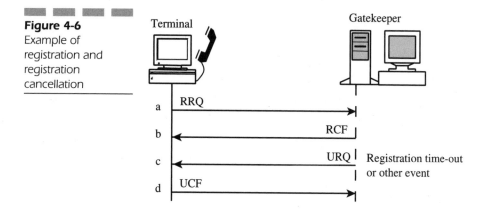

Figure 4-6
Example of registration and registration cancellation

NOTE: *This figure is only an example. A URQ message can be issued by an endpoint as well as by a gatekeeper.*

Endpoint Location

Endpoint location is a service that enables an endpoint or gatekeeper to request a real address when only an alias is available. In other words, endpoint location is a translation service.

An endpoint that wishes to obtain contact information for a given alias can send a *LocationRequest* (LRQ) message to a gatekeeper. The message can be sent to a particular gatekeeper or can be multicast to the gatekeeper discovery multicast address 224.0.1.41. The LRQ contains the aliases for which address information is required. In the multicast case, this situation is equivalent to asking the question, "Does anyone know who has alias XXX?"

Not only can an endpoint send the LRQ, but a gatekeeper can also send an LRQ to another gatekeeper to determine the address of an endpoint. This procedure would happen if the call signaling from an endpoint is to pass through the calling endpoint's gatekeeper. A gatekeeper that receives an LRQ from an endpoint, however, will not propagate the LRQ to another gatekeeper. A gatekeeper will send an LRQ to another gatekeeper only if the need for address translation originates at the first gatekeeper.

A *LocationConfirm* (LCF) message indicates a positive response to the LRQ. This message should be sent from the gatekeeper at which the endpoint is registered. Among some other optional parameters, this message contains a call-signaling address and an RAS signaling address for the located endpoint.

If a gatekeeper receives an LRQ message and the endpoint is not registered at that gatekeeper, then the gatekeeper should respond with a *LocationReject* (LRJ) message if the original LRQ had been received by the gatekeeper on its RAS channel. If the gatekeeper receives an LRQ by multicast on the gatekeeper discovery multicast address, then the gatekeeper should not send an LRJ.

Admission

Admission is the process by which an endpoint requests permission from a gatekeeper to participate in a call. The endpoint does by so sending an *AdmissionRequest* (ARQ) message to the gatekeeper. The endpoint will indicate the type of call in question (e.g., two-party or multi-party), the endpoint's own identifier, a call identifier (a unique string), a call-reference value (an integer value also used in Q.931 messages for the same call), and information regarding the other party or parties who will participate in the call. The information regarding other parties to the call includes one or more aliases and/or signaling addresses.

One of the most important mandatory parameters in the ARQ is the bandwidth parameter. This parameter specifies the amount of bandwidth required in units of 100 bits/s. Note that the endpoint should request the total media-stream bandwidth that is needed (excluding overhead). Thus, if a two-party call is necessary (with each party sending voice at 64kbps), then the bandwidth required is 128kbps, and the value carried in the bandwidth parameter is 1,280. The purpose of the bandwidth parameter is to enable the gatekeeper to reserve resources for the call. Through the use of the optional parameter transportQOS, however, the endpoint can indicate that it is willing to undertake resource reservation, rather than having the gatekeeper do it.

The gatekeeper indicates successful admission by responding to the endpoint with an *AdmissionConfirm* (ACF) message. This message includes many of the same parameters that are included in the ARQ. The difference is that when a given parameter is used in the ARQ, it is simply a request from the endpoint—whereas a given parameter value in the ACF is a firm order from the gatekeeper. For example, the ACF includes the bandwidth parameter, which might be a lower value than requested in the ARQ, in which case the endpoint must stay within the bandwidth limitations that the gatekeeper imposes. Equally, while the endpoint might have indicated that it will perform its own resource reservation, the gatekeeper might decide that it will take that responsibility.

Another parameter of particular interest in both the ARQ and the ACF is the `callModel` parameter, which is optional in the ARQ and mandatory in the ACF. In the ARQ, this parameter indicates whether the endpoint wishes to send call signaling directly to the other party or whether it prefers for call signaling to be passed via the gatekeeper. In the ACF, this parameter represents the gatekeeper's decision as to whether call signaling is to pass via the gatekeeper. The first scenario is known as direct call signaling. In that case, the calling endpoint sends an ARQ to its gatekeeper. After the ACF is returned from the gatekeeper, the endpoint sends call signaling directly to the remote endpoint. In the case of gatekeeper-routed call signaling, the calling endpoint's gatekeeper returns a `callModel` parameter in the ACF, indicating that call signaling should be sent via the gatekeeper. The endpoint then sends the call signaling to the gatekeeper, and the gatekeeper forwards the call signaling to the remote endpoint.

Figure 4-7 shows an example of direct call signaling, and Figure 4-8 shows an example of gatekeeper-routed call signaling. Both diagrams assume that the two endpoints are connected to the same gatekeeper, which might or might not be the case. In the event that the two endpoints are connected to different gatekeepers, each gatekeeper determines independently whether it would like to be in the path of the call signaling. Therefore, in the case of two gatekeepers, the signaling might pass via none, one, or both gatekeepers.

Figure 4-7
Direct call signaling

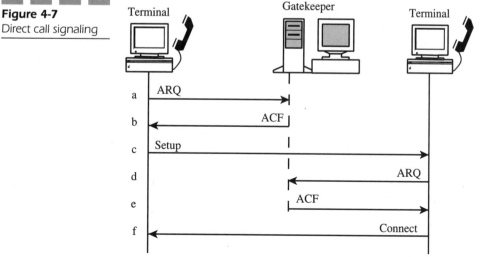

Figure 4-8
Gatekeeper-routed
call signaling

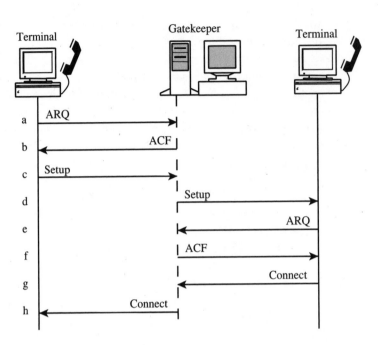

Figures 4-7 and 4-8 show Setup and Connect messages. These messages are part of call signaling (Q.931), as described later in this chapter. The Setup message is the first message in call establishment between endpoints, and the Connect message indicates that a call is accepted by the called endpoint. The message is sent when the call is answered.

Of course, a gatekeeper might decide to deny a particular admission request. In such an event, it responds to the ARQ with an *AdmissionReject* (ARJ) message, which includes a reason for the denial. Possible reasons include a lack of available bandwidth, an inability to translate a destination alias to a real address, or an endpoint not being registered.

Bandwidth Change

At any time during a call, an endpoint can request either an increase or decrease in allocated bandwidth. An endpoint itself can change the bit rate of any given channel without approval from the gatekeeper, provided that the total bandwidth does not exceed the limit specified by the gatekeeper. Consequently, an endpoint is not required to request a reduction in bandwidth, although it should do so as a means of informing the gatekeeper of a reduced bandwidth requirement—thereby enabling the gatekeeper to release certain resources for potential use in other calls. If a change is to occur in a call, however, and that change will lead to an aggregate bandwidth in excess of the limit specified by the gatekeeper, then the endpoint must first ask the gatekeeper for permission. The endpoint can increase the bandwidth only after receiving confirmation from the gatekeeper.

A change in bandwidth is requested through a *BandwidthRequest* (BRQ) message. The message indicates the new bandwidth requested and the call to which the bandwidth applies. The gatekeeper can either approve the new bandwidth with a *BandwidthConfirm* (BCF) message or deny the request with a *BandwithReject* (BRJ) message. If the request is denied, then the endpoint must live with whatever bandwidth has already been allocated—perhaps through the use of flow-control mechanisms.

Not only can an endpoint request a change in bandwidth, but the gatekeeper can also request an endpoint to change its bandwidth limit. For example, the gatekeeper can determine that network resource utilization is approaching capacity and can order an endpoint to reduce its bandwidth usage. If the gatekeeper requests a reduction in bandwidth, then the endpoint must comply with the request and must respond with a BCF message.

Note that changes in bandwidth are closely tied to H.245 signaling. Recall that H.245 specifies the signaling that occurs between endpoints that are involved in a call and relates to the actual media streams that the endpoints share. Also recall that H.245 defines the use of logical channels that carry the media between the endpoints. If there is to be a change in bandwidth, then there will be a direct impact to one or more logical channels. In fact, a reduction in bandwidth will require an existing logical channel to be closed and reopened with different parameter settings related to the new bandwidth limitation.

Figure 4-9 provides an example of the interaction between H.225.0 RAS signaling and H.245 control signaling for changes in bandwidth. In this example, an endpoint wishes to increase the bandwidth used on a call. The endpoint first uses the BRQ message to request permission from its gatekeeper. Assuming that the request is acceptable, the gatekeeper responds with a BCF message. The endpoint closes the existing logical channel to the remote endpoint and opens a new logical channel, which has a new bit rate. The remote endpoint sees that a higher bit rate is required, but before confirming the new bit rate, it must ask its gatekeeper for permission. Therefore, the endpoint also sends a BRQ and receives a BCF. Only upon receipt of the BCF from its gatekeeper does the remote endpoint confirm the opening of the new logical channel.

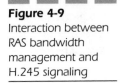

Figure 4-9
Interaction between
RAS bandwidth
management and
H.245 signaling

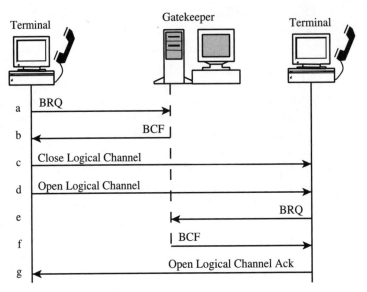

NOTE: *Figure 4-9 shows that the same gatekeeper controls the two endpoints. Different gatekeepers could just as easily control the two endpoints.*

H.245 signaling is described in greater detail later in this chapter.

Status

A gatekeeper should be informed of the status of an endpoint, simply to know whether the endpoint is still functioning and to obtain information regarding any call that the endpoint might currently have active.

Information to be provided by an endpoint to a gatekeeper is carried in an *InformationRequestResponse* (IRR) message. This message carries information about the endpoint itself, plus information about each call that the endpoint might currently have active. The per-call information includes items such as the call identifier, the call reference value, the call type (two-party or multi-party), the bandwidth in use, and RTP session information (CNAME, RTP address, RTCP address, etc.).

The gatekeeper can stimulate an endpoint to send an IRR in two ways. First, the gatekeeper can poll the endpoint by sending an *Information-*

Request (IRQ) message to the endpoint. The other way is for the gatekeeper to indicate within the ACF message that the endpoint should periodically send the information. This procedure is conducted by including the optional parameter `irrFrequency` in an ACF message. This parameter indicates the frequency (in seconds) at which an endpoint in a call should send an IRR message to the gatekeeper.

An IRR might or might not receive an acknowledgement. The gatekeeper and the endpoint determine whether the IRR receives an acknowledgement. First, the gatekeeper indicates whether or not it is willing to send acknowledgements to IRR messages and performs this action by setting the `BOOLEAN willRespondToIRR` parameter in the ACF or RCF message. If that parameter has been set, then an endpoint might indicate that a particular IRR should receive a response. The endpoint performs this action by setting the `BOOLEAN needResponse` parameter in the IRR.

The response to an IRR can be positive or negative. A positive response takes the form of an *InfoRequestAck* (IACK) message, and a negative response takes the form of an *InfoRequestNak* (INAK) message. The negative response should be returned only in cases in which a gatekeeper has received an IRR message in error, such as from an unregistered endpoint.

Disengage

At the end of a call, the various parties involved stop transmitting media to each other and terminate the session. At this stage, each endpoint should send a *DisengageRequest* (DRQ) message to its gatekeeper. The message contains, among other parameters, the call identifier, the call reference value, and a disengage reason. In the case that this call termination is normal, then the value of the reason will be "normalDrop."

A gatekeeper that receives the DRQ request normally responds with a *DisengageConfirm* (DCF) message. The gatekeeper might respond with a *DisengageReject* (DRJ) message under some unusual circumstances. Such scenarios could include an endpoint requesting to disengage when the endpoint is not registered with that gatekeeper or if the endpoint wishes to terminate a call that a different endpoint is handling.

Situations might exist in which a gatekeeper, rather than an endpoint, decides to terminate a call. In such a case, the gatekeeper sends a DRQ message to the endpoint. Upon receiving such a request, the endpoint must stop transmitting media and must bring the session to a close by using the appropriate H.245 control messages and Q.931 call-signaling messages. Once the session is terminated, the endpoint responds to the gatekeeper with a DCF message.

Resource Availability

H.225.0 defines two messages related to resource availability: the *ResourceAvailableIndicate* (RAI) message and the *ResourceAvailableConfirm* (RAC) message. The first message is sent from a gateway to a gatekeeper in order to inform the gatekeeper of the currently available call capacity and bandwidth for each protocol that the gateway supports. Of particular importance is a mandatory Boolean, almostOutofResources, parameter that, when set, indicates that the gateway is almost out of resources. The manufacturer chooses how close to capacity the gateway needs to be before this parameter is set. Equally, any action that is taken as a result of the parameter being set is also at the discretion of the system manufacturer. The gatekeeper acknowledges the RAI message with the RAC message.

Request in Progress

Finally, H.225.0 defines the *RequestInProgress* (RIP) message, which is a means of indicating that a given request might take longer than expected. H.225.0 specifies recommended timeout periods for various messages. If an entity cannot respond to a request within the applicable timeout period, then it should send a RIP message indicating the expected delay before it expects to send a response to the original request. If the entity can subsequently respond before the delay period expires, then it should do so.

If an entity expects a response to a RAS message and receives a RIP message instead, then it should wait for a response or wait for the RIP delay time to expire before resending the original message. Note that it is perfectly legal and sometimes necessary to retransmit a RAS message. After all, RAS messages are sent by using unreliable transport, as indicated in Figure 4-4. Therefore, the possibility exists that RAS messages can become lost.

Call Signaling

We already mentioned call signaling and that it is the signaling used between endpoints to enable the establishment and tear-down of calls. The messages used are Q.931 messages, as modified by recommendation H.225.0. At first reading, such a statement can be somewhat confusing. After all, Q.931 is the layer 3 signaling protocol for an ISDN user-network interface, and the various messages are defined in that recommendation. The fact is that H.225.0 takes advantage of the protocol defined in Q.931 and simply reuses the messages, with some modifications necessary for use in the overall H.323 architecture. H.225.0 also uses a Q.932 message. This technique is a pretty smart way of doing things, because it has avoided a great deal of protocol-development effort. H.225.0 does not use all of the messages defined in Q.931—just those that are necessary for the support of call-signaling functions in the H.323 architecture. The messages used are specified in Table 4-2.

H.225.0 specifies the modifications that are applicable to Q.931 messages when used for call signaling in an H.323 network. In general, relatively few changes are made to the body of the Q.931 messages. For example, H.225.0 does not define any exotic new information elements to be included in particular messages. On the other hand, H.225.0 does not take each Q.931 message as is, either. Instead, H.225.0 specifies a number of rules regarding the usage of information elements defined in Q.931. For example, H.225.0 does not enable the use of the Transit Network Selection information element, nor does it currently enable the use of high-layer compatibility and low-layer compatibility information elements. In fact, the changes specified by H.225.0 generally involve specifying that certain Q.931 mandatory information elements are either forbidden or optional when used in an H.323 network.

The big question, however, is how H.225.0's use of Q.931 messages enables the transfer of H.323-specific information. For example, a clear need exists to transport information regarding gatekeepers and H.245 addresses to be used for logical channels. These types of items are foreign to the usual ISDN environment of Q.931 signaling, which means that information must be added to the Q.931 messages. This task is performed through a rather clever use of the User-to-User information element. This information element in Q.931 ISDN messages is used to pass information transparently from one user to another by using the D channel. The network does not interpret the information. In H.225.0, this element is used to convey all of the extra information needed in an H.323 environment. Such information includes parameters such as a mandatory protocol identifier, endpoint aliases, H.245 addresses, etc. The ASN.1 syntax specified in H.225.0 provides the exact syntax to be used in the User-to-User information element of the various messages.

The following sections of this chapter provide a brief overview of some of the most important call-signaling messages. For complete details, the user can refer to the Q.931 and H.225.0 recommendations.

Table 4-2

Call signaling messages

Message	Function	Comment
Alerting	Sent by called endpoint to indicate that the called user is being alerted	Must be supported
Call Proceeding	An optional interim response sent by the called endpoint or gatekeeper prior to the sending of the Connect message	Should be sent if a called endpoint uses a gatekeeper
Connect	An indication that the called user has accepted the call	Must be supported
Progress	An optional message sent by the called endpoint prior to the Connect message	Can be used by a called gateway in the case of PSTN interworking
Setup	An initial message used to begin call establishment	Must be supported
Setup Acknowledge	An optional response to the Setup message	Can be forwarded from a gateway in the case of PSTN inter-working
Release Complete	Used to bring a call to an end	The Q.931 Release message is not used.
User Information	An optional message that is used to send additional call-establishment information	Can be used in overlap signaling
Notify	An optional message that can be used to provide information for presentation to the user	Can be used by the calling or called endpoint
Status	Sent in response to a Status Inquiry message or in response to an unknown message	Optional message; can be sent in either direction
Status Inquiry	A message used to query the remote end as to the current call status from its perspective	Can be sent in conjunction with RAS Status procedures
Facility (Q.932)	Used to redirect a call or to invoke a supplementary service	Can be sent by either end at any time; is useful for conveying information when no other message should be sent

Setup

The Setup message is the first call-signaling message sent from one endpoint to another in order to establish a call. If an endpoint uses the services of a gatekeeper, this message will be sent after the endpoint has received an ACF message from the gatekeeper providing permission for the endpoint to establish the call. The message must contain the Q.931 Protocol Discriminator, a Call Reference, a Bearer Capability, and the User-to-User information element.

Although the Bearer capability information element is mandatory, the concept of a bearer as used in the circuit-switched world does not map well to an IP network. For example, no B-channel exists in IP, and the actual agreement between endpoints regarding the bandwidth requirement is made as part of H.245 signaling, where RTP information such as payload type is exchanged. Consequently, many of the fields in the bearer capability information element, as defined in Q.931, are not used in H.225.0. Of those fields that are used in H.225.0, many are used only when the call has originated from outside the H.323 network and has been received at a gateway, where the gateway performs a mapping from the signaling received to the appropriate H.225.0 messages.

A number of parameters are included within the mandatory User-to-User information element, including the call identifier, the call type, a conference identifier, and information about the originating endpoint. Among the optional parameters, we might find a source alias, destination alias, H.245 address for subsequent H.245 messages, and a destination call-signaling address. Note that these parameters might supplement information that is already contained within the body of the Q.931 message. For example, the optional Calling Party number information element in the body of the message will contain the E.164 number of the source.

Call Proceeding

This optional message can be sent by the recipient of a Setup message to indicate that the Setup message has been received and that call-establishment procedures are underway.

As is the case for all call-signaling messages, the Protocol Discriminator, Call Reference, and Message Type elements are mandatory. The only other mandatory information element is User-to-User. Within that information element, the mandatory piece of information, in addition to the protocol identifier, is the destination information, which indicates the type of endpoint and the call identifier. Among the optional parameters is the H.245 address of the called end, which indicates where it would like H.245 messages to be sent.

Alerting

The called endpoint sends this message to indicate that the called user is being alerted. Besides the Protocol Discriminator, Call Reference, and Message Type, only the User-to-User information element is mandatory. The optional Signal information element might be returned if the called endpoint wishes to indicate a specific alerting tone to the calling party. The mandatory User-to-User information element in the Alerting message contains the same parameters as those that are defined for the Call Proceeding message.

Progress

The Progress message can be sent by a called gateway to indicate call progress, particularly in the case of interoperating with a circuit-switched network. The Cause information element, although optional, is used to convey information to supplement any in-band tones or announcements that might be provided. The User-to-User information element contains the same set of parameters as defined for the Call Proceeding and Alert messages.

Connect

The Connect message is sent from the called entity to the calling entity to indicate that the called party has accepted the call. While some of the messages from called party to calling party (such as Call Proceeding and Alerting) are optional, the Connect message must be sent if the call is to be completed. The User-to-User information element contains the same set of parameters as defined for the Call Proceeding, Progress, and Alert messages, with the addition of the Conference Identifier. This parameter is also used in a Setup message, and its use in the Connect message is to correlate this conference with that indicated in a Setup. Any H.245 address sent in a Connect message should match that sent in any earlier Call Proceeding, Alerting, or Progress message.

Release Complete

The Release Complete message is used to terminate a call. Unlike ISDN, Release Complete is not sent as a response to a Release message. In fact,

H.323 endpoints should never send a Release message. Instead, the Release Complete is all that is needed to bring the call to a close.

The optional Cause information element exists within the Release Complete message body. If this element is not used, then the User-to-User information element will contain a Release reason. On the other hand, if no Release reason is specified in the User-to-User information element, then the Cause information element should be included. One way or the other, the party who terminates the call should provide the other party with a reason for the termination.

Facility

The Facility message is actually defined in ITU-T recommendation Q.932. In H.225.0, this message is used in situations where a call should be redirected. This message might also be used to request certain supplementary services, as described in recommendation H.450.1. Within the User-to-User information element, one important item is the mandatory reason parameter, which provides extra information to the recipient of the Facility message. One important use is when a Setup message has been sent to an endpoint, and that endpoint's gatekeeper has indicated to the endpoint that it wishes to be involved in the call signaling. In such a case, the called endpoint would return a Facility message with the reason set to `routeCallToGatekeeper`. This action would cause the recipient of the Facility message to release the call and attempt to set up the call again via the called endpoint's gatekeeper.

Interaction between Call Signaling and H.245 Control Signaling

Before delving into some examples of call signaling, we should note a few points regarding the interplay between call signaling and H.245 control signaling. As mentioned previously, Q.931 call-signaling messages are used for the establishment and tear-down of calls between endpoints, and H.245 messages are used for the negotiation and establishment of media streams between call participants. Obviously, the two are closely tied together.

Using Q.931, call establishment begins with a Setup message, and the call is considered established upon the return of a Connect message from the called entity. In the interim, between the receipt of the Setup message and the sending of the Connect message, the called party might issue one or more progress-related messages (such as Call Proceeding or Alerting). Given that H.245 messaging must also take place, at which point during call signaling should it occur? In other words, which of the call-signaling messages should be used as a trigger to begin the exchange of H.245 messages?

The answer to these questions is equipment dependent. H.323 requires H.245 messages to be exchanged but does not mandate the exchange of these messages to occur at any specific point within the call-signaling process. Obviously, because the Setup message is the first call-signaling message to be sent, H.245 messages cannot be exchanged before the called endpoint has received a Setup message. The messages can be exchanged at any point after that, however. The called endpoint can initiate the exchange of H.245 messages upon the receipt of a Setup message, or the calling endpoint can initiate the exchange upon receipt of a Call Progress or Alerting message. If H.245 message exchange has not begun by the time a Connect message is sent, then the exchange should begin immediately after the Connect message is sent.

Call Scenarios

Now that we have described RAS signaling and call signaling, we will present a number of call scenarios to show how the different types of messages are used in the establishment and tear-down of calls. Granted, H.245 messaging has not yet been described, and H.245 messaging is a prerequisite for media exchange between the endpoints. We can assume in the following scenarios, however, that successful H.245 message exchange takes place immediately after a Connect message is sent.

Basic Call without Gatekeepers

A gatekeeper is an optional entity in H.323. In the absence of gatekeepers, call signaling occurs directly between endpoints. Figure 4-10 shows a typical call establishment and tear-down in the absence of gatekeepers. The call is initiated with a Setup message, "answered" with a Connect message, and released with a Release Complete message. Either party can send the Release Complete message. Note that no acknowledgement is necessary for the Release Complete message, because call signaling is sent by using reliable transport. The H.323 application does not have to worry about making sure that a given message reaches its destination. Lower layers, where retransmission can take place if necessary, handle this responsibility.

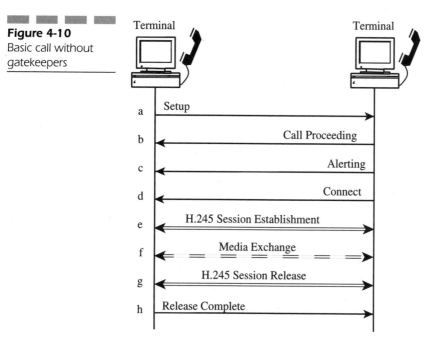

Figure 4-10
Basic call without
gatekeepers

Terminal

Terminal

a Setup

b Call Proceeding

c Alerting

d Connect

e H.245 Session Establishment

f Media Exchange

g H.245 Session Release

h Release Complete

A Basic Call with Gatekeepers and Direct-Endpoint Call Signaling

In the event that the endpoints are registered with a gatekeeper, prior to establishment of call signaling an endpoint must first gain permission from its gatekeeper. Furthermore, at the end of the call, an endpoint must notify the gatekeeper that the call has been disconnected.

When gatekeepers are used on the network, a given endpoint might send call signaling directly to the other endpoint, or it might send call signaling via its gatekeeper. Although an endpoint can indicate a preference, the gatekeeper chooses which path the call signaling will take.

Figure 4-11 shows a scenario in which different gatekeepers control two endpoints. Both gatekeepers decided that call signaling should be sent directly from endpoint to endpoint. We also assume that H.245 signaling goes from endpoint to endpoint. The originating endpoint first requests access permission from its gatekeeper. Only when permission is granted does the endpoint commence call signaling to the remote endpoint. Note that the remote endpoint immediately returns a Call Proceeding message prior to the endpoint asking its gatekeeper for permission to handle the call. By sending back the Call Proceeding message immediately, the endpoint signals

Figure 4-11
Direct-endpoint call
signaling

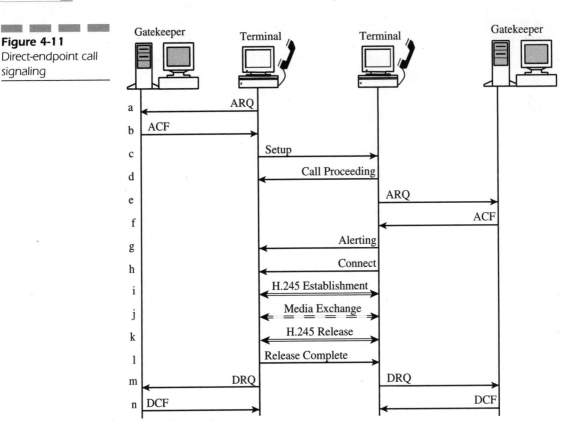

to the call originator that the Setup has been received successfully. Thus, the calling endpoint is not left in the dark while the called endpoint requests permission from its gatekeeper to handle the call.

In the example, we assume that H.245 message exchange begins immediately after the Connect message. These messages could optionally occur earlier.

A Basic Call with Gatekeeper/ Direct-Routed Call Signaling

A gatekeeper might choose for call signaling to be routed via the gatekeeper, rather than directly from endpoint to endpoint. In the event that the two endpoints are connected to different gatekeepers, one gatekeeper might choose to route the call signaling itself while the other gatekeeper

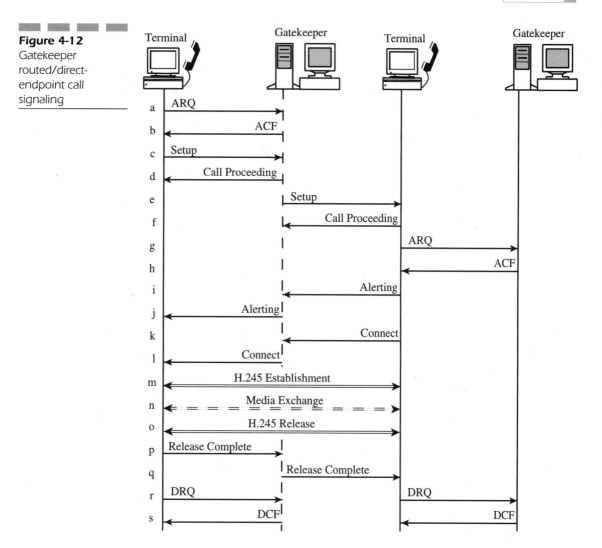

Figure 4-12
Gatekeeper routed/direct-endpoint call signaling

does not. Such a scenario appears in Figure 4-12, where the gatekeeper of the calling endpoint chooses to route the call signaling itself, and the gatekeeper of the called endpoint wishes not to be in the path of call-signaling messages.

In the first ACF message, the gatekeeper of the calling endpoint indicates that call signaling should be sent to the gatekeeper. The gatekeeper includes a call-signaling address of itself in the ACF. The endpoint sends a

Setup message to that address, and the gatekeeper responds immediately with Call Proceeding. The gatekeeper then sends a Setup message to the destination endpoint by using the call-signaling address of that endpoint.

If the called endpoint wishes to handle the call, it first responds with a Call Proceeding message and then sends an ARQ to its gatekeeper. In this case, the gatekeeper responds with an ACF indicating direct call signaling, as opposed to gatekeeper-routed call signaling.

The called endpoint alerts the user and sends an Alert message to the gatekeeper of the calling endpoint. That gatekeeper forwards the Alerting message to the calling endpoint. Once the called party accepts the call, the called endpoint sends a Connect message that follows the same path as the Alerting message.

At this point in the example, H.245 messages are exchanged in order to set up the media streams between the endpoints. Note that H.245 message exchange can take place earlier, depending on the options implemented in the two endpoints.

In the example, the H.245 messages are routed directly from endpoint to endpoint. Like call-signaling messages, however, they could be sent via one or both gatekeepers, depending on whether each gatekeeper wishes to be in the path of the H.245 messages. You must remember that the actual RTP streams will pass directly from endpoint to endpoint, regardless of whether a gatekeeper is in the path of call signaling and/or H.245 control signaling. No requirement exists for the media streams to follow the same path as the H.245 control messages.

A Basic Call with Gatekeeper-Routed Call Signaling

The signaling process is slightly more complex when the gatekeeper that controls the called endpoint wishes to be in the path of call-signaling messages. Such a scenario appears in Figure 4-13, where both gatekeepers are involved in call signaling. Again, in the example, we assume that H.245 control messages are passed directly from endpoint to endpoint, rather than via one or more gatekeepers.

Upon receiving the Setup message from the calling endpoint's gatekeeper, the called endpoint requests permission from its own gatekeeper to handle the call. Because the gatekeeper wants to handle the call signaling, it returns an ARJ message with a cause code of `routeCallToGatekeeper`.

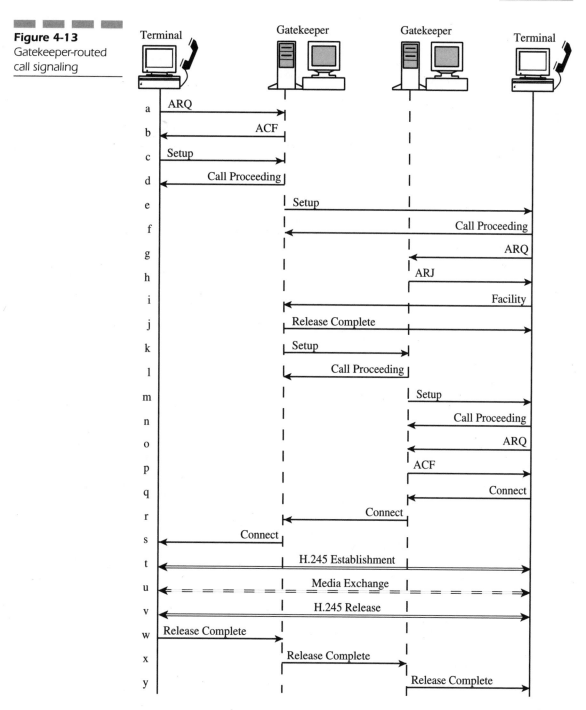

Figure 4-13
Gatekeeper-routed
call signaling

This action causes the called endpoint to send a Facility message to the gatekeeper of the calling endpoint. The Facility reason indicates that the call should be routed to the gatekeeper, and the call-signaling address of the gatekeeper is also included so that the calling gatekeeper knows where to send the call signaling.

The gatekeeper on the originating side releases the call to the endpoint and attempts to establish the call to the gatekeeper on the terminating side instead. Upon receiving the Setup message, the gatekeeper on the terminating side returns a Call Proceeding message and forwards the Setup message to the called endpoint. The called endpoint immediately returns a Call Proceeding message, then requests permission to handle the call (ARQ) and receives permission from the gatekeeper (ACF).

Once the user is alerted, the Alerting message (not shown) is passed from the called endpoint via the two gatekeepers to the calling endpoint. The Connect message follows the same path when the called user answers.

Once media have been exchanged, the Release Complete message is passed from one endpoint to the other via the two gatekeepers. Each endpoint then disengages by using the DRQ/DCF message exchange with its gatekeeper (not shown).

Optional Called-Endpoint Signaling

The previous gatekeeper-routed example assumes that the gatekeeper of the calling endpoint knows the call-signaling address of the called endpoint. As you can see in Figure 4-13, such a situation can lead to a rather cumbersome signaling exchange if the gatekeeper of the called endpoint wishes to handle call signaling. The exchange does not have to be so complicated, however.

Given that different gatekeepers control the two endpoints in our examples, the possibility exists that the gatekeeper of the calling endpoint does not know the call-signaling address of the endpoint to be called. The gatekeeper might only have an alias for that endpoint, in which case it will have to issue a *Location Request* (LRQ) to the gatekeeper of the endpoint to be called—perhaps by multicast—in order to determine the call-signaling address of the endpoint.

If the gatekeeper of the terminating endpoint does not wish to handle call signaling, then it will return the call-signaling address of the endpoint in the *Location Confirm* (LCF). At that point, the call will proceed largely according to the example in Figure 4-12. On the other hand, if the gatekeeper wishes to handle call signaling, it can return its own call-signaling address in the LRQ. In such an event, the sequence of events would be

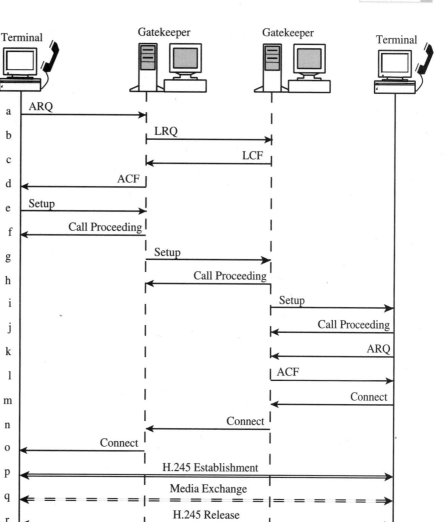

Figure 4-14
Optional called-endpoint signaling

similar to Figure 4-14. Again, the Alerting message and the RAS Disengage sequence are not shown. As we can see, the setup from the beginning is in accordance with the desires of the remote gatekeeper. Therefore, the call establishment is much simpler, and call redirection using the Facility message is not necessary.

H.245 Control Signaling

We have mentioned the H.245 control protocol on several occasions in this chapter already. As described, H.245 is the protocol that is used between session participants to establish and control media streams. For a straightforward, two-party voice call, this protocol takes care of ensuring that participants agree on the media formats to be sent and received and the bandwidth requirements. For more complex multimedia calls, this protocol takes care of mutiplexing multiple media streams for functions such as lip synchronization between audio and video.

You should note that H.245 is not responsible for carrying the actual media. For example, an H.245 packet containing a sample of coded voice does not exist. That is the job of RTP. Instead, H.245 is a control protocol that manages the media sessions.

H.245 is not dedicated to use for VoIP; rather, it is a more generic protocol used for the control of media streams and is designed to be used with a large number of applications. Consequently, the specification is a large document. The intent in this book is not to describe H.245 in great detail, but rather to provide a general understanding of how it works in an H.323 environment and to provide descriptions of the most common messages and procedures.

H.245 Message Groupings

H.245 involves the sending of messages from one endpoint to another. These messages fall into four groups:[1]

Requests   Requests are messages that require the recipient to perform some action and to send an immediate response.

Responses   These messages are sent in reply to Requests.

Commands   These messages require the recipient to perform some action; however, no explicit response is necessary.

Indications   These messages are of an informational nature only. The recipient of the message is not required to perform any specific action, and no response is expected.

Only a selection of the most common H.245 messages is described in this chapter.

---

[1]The choice of names for these groups is a little unfortunate. At first glance, they have an uncanny resemblance to the service primitives request, indication, response, and confirm as used in some SS7 applications. They are not the same, however.

The Concept of Logical Channels

H.245 handles media streams through the use of logical channels. A logical channel is a unidirectional media path between two endpoints. Generally, in an IP environment, we can view such a logical channel as simply an IP address and port number supporting a particular type of media (such as G.729B encoded voice). Each logical channel has a number that is specified by the sending entity.

Logical channels are unidirectional from sender to receiver. Therefore, in a two-party conversation, two logical channels exist. This separation enables one terminal to potentially send voice in one format and to receive voice in another format. Although logical channels are unidirectional, a perusal of the H.245 recommendation will find mention of a bidirectional channel. Such an entity still consists of two logical channels, although they are associated with each other.

To establish a media stream from one endpoint to another, the endpoint that wishes to transmit information opens a logical channel. The endpoint does so by sending a message to the far end, indicating the logical channel number and information about the media to be sent, such as RTP payload type. The message is called Open Logical Channel. If the far endpoint wishes to receive the media, then it responds with a positive acknowledgement (Open Logical Channel Ack). Included in the acknowledgement is an RTP port to which the media stream should be sent.

The H.245 messages are carried on the H.245 control channel. Each endpoint or gatekeeper will establish one H.245 control channel for each call in which it is participating. The H.245 control channel is carried on a special logical channel (channel number 0). This logical channel is special in that it is not opened and closed like other logical channels. Instead, this channel is considered permanently open as long as the endpoint is involved in a call.

H.245 Procedures

H.245 includes a number of procedures. For example, we briefly discussed the concept of opening logical channels. Several other procedures exist in H.245, however, and some of them we discuss here briefly.

Before a logical channel can be opened, the sending endpoint must understand the capabilities of the receiving endpoint. Trying to send voice in a format that the receiving entity cannot handle is fruitless. Consequently, H.245 provides a procedure known as Capabilities Exchange, which involves a set of messages that can be exchanged between endpoints to inform each other of their respective capabilities. These capabilities include sending and/or receiving media, determining the media formats that are supported, and determining what formats can be supported simultaneously (e.g., particular combinations of voice and video).

In conference calls where two or more endpoints also have an MC, a potential problem exists when trying to identify who is in control of the conference. Equally, imagine a situation in which two endpoints have an urgent need to communicate with each other and each tries to start a session with the other at the same time. To resolve these situations, H.245 has a procedure for master-slave determination.

Given the choice, most people prefer to eat French fries rather than Brussels sprouts. Similarly, just because an endpoint can handle media in a particular format, it might have a preference for one format over another. While Capability Exchange enables one endpoint to indicate to another what types of media it can handle, the Request Mode procedure enables an endpoint to rank formats in order of preference.

Capabilities Exchange Capabilities Exchange enables endpoints to share information regarding their receive and transmit capabilities. When a session is to be established between endpoints, the Capabilities Exchange procedure starts first. After all, trying to open logical channels between two endpoints if they do not even speak the same language is useless. We must first determine what each endpoint can support.

Receive capabilities indicate which media formats an endpoint can receive and process. Other endpoints should limit what they transmit to what the endpoint has indicated that it can handle. Transmit capabilities indicate which media formats an endpoint can send. A remote endpoint can treat this information like a menu of possible media formats from which it can choose. The endpoint can subsequently indicate a preference for one format over another. If an endpoint does not indicate any transmit capabilities, then we assume that the endpoint cannot and will not transmit any media.

An endpoint indicates its capabilities in a TerminalCapabilitySet message, which is a request message (and therefore requires a response). The message indicates a sequence number plus the types of audio and video formats that the endpoint can send and receive. The send and receive formats are indicated separately. Furthermore, the capabilities indicate what media can be handled simultaneously—for example, a particular audio and video combination.

The TerminalCapabilitySetAck message acknowledges the TerminalCapabilitySet message. This message is empty, except for a sequence number that matches the sequence number received in the original request. If the endpoint receiving the request finds a problem with the request, then it responds with a TerminalCapabilitySetReject message indicating the reason for rejection. If there is no response within a timeout period, then the TerminalCapabilitySetRelease message is sent. This indication message does not require a response.

As well as the capability for an endpoint to unilaterally permit another endpoint to know about its capabilities, an endpoint can also request another endpoint to send information about its capabilities. This procedure occurs by using the message SendTerminalCapabilitySet. Strictly speaking, this message is a command message—and according to the definition of a command message, it does not require a specific response. The endpoint that receives the message, however, should subsequently send a Terminal-CapabilitySet message.

The SendTerminalCapabilitySet message contains two options. The sender can request the far endpoint to indicate all of its capabilities. Alternatively, the sending endpoint can request confirmation about specific capabilities that the far endpoint already described. This feature could be used, for example, in a situation where there has been a break in communication and one endpoint wants to make sure that it still has the latest information about the other endpoint.

Master-Slave Determination   Within any conference, one of the endpoints needs to be the master. Who the master is has particular importance when a multi-party conference is to take place and several of the entities involved could control the conference. Deciding who is master is accomplished through the master-slave determination procedure.

The master-slave determination procedure involves two pieces of information at each entity. The first is a terminal type value, and the second is a random number between 1 and 16,777,215. A master is created after comparing the values of terminal type. Whichever endpoint has the larger value of terminal type is automatically the master. H.323 specifies the values of terminal type that should be applied to different types of endpoints. For example, a terminal that does not have an MC has a terminal type value of 50. A gateway that does not contain an MC has a terminal type value of 60. An MCU that supports audio, video, and data conferencing has a terminal type value of 190, and an MCU that is currently managing a conference has the highest possible value: 240. Only in the case of two terminals having the same value of terminal type is the random number considered, and the terminal that chooses the highest random number value is the master.

In order to determine the master, the endpoints must share their values of terminal type and random numbers. This action is performed through the use of the Master-Slave Determination message and the Master-Slave Determination Ack message. Either entity can begin the master-slave determination procedure by sending the Master-Slave Determination message, which contains a value of terminal type and a random number. The receiving entity compares the value of terminal type with its own value, and if necessary, compares the received random number with a random number that it chose itself. Based on these comparisons, the entity makes the determination of who is the master and returns a Master-Slave Determination Ack message.

The Master-Slave Determination Ack message indicates the sender's view of the other party's ranking. If the message contains a "master" indication, the receiver of the message is the master. If it contains a "slave" indication, the receiver of the message is the slave.

Establishing and Releasing Media Streams   One or more logical channels carry media streams between participants. These channels need to be established before media can be exchanged and need to be closed at the end of a call.

Opening Unidirectional Logical Channels   Sending an Open Logical Channel request message opens a logical channel. This message contains a mandatory parameter called `forwardLogicalChannelParameters`, which relates to the media to be sent in the forward direction (i.e., from the endpoint issuing this request). This parameter contains information regarding the type of data to be sent (e.g., G.728 audio), an RTP session ID, an RTP payload type, an indication as to whether silence suppression is to be used, a description of redundancy encoding (if used), etc.

If the recipient of the message wishes to accept the media to be sent, then it will return an OpenLogicalChannelAck message containing the same logical channel number as received in the request and a transport address to which the media stream should be sent.

The process is shown in Figure 4-15, where a single logical channel is opened. The initiator of the logical channel has not indicated that it wishes to receive media. We are not saying that the initiator is completely unwilling to receive media, however. If the called endpoint also wishes to send media, then after the first logical channel is opened, it can choose to open a logical channel in the reverse direction.

If an endpoint does not wish to accept a request to open a channel, it can reply with the OpenLogicalChannelReject message. This message contains the logical channel number received in the request and a reason for the rejection, such as an inability to handle the proposed media format.

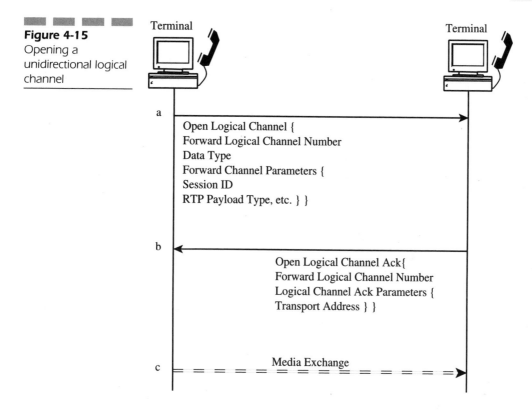

Figure 4-15
Opening a
unidirectional logical
channel

a

Open Logical Channel {
Forward Logical Channel Number
Data Type
Forward Channel Parameters {
Session ID
RTP Payload Type, etc. } }

b

Open Logical Channel Ack{
Forward Logical Channel Number
Logical Channel Ack Parameters {
Transport Address } }

c   Media Exchange

Opening Bidirectional Logical Channels The foregoing discussion regarding the opening of unidirectional logical channels mentions the fact that two endpoints can send media to each other by each opening a logical channel in each direction. This process requires four messages: two requests and two responses. If an endpoint wishes to send media and is expecting to receive media in return, however, as would be the case for a typical voice conversation, then it can attempt to establish a bidirectional logical channel. This procedure is a means of establishing two logical channels, one in each direction, in a slightly more efficient manner.

The process is depicted in Figure 4-16 and begins with the sending of an Open Logical Channel request message. In addition to forward logical channel parameters, however, the message also contains reverse logical channel parameters that describe the type of media that the endpoint is willing to receive and to where that media should be sent.

Upon receiving the request, the far endpoint can send an Open Logical Channel Ack message containing the same logical channel number for the forward logical channel, a logical channel number for the reverse logical

Figure 4-16
Opening bidirectional
logical channels

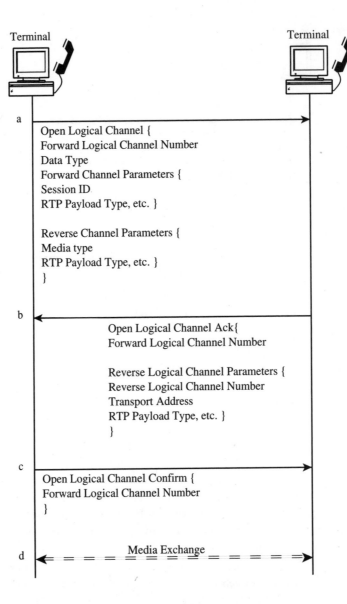

a
> Open Logical Channel {
> Forward Logical Channel Number
> Data Type
> Forward Channel Parameters {
> Session ID
> RTP Payload Type, etc. }
>
> Reverse Channel Parameters {
> Media type
> RTP Payload Type, etc. }
> }

b
> Open Logical Channel Ack{
> Forward Logical Channel Number
>
> Reverse Logical Channel Parameters {
> Reverse Logical Channel Number
> Transport Address
> RTP Payload Type, etc. }
> }

c
> Open Logical Channel Confirm {
> Forward Logical Channel Number
> }

d
Media Exchange

channel, and descriptions that are related to the media formats it is willing to send. These media formats should be chosen from the options originally received in the request, thereby ensuring that the called end will only send media that the calling end supports.

Upon receipt of the Open Logical Channel Ack message, the originating endpoint responds with an Open Logical Channel Confirm message to indicate that all is well. Media flow can now begin in both directions.

Closing Logical Channels and Ending a Session   The closing of a logical channel involves the sending of a CloseLogicalChannel message. In the case of a successful closure, the far end should send the response message CloseLogicalChannelAck.

In general, a logical channel can only be closed by the entity that created it in the first place. For example, in the case of a unidirectional channel, only the sending entity can close the channel. The receiving endpoint in a unidirectional channel can humbly request the sending endpoint to close the channel, however. The endpoint does so by sending the RequestChannel-Close message, indicating the channel that the endpoint would like to have closed. If the sending entity is willing to grant the request, it responds with a positive acknowledgment and then proceeds to close the channel.

One exception should be noted regarding the closure of a bidirectional channel. When an endpoint closes a forward logical channel, it also closes the reverse logical channel when the two are part of a bidirectional channel.

Once all logical channels in a session are closed, then the session is terminated when an endpoint sends an EndSession command message. The receiving endpoint responds with an EndSession command message. Once an entity has sent this message, it must not send any more H.245 messages related to the session.

Figure 4-17 provides an example of channel closure where one entity requests that the other close the channel, followed by the channel closure.

Fast-Connect Procedure

Having reviewed RAS, Q.931, and H.245 signaling, we can construct a call flow that incorporates more of the signaling that is required for call establishment between two endpoints. Figure 4-18 shows an example of such a call establishment and can be considered a more complete version of the example depicted in Figure 4-11. One can see that many messages are required even for such a simple call (hence the term "slow start" in the figure caption). Furthermore, when we consider that the figure does not show capability exchange or master-slave determination, we see that even more messaging is required. If one imagines a scenario with gatekeeper-routed call signaling and the need for the Facility message, as shown in Figure 4-13, plus the possibility of gatekeeper-routed H.245 control signaling, the number of messages required can clearly be greater still. Call establishment becomes cumbersome and is certainly not conducive to fast call setup. Considering that fast call setup is a requirement for carrier-grade operation, the

Figure 4-17
Closing a logical
channel

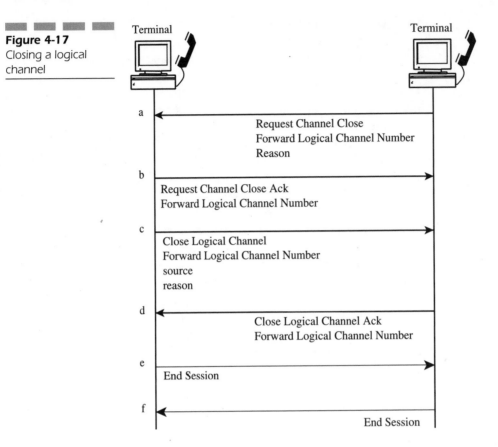

a
Request Channel Close
Forward Logical Channel Number
Reason

b
Request Channel Close Ack
Forward Logical Channel Number

c
Close Logical Channel
Forward Logical Channel Number
source
reason

d
Close Logical Channel Ack
Forward Logical Channel Number

e
End Session

f
End Session

situation is lamentable. Thankfully, the designers of H.323 recognized these issues and defined a procedure that reduced signaling overhead and speeded things considerably. Appropriately, this procedure is called the Fast-Connect procedure.

The Fast-Connect procedure involves setting up media streams as quickly as possible. To achieve this goal, the Setup message can contain a faststart element within the User-to-User information element. The fast-start element is actually one or more OpenLogicalChannel request messages that contain all of the information that would normally be contained in such a request. This element includes reverse logical-channel parameters if the calling endpoint expects to receive media from the called endpoint.

If the called endpoint also supports the procedure, it can return a fast-start element in one of the Call Proceeding, Alerting, Progress, or Connect messages. That faststart element is basically another OpenLogicalChannel message that appears similar to a request to open a bidirectional logical

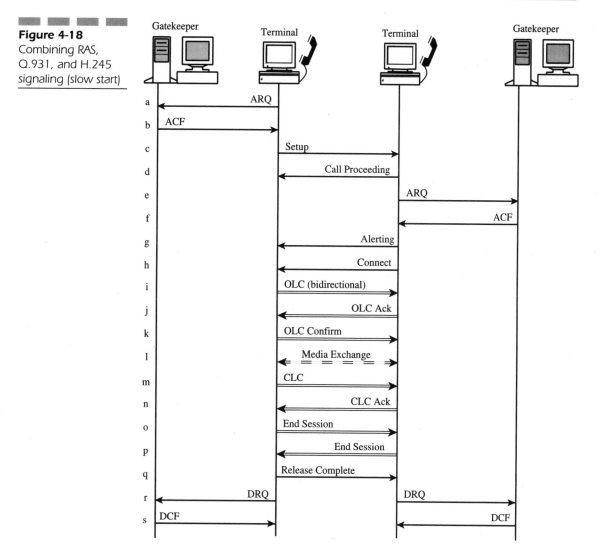

Figure 4-18
Combining RAS,
Q.931, and H.245
signaling (slow start)

channel. The included choices of media formats to send and receive are chosen from those offered in the faststart element of the incoming Setup message. The calling endpoint has effectively offered the called endpoint a number of choices for forward and reverse logical channels, and the called endpoint has indicated those choices that it prefers. The logical channels are now considered open as if they had been opened according to the procedures of H.245. The use of faststart is depicted in Figure 4-19.

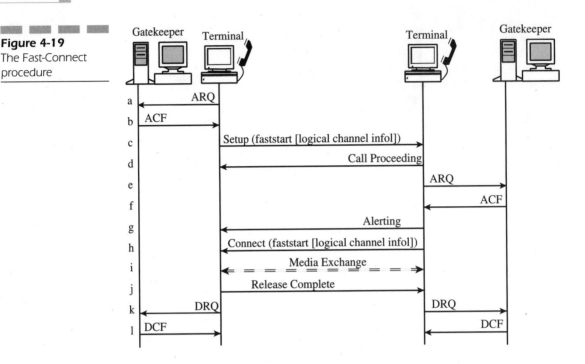

Figure 4-19
The Fast-Connect
procedure

> **NOTE:** *Note that the faststart element from called party to calling party can be sent in any message up to and including the Connect message. If this element has not been included in any of the messages, then the calling endpoint will assume that the called endpoint either cannot or does not wish to support faststart. In such a case, the standard H.245 methods must be used.*

The use of the Fast Connect procedure means that H.245 information is carried within the Q.931 messages, and there is no separate H.245 control channel. Therefore, bringing a call to a conclusion is also faster. The call is released simply by sending the Q.931 Release Complete message. When used with the Fast-Connect procedure, this procedure has the effect of closing all of the logical channels associated with the call and is equivalent to using the procedures of H.245 to close the logical channels.

H.245 Message Encapsulation

In addition to the Fast-Connect procedure, H.323 also enables H.245 messages to be encapsulated with Q.931 messages as octet strings. An endpoint that wishes to encapsulate such messages with Q.931 messages sets the element h245Tunneling to true in the first Q.931 message that it sends and keeps the element set to true in every other message that it sends, as long as it wishes to support such encapsulation. For a calling endpoint, the h245Tunneling element should be set to true in the Setup message. Note that a called endpoint can use encapsulation only if the calling endpoint indicated support for encapsulation in the Setup message.

The encapsulated data is actually contained with the h245Control element—which, like all H.323 information, is included in the User-to-User information element of the Q.931 message.

Note that there is a certain conflict between encapsulation and faststart. After all, the faststart element itself is a type of encapsulated H.245 message. Consequently, a calling endpoint should not include both a faststart element and an encapsulated H.245 message in a Setup message. If the endpoint wishes to support encapsulation with later commands, however, then it should still set the h245Tunneling element to true so that the called endpoint knows that the calling endpoint can support encapsulation.

As is the case for the Fast-Connect procedure, the use of encapsulation means that no separate H.245 control channel exists, because H.245 messages are carried on the call-signaling channel. What happens if an endpoint needs to send an H.245 message at a time when it does not particularly need to send a call-signaling message? We can handle this situation in two ways. First, an endpoint can send a Facility message with an encapsulated H.245 message. Alternatively, either endpoint can choose to change to a separate H.245 control channel.

An endpoint indicates a desire to switch to a separate H.245 control channel by populating the h245address element in a Q.931 message. In such a case, any existing logical channels established by the Fast-Connect procedure or through encapsulation will be inherited by the H.245 control channel as if they had been established through normal H.245 procedures.

If the endpoint wishes to change to a separate H.245 channel and has no need to send a Q.931 message, then it has no means of indicating an H.245 address to the other endpoint. Again, the Facility message comes to the rescue, because this message can be sent at any time. All that is necessary is to send a Facility message with a FacilityReason of startH245 and with the h245address element populated with the endpoint's own H.245 address. This action effectively establishes a separate H.245 control channel. Any subsequent Q.931 messages must have the h245Tunneling element set to false, because anything else would be inconsistent with a separate H.245 control channel.

Conference Calls

The previous descriptions and examples have focused on the signaling that is necessary for the establishment and release of two-party calls. Situations will occur, however, in which there are several parties involved in a multipoint conference. In fact, H.323 defines the *Multipoint Controller* (MC) for the purpose of managing multipoint conferences. A conference call can be established in two ways. One way is to establish a pre-arranged conference in which participants call in to a separate MCU that controls the conference. The other technique is an ad-hoc conference, where a two-party call needs to be expanded to include other parties.

A Pre-Arranged Conference

In a pre-arranged conference, the individual participants are connected to the conference through the establishment of a call with the MCU. In other words, connections to the MCU are established through Q.931 call signaling. The sharing of media streams can be accomplished in a centralized manner via the MCU or in a decentralized multicast manner. In the centralized scenario, both signaling and media pass via the MCU. In the decentralized scenario, only the signaling passes via the MCU while the media streams are multicast from each endpoint to each other endpoint. The MCU determines the mode of the conference.

The MCU specifies the conference mode through the use of the H.245 command message Communication Mode command. This command specifies all of the sessions in the conference (and for each session, various session data plus a unicast or multicast address). Note that the command only specifies the transmit requirements of each endpoint, not the receive requirements. The reason is because the receive requirements will be specified in separate Open Logical Channel commands that will be subsequently sent from the MCU to the various endpoints.

An Ad-Hoc Conference

Quite often, the need will exist to expand an existing two-party call to a conference involving three or more participants. Such an occurrence might not be foreseen at the beginning of the two-party call, however. Given that most PBXs and even residential phone service enable multi-party calls and the expansion of a call to include other parties, the same capability needs to be supported in a VoIP environment.

Many H.323 endpoints or gatekeepers include an MC function. This function is not a callable entity like an MCU that is used for a call-in conference, but it can be utilized in the event that a two-party call needs to be expanded to three or more parties. In order for a two-party call to be expanded to an ad-hoc conference, one of the endpoints must contain an MC, or one of the endpoints' gatekeepers must contain an MC (in which case gatekeeper-routed call signaling must be used).

In the case that these requirements are met, then the two-party call can be established as a conference call between two entities. This process requires the Setup message from endpoint 1 to endpoint 2 to contain a unique *Conference ID* (CID). If endpoint 2 wishes to accept the call, then the Connect message that is returned should contain the same value of CID. H.245 capability exchange and master-slave determination procedures are performed, and the call is established (as shown in Figure 4-20).

Subsequently, if the master endpoint decides to invite another party (endpoint 3) to join the conference, it sends a Setup message to endpoint 3 with the CID value of the conference and the conference goal parameter set to `invite`. Assuming that endpoint 3 wishes to join the conference, it includes the same CID in the Connect message. Then, capability exchange and master-slave determination is performed between endpoint 1 and endpoint 3. Finally, the MC at endpoint 1 sends the H.245 indication Multipoint Conference to each of the participants.

Figure 4-20
An ad-hoc
conference

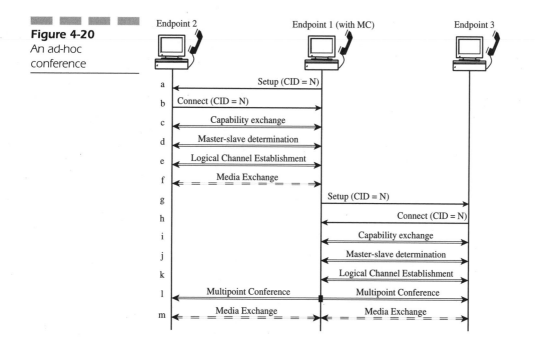

The Session Initiation Protocol (SIP)

Introduction

Many consider the *Session Initiation Protocol* (SIP) a powerful alternative to H.323. They say SIP is a more flexible solution, simpler than H.323, easier to implement, better suited to the support of intelligent user devices, and better suited to the implementation of advanced features. These factors are of major importance to any equipment vendor or network operator. Simplicity means that products and advanced services can be developed faster and made available to subscribers more quickly. The features themselves mean that operators are better able to attract and retain customers and also to offer the potential for new revenue streams.

SIP is designed to be a part of the overall *Internet Engineering Task Force* (IETF) multimedia data and control architecture. As such, SIP is used in conjunction with several other IETF protocols, such as the *Session Description Protocol* (SDP), the *Real-Time Streaming Protocol* (RTSP), and the *Session Announcement Protocol* (SAP).

The Popularity of SIP

Originally developed in the *Multiparty Multimedia Session Control* (MMU-SIC) working group of the IETF, the great interest that SIP has engendered has led to the creation of a separate SIP working group. Although RFC 2543 defines SIP, the SIP working group is proceeding at a brisk pace in order to include numerous enhancements to the protocol. Nevertheless, many developers are including SIP in their products without waiting for those enhancements, simply because SIP is seen as the way of the future for VoIP signaling. In fact, many believe that SIP, in conjunction with the MGCP or MEGACO (as described in Chapter 6, "Media Gateway Control and the Softswitch Architecture") will be the dominant *Voice over IP* (VoIP) signaling architecture in the future.

SIP's development and its implementation by system developers has involved a number of slightly unusual events known as "bake-offs." During these events, various vendors come together and test their products against each other to ensure that they have implemented the specification correctly and to ensure compatibility with other implementations. To give an indication of SIP's popularity, Figure 5-1 shows how many companies have been involved in the various SIP bake-offs that have been held as of this writing. We expect that there will be yet a greater number of companies participating in the next SIP bake-off, which will occur in August 2000. In addition, ETSI volunteered to host a future bake-off. The increasing number of participating

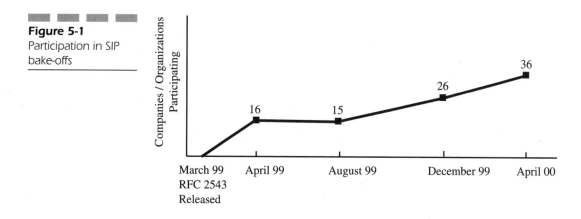

Figure 5-1
Participation in SIP bake-offs

companies indicates the growing support for SIP within the industry. The fact that ETSI wishes to host a bake-off attests to the growing international appeal of SIP.

SIP Architecture

SIP is a signaling protocol that handles the setup, modification, and teardown of multimedia sessions. SIP, in combination with other protocols, describes the session characteristics to potential session participants. Although strictly speaking, SIP is written such that the media to be used in a given session could use any transport protocol, the media will normally be exchanged by using RTP as the transport protocol.

Likely, SIP messages will pass through some of the same physical facilities as the media to be exchanged. SIP signaling should be considered separately from the media itself, however. Figure 5-2 shows the logical separation between signaling and session data. This separation is important, because the signaling can pass via one or more proxy or redirect servers while the

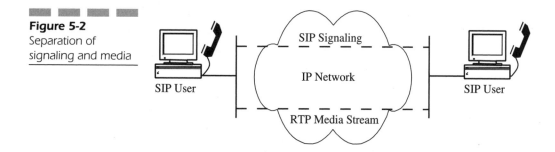

Figure 5-2
Separation of signaling and media

media stream takes a more direct path. This process can be considered somewhat analogous to the separation of signaling and media already described for H.323.

SIP Network Entities

SIP defines two basic classes of network entities: clients and servers. Strictly speaking, a client (also known as a user agent client) is an application program that sends SIP requests. A server is an entity that responds to those requests. Thus, SIP is a client-server protocol.

VoIP calls using SIP originate at a client and terminate at a server. A client might be found within a user's device, which could be a PC with a headset attachment or a SIP phone, for example. Clients might also be found within the same platform as a server. For example, SIP enables the use of proxies, which act as both clients and servers.

Four different types of servers exist: proxy server, redirect server, user agent server, and registrar.

A proxy server acts similarly to a proxy server that is used for Web access from a corporate *Local-Area Network* (LAN). Clients send requests to the proxy, and the proxy either handles those requests itself or forwards them to other servers, perhaps after performing some translation. To those other servers, it appears as though the message is coming from the proxy rather than from some entity hidden behind the proxy. Given that a proxy receives requests and sends requests, it incorporates both server and client functionality. Figure 5-3 shows an example of the operation of a proxy server. You do not need much imagination to realize how this type of functionality

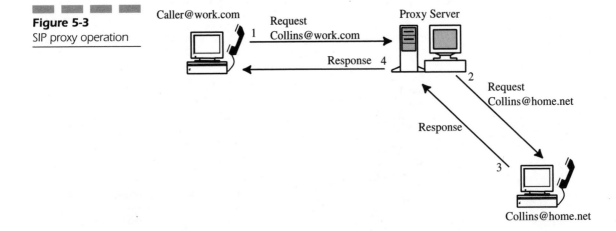

Figure 5-3
SIP proxy operation

can be used for call forwarding/follow-me services. For example, if the message from the caller to Collins in Figure 5-3 offers an invitation to participate in a call, the net effect is that the call will be forwarded to Collins at home. Of course, the proxy must be aware that Collins happens to be at home instead of at work.

A redirect server is a server that accepts SIP requests, maps the destination address to zero or more new addresses, and returns the translated address to the originator of the request. Thereafter, the originator of the request can send requests to the address(es) returned by the redirect server. A redirect server does not initiate any SIP requests of its own. Figure 5-4 shows an example of the operation of a redirect server. This process can be another means of providing the call forwarding/follow-me service that a proxy server can offer. The difference is, however, that in the case of a redirect sever, the originating client actually forwards the call. The redirect server simply provides the information necessary to enable the originating client to perform this task, after which the redirect server is no longer involved.

A user agent server accepts SIP requests and contacts the user. A response from the user to the user agent server results in a SIP response on behalf of the user. In reality, a SIP device (such as a SIP-enabled telephone) will function as both a user agent client and as a user agent server. Acting as a user agent client, the SIP device can initiate SIP requests. Acting as a user-agent server, the device can receive and respond to SIP requests. In practical terms, the device can initiate and receive calls, which enables SIP (a client-server protocol) to be used for peer-to-peer communication.

Figure 5-4
SIP redirect operation

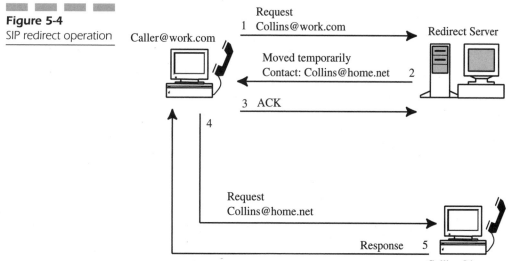

A registrar is a server that accepts SIP REGISTER requests. SIP includes the concept of user registration, whereby a user indicates to the network that he or she is available at a particular address. Such registration occurs through the issuance of a REGISTER request from the user to the registrar. Typically, a registrar is combined with a proxy or redirect server.

Given that practical implementations will involve combining the user agent client and user agent server and combining registrars with either proxy servers or redirection servers, a real network might well involve only user agents and redirection or proxy servers.

SIP Call Establishment

At a high level, SIP call establishment is simple, as shown in Figure 5-5. The process starts with a SIP INVITE message, which is used from the calling party to the called party. The message invites the called party to participate in a session—i.e., a call. A number of interim responses to the INVITE might exist prior to the called party accepting the call. For example, the caller might be informed that the call is queued and/or that the called party is being alerted (i.e., the phone is ringing). Subsequently, the called party answers the call, which generates an OK response back to the caller. The calling client acknowledges that the called party has answered by issuing an ACK message. At this point, media are exchanged. This media

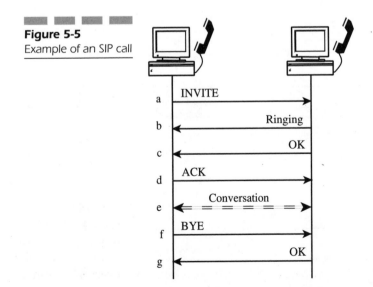

Figure 5-5
Example of an SIP call

will most often be in the form of regular speech but could also be other media, such as video. Finally, one of the parties hangs up, which causes a BYE message to be sent. The party receiving the BYE message sends OK to confirm receipt of the message. At that point, the call is over.

SIP Advantages over Other Signaling Protocols

Given that the call establishment and release is pretty straightforward, one might wonder what is so special about SIP. After all, any call-signaling protocol must have a means for one party to call another, a means to indicate acceptance of the call, and a means to release the call. SIP performs these actions and more but does not go far beyond these basic requirements. In other words, SIP attempts to keep the signaling as simple as possible. Contrast the number of messages in Figure 5-5 with the number of messages necessary to establish an equivalent H.323 call—particularly if the H.323 uses gatekeeper-routed signaling and does not use the Fast-Connect procedure. The fact that SIP involves less signaling means that calls can be established faster, and rapid call setup is a key requirement of a high-quality carrier-grade service.

Furthermore, not only do the SIP messages themselves make SIP powerful, but also the various pieces of information that can be included within messages and responses make SIP a useful protocol. Not only does SIP enable a range of standard information to be included in requests and responses, but it also enables lots of non-standard information to be included. By enabling useful information to be included, SIP permits the user devices and the users themselves to make various intelligent decisions about call handling. Consequently, calling or called parties can invoke various services.

Imagine, for example, that a call is directed to a user who is currently not available. Obviously, the SIP response will indicate that the user is unavailable (no big surprise). The response could, however, also include an indication that the user expects to be available again at a specific time, such as 4 P.M. In such a case, the calling terminal could perform two actions. First, the terminal could tell the calling party that the called user expects to be available at 4 P.M. Secondly, at 4 P.M., the terminal could ask the caller whether he or she wants to make the call again, and if so, automatically set up the call. Such a scenario appears in Figure 5-6. This mechanism is a clever and simple IP version of a call-completion service, which is similar to the *Call Completion to Busy Subscriber* (CCBS) of the telephony world. This mechanism is also much more efficient than calling back every 15 minutes

Figure 5-6
Example of an SIP-
enabled service

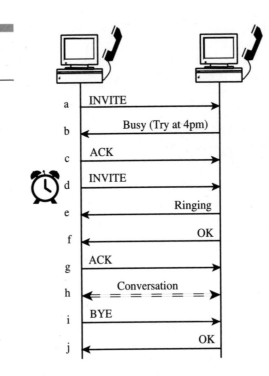

in the hopes that the person will finally answer or paying the telephone company to keep trying to connect the call.

This function is just one simple example of a service that SIP provides. Because SIP provides many pieces of information for inclusion in messages, and because additional, non-standard information can also be included, the opportunity exists to offer numerous intelligent features to subscribers. Furthermore, the customer has control of those features. No longer does a customer need to subscribe to a particular feature and have the network operator actually control that feature.

The following sections of this chapter are devoted to describing the fundamentals of SIP. From this description, the reader should gain a good understanding of how SIP works, how it compares to other standards (such as H.323), and how SIP can be used to provide new and advanced features to users.

Overview of SIP Messaging Syntax

As mentioned, SIP is a signaling protocol—and as such, it has a particular syntax. In the case of SIP, the syntax is text-based (uses the ISO 10646 character set) and has a similar look and feel to the *Hypertext Transfer Protocol* (HTTP). One advantage of this approach is that programs designed for parsing of HTTP can be adapted relatively easily for use with SIP. One obvious disadvantage compared to binary encoding, however, is that the messages themselves consume more bandwidth.

SIP messages are either requests from a client to a server or responses (which are also known as status messages) from a server to a client. Each message, whether a request or a response, contains a start line followed by zero or more headers and is optionally followed by a message body:

```
message = start-line
                *message-header
            CRLF
            [message-body]
```

Given that SIP defines only request and status messages, then the start line is

```
Start-line = request-line | status-line
```

The request line specifies the type of request being issued, while the response line indicates the success or failure of a given request. In the case of failure, the line indicates the type of failure or the reason for failure.

Message headers provide additional information regarding the request or response. Information that is obviously required includes the message originator and the intended message recipient. Message headers also offer the means for carrying additional information, however. For example, the Retry-after header indicates when a request should be attempted again. In Figure 5-6, this header would convey the fact that the called user expects to be available again at 4 P.M. Another example of a useful header is the Subject header, which enables a caller to indicate the reason for a call. The called party can choose to accept or reject a call, depending on the subject in question.

The message body normally describes the type of session to be established, including a description of the media to be exchanged. Thus, for a given call, the message body might indicate that the caller wishes to communicate by using voice that is coded according to G.711 A-law. Note, however, that SIP does not define the structure or content of the message body. This structure and content is described by using a different protocol. The most common structure for the message body uses the *Session Description Protocol* (SDP) described later in this chapter. The message body, however, could contain information that is coded according to another standard. In fact, the message body could actually include an *ISDN User Part* (ISUP) message in binary format, thereby enabling SIP to carry ISUP information.

This function would be used, for example, in a scenario where the SIP network is interoperating with the *Public-Switched Telephone Network* (PSTN) by using *Signaling System 7* (SS7). SIP does not greatly care about what the message body happens to say; rather, it is concerned only with making sure that the message body is carried from one party to the other. The message body is examined only at the two ends. The message body can be considered a sealed envelope. SIP carries the message from one end to the other but does not look inside the envelope.

SIP Requests

A SIP request begins with a request line, which comprises a method token, a REQUEST-URI, and an indication of the SIP version. The method token identifies the specific request being issued, and the Request-URI is the address of the entity to which the request is being sent. The three components of the request line are separated by spaces, and the line itself is terminated by a CRLF character. Thus, the syntax is as follows:

```
request-line = method SP Request-URI SP SIP-Version CRLF
```

RFC 2543 defines six different methods (and hence, six different types of requests): INVITE, ACK, OPTIONS, BYE, CANCEL, and REGISTER.

The INVITE method initiates a session. For a simple call between two parties, INVITE is used to initiate the call, with the message including information regarding the calling and called parties and the type of media to be exchanged. For those who are familiar with ISUP, the INVITE request is akin to the *Initial Address Message* (IAM). As well as the means for initiating a simple two-party call, INVITE also offers the capability to initiate a multi-party conference call.

Once it has received a final response to an INVITE, the client that initiated the INVITE sends an ACK. This method confirms that the final response has been received. For example, if the response to an INVITE is that the called user is busy and the call cannot be completed at that time, then the calling client will send an ACK. On the other hand, if the response to the INVITE indicates that the called user is being alerted or that the call is being forwarded, then the client does not send an ACK, because such responses are not considered final.

The BYE method terminates a session and can be issued by either the calling or called party. This method is used when the party in question hangs up.

The OPTIONS method queries a server as to its capabilities. This method could be used, for example, to determine whether a called user agent can support a particular type of media or to determine how a called user agent would respond if sent an INVITE. In such a case, the response might indicate that the user can support certain types of media, or perhaps that the user is currently unavailable.

The CANCEL method terminates a pending request. For example, CANCEL could be used to terminate a session in which an INVITE has been sent but a final response has not yet been received. As discussed later in this chapter, we can initiate a parallel search to multiple destinations (e.g., if a user is registered at multiple locations). If such a case exists, and a final response has been received from one of the destinations, CANCEL can be used to terminate the pending requests at the other destinations.

A user agent client uses the REGISTER method to log in and register its address with a SIP server, thereby letting the registrar know the address at which the user is located. The user agent client might register with a local SIP server at startup, with a known registrar server whose address is configured within the user agent, or with multicast to the "all SIP Servers" multicast address (224.0.1.175). A client can register with multiple servers, and a given user can have several registrations listed at the same registrar. This situation would occur when a user has logged in at several terminals or devices. In the case where a user has multiple active registrations, calls to the user can be sent to all registered destinations. This procedure can enable a "one-number" service in which a user publishes just a single number, but when that number is called, the user's office phone, home office phone, and wireless phone all ring. Figure 5-7 shows an example of how this process would work and also shows the usage of the CANCEL message in such a scenario.

SIP Responses

The start line of a SIP response is a status line. This line contains a status code, which is a three-digit number indicating the outcome of the request. The start line will also contain a reason phrase, which provides a textual description of the outcome. The client software will interpret the reason code and will act upon it, while the reason phrase could be presented to the human user to aid in understanding the response. The syntax of the status line is as follows:

```
status-line = SIP version SP status code SP reason-phrase CRLF
```

Figure 5-7
Multiple registrations
enabling a "one-
number" service

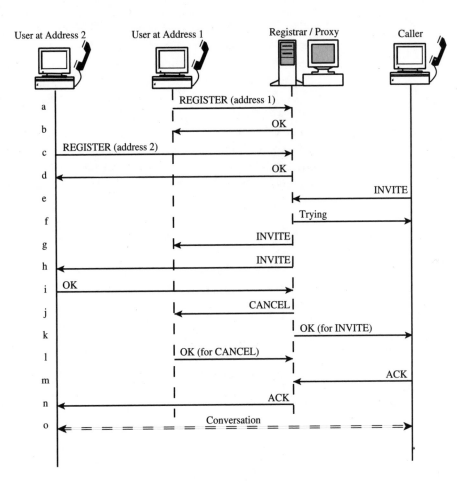

Status codes defined in RFC 2543 have values between 100 and 699, with the first digit of the reason code indicating the class of response. Thus, all status codes between 100 and 199 belong to the same class. The different classes are as follows:

- *1XX* Informational (e.g., 181 indicates that the call is being forwarded)

- *2XX* Success (only code 200 is defined, which means that the request has been understood and has been performed. In the case of an INVITE, a response of 200 is used to indicate that the called party has accepted the call)

- *3XX* Redirection (e.g., 302, which indicates that the called party is not available at the address used in the request and that the request should be reissued to a new address included with the response)

- *4XX* Request Failure (e.g., 401, which indicates that the client is not authorized to make the request)
- *5XX* Server Failure (e.g., 505, which indicates that the server does not support the SIP version specified in the request)
- *6XX* Global Failure (e.g., 604, indicating that the called user does not exist anywhere)

All responses, except for 1XX responses, are considered final and should be acknowledged with an ACK message if the original message happened to be an INVITE. RFC 2543 specifies that 1XX responses are provisional and do not need to be acknowledged. Table 5-1 gives a complete listing of the status codes defined in RFC 2543, along with proposed reason phrases.

Table 5-1

Status codes and proposed reason phrases

Category	Status Code	Reason Phrase
Informational[1]	100	Trying
	180	Ringing
	181	Call is being forwarded
	182	Queued
Success	200	OK
Redirection	300	Multiple choices
	301	Moved permanently
	302	Moved temporarily
	305	Use proxy
	380	Alternative service
Request Failure[2]	400	Bad request
	401	Unauthorized
	402	Payment required
	403	Forbidden
	404	Not found
	405	Method not allowed
	406	Not acceptable

(continues)

Table 5-1

Continued

Category	Status Code	Reason Phrase
	407	Proxy Authentication required
	408	Request timeout
	409	Conflict
	410	Gone
	411	Length required
	413	Request entity too large
	414	Request-URI too long
	415	Unsupported media type
	420	Bad extension
	480	Temporarily not available
	481	Call leg/transaction does not exist
	482	Loop detected
	483	Too many hops
	484	Address incomplete
	485	Ambiguous
	486	Busy here
Server Failure	500	Server internal error
	501	Not implemented
	502	Bad gateway
	503	Service unavailable
	504	Gateway timeout
	505	SIP version not supported
Global-Failure	600	Busy everywhere
	603	Decline
	604	Does not exist anywhere
	606	Not acceptable

[1]A new informational response, 183 (Session Progress) has been proposed as an extension to SIP.

[2]Two new Request-Failure responses, 487 (Request Terminated) and 488 (Not acceptable here) have been proposed for inclusion in a revised version of SIP.

SIP Addressing

As with any signaling protocol, requests and responses are sent to particular addresses. In SIP, these addresses are known as SIP URLs (Uniform Resource Locators). These addresses take the form of user@host, which is similar to an e-mail address. In many cases, a given user's SIP address can, in fact, be guessed from the user's e-mail address. Although they might look similar, they are different. Whereas an e-mail address uses a mailto URL (such as mailto:collins@home.net), a SIP URL has the syntax sip:collins@home.net.

SIP deals with multimedia sessions, which could include voice. Furthermore, we can interwork with a traditional circuit-switched network. For these reasons, SIP enables the user portion of the SIP address to be a telephone number. Thus, we could have a SIP address such as sip:3344556789@telco.net. In a given network, such a SIP address could be used to cause routing of media to a gateway that interfaces with the traditional telephone company. A number of parameters that provide more information can also supplement a SIP URL. For example, to clearly indicate that a call is to a telephone number, we could supplement the URL with the term user=phone. In such a case, the URL would have the following format: sip:3344556789@telco.net;user=phone.

Message Headers

RFC 2543 defines a number of different message headers. These items of information are included in a request or response in order to provide further information about the message or to enable the appropriate handling of the message. In that respect, these headers are akin to the message parameters or information elements included in any typical signaling protocol, such as ISUP, Q.931, etc. For example, the To: header in an INVITE message indicates the called party, and the From: header indicates the calling party. The intention in this chapter is to describe the usage of headers and to give an understanding of the most common and useful headers. For complete definitions, please refer to RFC 2543.

Four main categories of headers exist: general headers, request headers, response headers, and entity headers. These headers are listed in Table 5-2 according to those four categories. Depending on the request or response, certain headers are mandatory, certain headers are optional, and certain headers are not applicable.

Table 5-2

Four categories of
message headers

General Headers	Request Headers	Response Headers	Entity Headers
Accept	Hide	Proxy-Authenticate	Allow
Accept-Encoding	In-Reply-To*	Retry-After	Content-Disposition*
Accept-Language	Max-Forwards	Server	Content-Encoding
Authorization[3]	Priority	Unsupported	Content-Language*
Call-ID	Proxy-Authorization	Warning	Content-Length
Contact	Proxy-Require	WWW-Authenticate	Content-Type
CSeq	Route		Expires
Date	Response-Key		
Encryption	Subject		
From			
Organization			
Record-Route			
Require			
Supported*			
Timestamp			
To			
User-Agent			
Via			
MIME-Version*			

*These header fields are not specified in RFC 2453 but have been included in a draft update to the specification.

[3]RFC 2543 defines this header as a request header. It can, however, also be used in responses.

Table 5-3 provides a mapping between requests and headers, where "M" means mandatory, "O" means optional, and "N/A" means not applicable[4]. If a header is listed in Table 5-2 but not on Table 5-3, then the header is not used in any requests.

Table 5-3

Mapping between
requests and
headers

Header	ACK	BYE	CAN	INV	OPT	REG
Accept	N/A	O	O	O	O	O
Accept-Encoding	N/A	O	O	O	O	O
Accept-Language	N/A	O	O	O	O	O
Authorization	O	O	O	O	O	O
Call-ID	M	M	M	M	M	M
Contact	O	N/A	N/A	O	O	O
Content-Disposition	O	O	N/A	O	O	O
Content-Encoding	O	O	N/A	O	O	O
Content-Language	O	O	N/A	O	O	O
Content-Length	M	M	N/A	M	M	M
Content-Type	O	O	N/A	O	O	O
CSeq	M	M	M	M	M	M
Date	O	O	O	O	O	O
Encryption	O	O	O	O	O	O
Expires	N/A	N/A	N/A	O	N/A	O
From	M	M	M	M	M	M
Hide	O	O	O	O	O	O
In-Reply-To	N/A	N/A	N/A	O	N/A	N/A
Max-Forwards	O	O	O	O	O	O
MIME-Version	O	O	O	O	O	O
Organization	N/A	N/A	N/A	O	O	O
Priority	N/A	N/A	N/A	O	N/A	N/A
Proxy-Authorization	O	O	O	O	O	O
Proxy-Require	O	O	O	O	O	O
Record-Route	O	O	O	O	O	O
Require	O	O	O	O	O	O
Retry-After	N/A	N/A	N/A	N/A	N/A	O

(continues)

Table 5-3

Continued

Header	ACK	BYE	CAN	INV	OPT	REG
Response-Key	N/A	O	O	O	O	O
Route	O	O	O	O	O	O
Subject	N/A	N/A	N/A	O	N/A	N/A
Supported	N/A	O	O	O	O	O
Timestamp	O	O	O	O	O	O
To	M	M	M	M	M	M
User-Agent	O	O	O	O	O	O
Via	M	M	M	M	M	M

[4]Some differences exist between Tables 5-3 and 5-4 and the information contained in RFC 2543, due to the fact that the authors have proposed certain changes to RFC2543. These changes are included in a new Internet draft.

Table 5-4 provides a mapping between headers and responses. You should note that the inclusion of a particular header in a response depends upon both the status code of the response and the request that led to the response. The "Status Code" column indicates the status codes for which a given header can be included in the response. In some cases, a given header can be used only with certain status codes. In other cases, a given header can be used with all status codes. The six columns corresponding to the six request methods indicate whether a given header can be used in a response to that particular type of request. For example, the Allow header can be used with a 200 or 405 status code; however, it can only be used with a 200 response if the response is sent as a result of an OPTIONS request.

The indication (c) means that the value of a header is copied from the request to the response.

General Headers General headers can be used within both requests and responses. These headers contain basic information that is needed for the handling of requests and responses. Examples include the "To:" header field, which indicates the recipient of the request, the "From:" header field, which indicates the originator of the request, and the "Call-ID:" header field, which uniquely identifies a specific invitation to a session.

One of the most useful general headers is the "Contact:" header, which provides a URL for use in future communication regarding a particular session. In a SIP INVITE, the Contact: header might be different from the From: header. In other words, the initiator of a SIP session does not need

Table 5-4

Mapping between headers and responses

Header	Status Code	ACK	BYE	Requests CAN	INV	OPT	REG
Accept	All but 415	N/A	N/A	N/A	N/A	O	O
Accept	415	N/A	O	O	O	O	O
Accept-Encoding	415	N/A	O	O	O	O	O
Accept-Language	415	N/A	O	O	O	O	O
Allow	200	N/A	N/A	N/A	O	O	O
Allow	405	M	M	M	M	M	M
Authorization	All	O	O	O	O	O	O
Call-ID	All (c)	M	M	M	M	M	M
Contact	1XX	N/A	N/A	N/A	O	O	N/A
Contact	2XX	N/A	N/A	N/A	O	O	O
Contact	3XX, 485	N/A	O	N/A	O	O	O
Content-Disposition	All	O	O	N/A	O	O	O
Content-Encoding	All	O	O	N/A	O	O	O
Content-Language	All	O	O	N/A	O	O	O
Content-Length	All	M	M	M	M	M	M
Content-Type	All	O	O	N/A	O	O	O
CSeq	All (c)	M	M	M	M	M	M
Date	All	O	O	O	O	O	O
Encryption	All	O	O	O	O	O	O
Expires	All	N/A	N/A	N/A	O	N/A	O
From	All (c)	M	M	M	M	M	M
MIME-Version	All	O	O	O	O	O	O
Organization	All	N/A	N/A	N/A	O	O	O
Proxy-Authenticate	401, 407	O	O	O	O	O	O
Record-Route	2XX, 401, 484	O	O	O	O	O	O

(continues)

Table 5-4

Continued

Header	Status Code	ACK	BYE	CAN	INV	OPT	REG
				Requests			
Require	All	O	O	O	O	O	O
Retry-After	404, 413, 480, 486, 500, 503, 600, 603	O	O	O	O	O	O
Server	All	O	O	O	O	O	O
Supported	All	N/A	O	O	O	O	O
Timestamp	All	O	O	O	O	O	O
To	All (c)	M	M	M	M	M	M
Unsupported	420	O	O	O	O	O	O
User-Agent	All	O	O	O	O	O	O
Via	All (c)	M	M	M	M	M	M
Warning	All	O	O	O	O	O	O
WWW-Authenticate	401	O	O	O	O	O	O

to be a participant in the session. An example of such usage would be a case in which a multi-party session is organized and initiated by an administrator who does not participate in the session itself. Used in responses, the Contact: header is useful for directing further requests (e.g., ACK) directly to the called user in the case where the original request passed through one or more proxies. This header can also be used to indicate a more appropriate address in the case where an INVITE issued to a given *Uniform Resource Identifier* (URI) failed to reach the user. An example of such usage would be with a 302 response (moved temporarily), where the Contact: header in the 302 response gives the current URI of the user.

Request Headers   Request headers apply only to SIP requests and are used to provide additional information to the server regarding the request itself or regarding the client. Examples include the "Subject:" header field, which can be used to provide a textual description of the topic of the session; the "Priority:" header field, which is used to indicate urgency of the request (emergency, urgent, normal, or non-urgent); and the "Authorization:" header field, which enables authentication of the request originator.

Response Headers   Response header fields apply only to response (status) messages. They are used to provide further information about the response—information that cannot be included in the status line. Examples of response header fields include the "Unsupported:" header field used to identify those features not supported by the server and the "Retry-After" header field, which can indicate when a called user will be available (in the case where the user is currently busy or unavailable).

Entity Headers   In SIP, the message body contains information about the session or information to be presented to the user. In the case of information regarding a session, the session description is most often specified according to SDP, in order to indicate information such as the RTP payload type and an address and port to which media should be sent. The purpose of the entity headers is to indicate the type and format of information included in the message body, so that the appropriate application can be called upon to act on the information within the message body. The entity header fields are Content-Length, Content-Type and Content-Encoding, Content Disposition, Content-Language, Allow, and Expires.

The Content-Length header field specifies the length of the message body in octets. The Content-Type header field indicates the media type of the message body. For VoIP, this header will usually indicate SDP, in which case the header field will appear as `Content-Type: application/sdp`.

The Content-Encoding header field is used to indicate any additional codings that have been applied to the message body (and hence, which decoding actions the recipient needs to take in order to obtain the media type indicated by the Content-Type header field). For example, you can legally compress the content of the message body. In such a case, the Content-Encoding header field would be used to pass information related to the compression scheme used.

The Allow header field does not need to be associated with a session description; rather, it is used in a Request to indicate the set of methods supported. This header field can be used in a response to indicate what a called server supports. For example, if a server is to respond with the status code 405 (Method Not Allowed), it must also include the Allow header to indicate what it does support.

The Expires header field gives the date and time after which the content of the message is no longer valid. For example, this header field can be used to limit the amount of time that a registration is active at a registrar.

Examples of SIP Message Sequences

Building upon the information provided already in this chapter, we provide a number of examples to demonstrate how SIP messages might appear in various situations. We do not explain all possible options in these examples, just some of the more common occurrences. For details about all available options, refer to RFC 2543.

Registration

The first request that a client issues is likely to be REGISTER, because this request provides the server with an address at which the user can be reached for SIP sessions. This request is somewhat similar to the Registration Request between a terminal and a gatekeeper in H.323.

Figure 5-8 shows a typical registration scenario. In this scenario, Collins has logged in at host `station1.work.com`. This action causes a REGISTER request to be sent to the local registrar. The Via: header field contains the path taken by the request so far, which requires the originating client to insert its own address in this field. Note the format of the Via: header—in particular, the fact that this header specifies the transport being used. The default is UDP.

The From: header field indicates the address of the individual who has initiated the registration. The To: header field indicates the "address of record" of the user being registered and is the address that the registrar will store for that user. The From: and To: fields will be identical in the case where a user is registering himself. These fields do not need to be identical, however, which means that one individual can perform a registration on behalf of another. Note that the To: header field is not used to contain the address of the registrar. That address is indicated in the first line of the request.

The originating client sets the Call-ID: header field. All REGISTER requests for an individual client should use the same value of Call-ID. In order to avoid the possibility that different clients might choose the same Call-ID, the recommended syntax for the Call-ID is `local-id@host`, thereby making the Call-ID host-specific.

The REGISTER request does not contain a message body, because the message is not used to describe a session of any kind. Therefore, the Content-Length: field is set to zero.

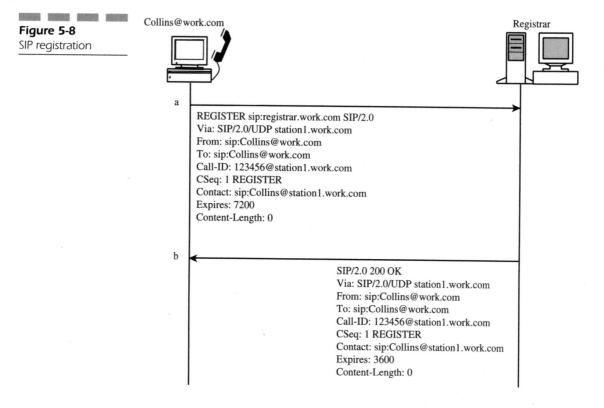

Figure 5-8
SIP registration

Collins@work.com

Registrar

a

REGISTER sip:registrar.work.com SIP/2.0
Via: SIP/2.0/UDP station1.work.com
From: sip:Collins@work.com
To: sip:Collins@work.com
Call-ID: 123456@station1.work.com
CSeq: 1 REGISTER
Contact: sip:Collins@station1.work.com
Expires: 7200
Content-Length: 0

b

SIP/2.0 200 OK
Via: SIP/2.0/UDP station1.work.com
From: sip:Collins@work.com
To: sip:Collins@work.com
Call-ID: 123456@station1.work.com
CSeq: 1 REGISTER
Contact: sip:Collins@station1.work.com
Expires: 3600
Content-Length: 0

The response to the REGISTER request is positive, as indicated by the 200 (OK) in the response line. Note that the content of the Via: header field is copied from request to response. This situation is also the case for the CSeq: header field. The CSeq: header (command sequence) is a field that indicates the method used in the request and an integer number. Consecutive requests for the same Call-ID must use contiguous increasing integer numbers. While not as critical for REGISTER requests, the use of the CSeq is important in other requests where different requests can be issued for the same Call-ID, in which case the CSeq is used to avoid ambiguity. A response to any request must use the same value of CSeq as used in the request itself.

In Figure 5-8, Collins indicated that future SIP messages should be routed to `sip:collins@station1.work.com`. Through the use of the Expires: header, Collins requested for the registration to be effective for two hours. The registrar chose to override this aspect of the request and limited the duration of the request to one hour. A registrar can change the length of time for which a given registration is effective, but if a registrar chooses to do so, it will normally set a lower registration interval than requested. The registrar will not set a higher interval. The Expires: header can be specified as a duration in seconds or as a specific date and time. In the case that it is specified as a number of seconds, the maximum value is such that a registration can be active for up to approximately 136 years, which should be long enough for anyone.

Having registered as shown in Figure 5-8, Collins could subsequently register at another terminal. In such a case, both registrations would become active, and subsequent invitations destined for Collins would be routed to both terminals. Of course, before registering at the second location, Collins could cancel the existing registration. This procedure requires sending another REGISTER request for the same address of record and Contact and specifying a registration interval of 0. In this case, the REGISTER request would be identical to the first REGISTER request, with the exceptions that the CSeq integer is incremented and the Expires header field contains the value 0. If Collins wished to cancel all existing registrations, then he would send a REGISTER message with the Expires header field set to 0 and the Contact header field populated with the wildcard character * (asterisk). The Expires header value of 0 indicates the cancellation of a registration, and the Contact: header value of * indicates that the request is applicable to all contact information for Collins.

Invitation

The INVITE request is the most fundamental and important SIP request, because it is the request used to initiate a session (i.e., establish a call).

Figure 5-9 shows a two-party call in which Collins initiates the session. The INVITE request is issued to `manager@work.com`, as seen from the Request-URI. The To: header field has the same value as the Request-URI, because in this case, we are not traversing any proxies. (As described later in this chapter, the Request-URI can be changed when a message passes through a proxy.) The From: header field indicates that the call is from `collins@work.com`. As you can see in Figure 5-9, SIP enables a display name to be used with the SIP URL. Thus, the called user's terminal could display the name Daniel when alerting the called user, rather than displaying the SIP URL. The optional header field Subject: is used to indicate the nature of the call, and the type of media that Daniel wishes to use is described within the message body. The Content-Type: entity header field indicates that this particular message body is described according to SDP. In the example, the length of xxx is used to

Figure 5-9

SIP establishment of a two-party call

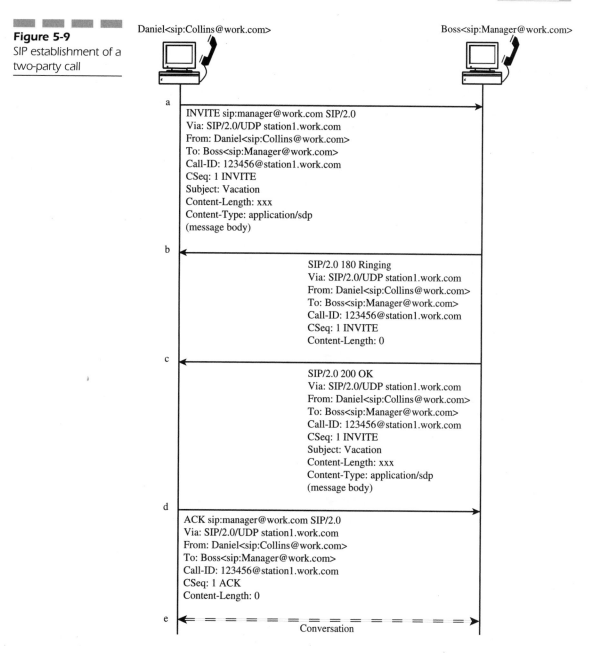

Daniel<sip:Collins@work.com>

Boss<sip:Manager@work.com>

a

INVITE sip:manager@work.com SIP/2.0
Via: SIP/2.0/UDP station1.work.com
From: Daniel<sip:Collins@work.com>
To: Boss<sip:Manager@work.com>
Call-ID: 123456@station1.work.com
CSeq: 1 INVITE
Subject: Vacation
Content-Length: xxx
Content-Type: application/sdp
(message body)

b

SIP/2.0 180 Ringing
Via: SIP/2.0/UDP station1.work.com
From: Daniel<sip:Collins@work.com>
To: Boss<sip:Manager@work.com>
Call-ID: 123456@station1.work.com
CSeq: 1 INVITE
Content-Length: 0

c

SIP/2.0 200 OK
Via: SIP/2.0/UDP station1.work.com
From: Daniel<sip:Collins@work.com>
To: Boss<sip:Manager@work.com>
Call-ID: 123456@station1.work.com
CSeq: 1 INVITE
Subject: Vacation
Content-Length: xxx
Content-Type: application/sdp
(message body)

d

ACK sip:manager@work.com SIP/2.0
Via: SIP/2.0/UDP station1.work.com
From: Daniel<sip:Collins@work.com>
To: Boss<sip:Manager@work.com>
Call-ID: 123456@station1.work.com
CSeq: 1 ACK
Content-Length: 0

e

Conversation

indicate some non-zero value. The actual length depends upon the content of the message body.

In Figure 5-9, the first response received is an indication that the user is being alerted. This information is conveyed through the use of the 180 status code.

NOTE: *Note that most of the header fields are copied from the request to the response, with the exception of the Content-Length and Content-Type, because this response does not contain a message body.*

Subsequently, the called user answers, and a 200 (OK) response appears. This response to an INVITE includes a number of header fields that have been copied from the original request. In addition, this response contains a message body describing the media that the called party wishes to use.

Finally, the caller sends an ACK to confirm receipt of the response. Note that the content of the CSeq: header field has changed to reflect the new request. For ACK, however, the integer part of CSeq: is not incremented. Also note that while most requests are answered with a 200 (OK) response if the request has been received and handled correctly, this situation is not the case for an ACK request. Once the ACK has been sent, the parties can exchange media.

Termination of a Call

A user agent who wishes to terminate a call does so through the issuance of a BYE request. Any party to the call can initiate this request. Upon receipt of a BYE request, the recipient of the request should immediately stop transmitting all media directed at the party who has issued the BYE request.

Figure 5-10 shows a call termination corresponding to the call origination shown in Figure 5-9. As you can see, the BYE request from Daniel has many of the same header values as in the original INVITE request, with the important exception that the command sequence (CSeq) has changed to reflect the BYE request, as opposed to the INVITE request. As with any successfully completed request (besides ACK), a 200 response appears.

Redirect and Proxy Servers

The examples shown so far have involved one user agent communicating directly with another. While that is certainly possible, a more likely scenario will involve the use of one or more proxy or redirect servers.

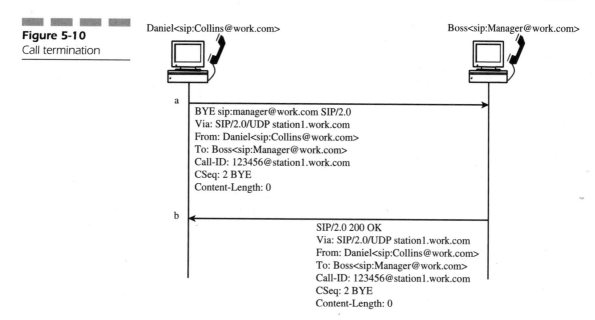

Figure 5-10
Call termination

Daniel<sip:Collins@work.com>   Boss<sip:Manager@work.com>

a

```
BYE sip:manager@work.com SIP/2.0
Via: SIP/2.0/UDP station1.work.com
From: Daniel<sip:Collins@work.com>
To: Boss<sip:Manager@work.com>
Call-ID: 123456@station1.work.com
CSeq: 2 BYE
Content-Length: 0
```

b

```
SIP/2.0 200 OK
Via: SIP/2.0/UDP station1.work.com
From: Daniel<sip:Collins@work.com>
To: Boss<sip:Manager@work.com>
Call-ID: 123456@station1.work.com
CSeq: 2 BYE
Content-Length: 0
```

Redirect Servers

A redirect server normally responds to a request with an alternative address to which the request should be directed. The exceptions are cases where the server has received a request that cannot be supported, in which case it will return a 4XX, 5XX, or 6XX response, or when the server receives a CANCEL request—to which it should respond with a 200 response.

Consider again the example call from Daniel to Boss, as depicted in Figure 5-9. Instead of the call-completion scenario depicted in Figure 5-9, let's assume that Boss is out of the office and has registered at a different location. Using a redirect server, we would see the call scenario as shown in Figure 5-11. The initial INVITE is sent to the Boss at the redirect server. To avoid clutter, we do not show the Subject: header field, the entity header fields, and the message body—although they are normally included in an INVITE.

The server responds with the status code 302 (Moved temporarily), and within the response, the message includes the Contact: header field. The content of this field is the address that the caller should try as an alternative.

The client then creates a new INVITE. This new INVITE uses the address received from the redirect server as the Request-URI, while the To:

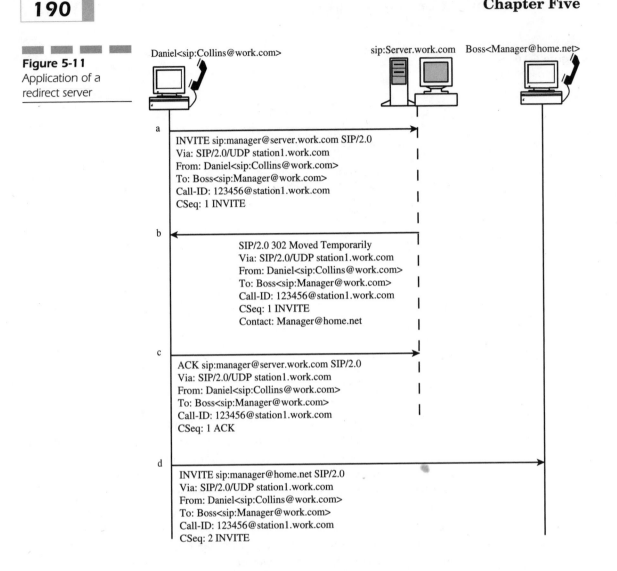

Figure 5-11
Application of a
redirect server

a
> INVITE sip:manager@server.work.com SIP/2.0
> Via: SIP/2.0/UDP station1.work.com
> From: Daniel<sip:Collins@work.com>
> To: Boss<sip:Manager@work.com>
> Call-ID: 123456@station1.work.com
> CSeq: 1 INVITE

b
> SIP/2.0 302 Moved Temporarily
> Via: SIP/2.0/UDP station1.work.com
> From: Daniel<sip:Collins@work.com>
> To: Boss<sip:Manager@work.com>
> Call-ID: 123456@station1.work.com
> CSeq: 1 INVITE
> Contact: Manager@home.net

c
> ACK sip:manager@server.work.com SIP/2.0
> Via: SIP/2.0/UDP station1.work.com
> From: Daniel<sip:Collins@work.com>
> To: Boss<sip:Manager@work.com>
> Call-ID: 123456@station1.work.com
> CSeq: 1 ACK

d
> INVITE sip:manager@home.net SIP/2.0
> Via: SIP/2.0/UDP station1.work.com
> From: Daniel<sip:Collins@work.com>
> To: Boss<sip:Manager@work.com>
> Call-ID: 123456@station1.work.com
> CSeq: 2 INVITE

header remains the same. The call-ID also remains the same, but the command sequence number (CSeq) is incremented. This number is always incremented when a second request of the same type is made for a given call. In this case, this request is the second INVITE for the call. The fact that the To: header is the same enables Boss (at home) to recognize that the call was originally made to his or her number at work. Based on this information (plus the identity of the caller and the subject of the call), Boss might or might not decide to accept the call. This example shows how SIP provides information to enable a user to make intelligent call-handling decisions.

Proxy Servers

A proxy server sits between a user agent client and the far-end user agent server. A proxy accepts requests from the client and forwards them, perhaps after some translation. Let's assume that Boss (at home) calls Collins (at work). If we assume that the call will pass via a proxy, as might be expected, then the sequence of messages would appear as in Figure 5-12. Again, the Subject: and entity header fields are omitted in the figure but are normally included.

The proxy receives the INVITE and forwards it. In doing so, the proxy changes the Request-URI. In general, a proxy will respond to an INVITE request with a provisional response code of 100 (Trying). This action is performed in parallel with forwarding the INVITE to the called user. The requirement to issue the provisional 100 response is mandatory in some cases and optional in others. In general, a server should provide this provisional response in all cases.

Of major importance for proxy servers is the Via: header field. The Via: header is used to indicate the path taken by a request so far. When a request is generated, the originating client inserts its own address into a Via: header field. Each proxy along the way also inserts its address into a new Via: header field (placed in front of any existing Via: headers, as seen in Figure 5-12). Therefore, the collection of Via: headers provides a map of the path taken through the network by a given request. When a proxy receives a request, it first checks whether its own address is already included in one if the header fields. If so, then the request has already been at that proxy, which indicates that a loop has been created. In such an event, the proxy will then respond to the request with status code 482 (loop detected). Otherwise, the proxy will pass the request onward after inserting its own address into a new Via: header field.

Responses also include Via: header fields, which are used to send a response back through the network along the same path that the request used (in reverse, of course). When a server responds to a request, the response contains the list of Via: headers exactly as received in the request. When a proxy receives a response, the first Via: header should refer to itself. If not, a problem exists, and the message is discarded. Assuming that the first Via: header field indicates the proxy itself, then the proxy removes the header and checks whether a second Via: header exists. If not, then the message is destined for the proxy itself. If a second Via: header field exists, then the proxy passes the response to the address in that field. In this way, the response finds its way back to the originator of the request along the path that the request originally took.

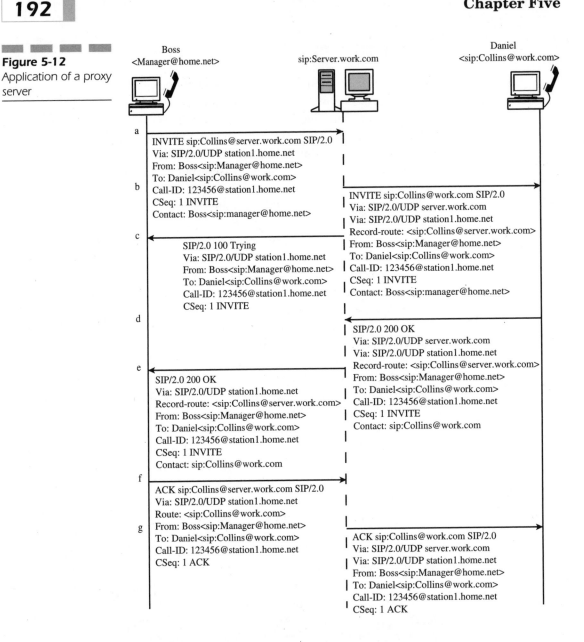

Figure 5-12
Application of a proxy server

Boss
<Manager@home.net>

sip:Server.work.com

Daniel
<sip:Collins@work.com>

a
INVITE sip:Collins@server.work.com SIP/2.0
Via: SIP/2.0/UDP station1.home.net
From: Boss<sip:Manager@home.net>
To: Daniel<sip:Collins@work.com>
Call-ID: 123456@station1.home.net
CSeq: 1 INVITE
Contact: Boss<sip:manager@home.net>

b
INVITE sip:Collins@work.com SIP/2.0
Via: SIP/2.0/UDP server.work.com
Via: SIP/2.0/UDP station1.home.net
Record-route: <sip:Collins@server.work.com>
From: Boss<sip:Manager@home.net>
To: Daniel<sip:Collins@work.com>
Call-ID: 123456@station1.home.net
CSeq: 1 INVITE
Contact: Boss<sip:manager@home.net>

c
SIP/2.0 100 Trying
Via: SIP/2.0/UDP station1.home.net
From: Boss<sip:Manager@home.net>
To: Daniel<sip:Collins@work.com>
Call-ID: 123456@station1.home.net
CSeq: 1 INVITE

d
SIP/2.0 200 OK
Via: SIP/2.0/UDP server.work.com
Via: SIP/2.0/UDP station1.home.net
Record-route: <sip:Collins@server.work.com>
From: Boss<sip:Manager@home.net>
To: Daniel<sip:Collins@work.com>
Call-ID: 123456@station1.home.net
CSeq: 1 INVITE
Contact: sip:Collins@work.com

e
SIP/2.0 200 OK
Via: SIP/2.0/UDP station1.home.net
Record-route: <sip:Collins@server.work.com>
From: Boss<sip:Manager@home.net>
To: Daniel<sip:Collins@work.com>
Call-ID: 123456@station1.home.net
CSeq: 1 INVITE
Contact: sip:Collins@work.com

f
ACK sip:Collins@server.work.com SIP/2.0
Via: SIP/2.0/UDP station1.home.net
Route: <sip:Collins@work.com>
From: Boss<sip:Manager@home.net>
To: Daniel<sip:Collins@work.com>
Call-ID: 123456@station1.home.net
CSeq: 1 ACK

g
ACK sip:Collins@work.com SIP/2.0
Via: SIP/2.0/UDP server.work.com
Via: SIP/2.0/UDP station1.home.net
From: Boss<sip:Manager@home.net>
To: Daniel<sip:Collins@work.com>
Call-ID: 123456@station1.home.net
CSeq: 1 ACK

Proxy State Strictly speaking, a proxy can be either stateless or stateful. If stateless, then the proxy takes an incoming request, performs whatever translation is necessary, forwards the corresponding outgoing request, and forgets that anything ever happened. If stateful, then the proxy remembers incoming requests and corresponding outgoing requests and can act more intelligently on subsequent requests and responses related to the same session.

Two important headers in Figure 5-12 are the Record-Route: and the Route: headers. In the example, all messages and responses pass through the same proxy. This situation does not have to be the case, however. While the response to a given request must return along the same route as used by the request, two consecutive requests do not need to follow the same path. Given that an initial INVITE can contain a Contact: header and responses can also contain a Contact: header, the two ends of the call might have the addressing information necessary to communicate directly. Hence, after the initial INVITE request and response, subsequent requests and responses can be sent end-to-end. A proxy might require that it remain in the signaling path for all subsequent requests, however. In particular, a stateful proxy must stay in the signaling path in order to maintain state regarding the session. A proxy can ensure that it remains in the signaling path through the use of the Record-Route: header.

For stateful proxies, each proxy along the path of the INVITE inserts its address into the Record-Route: header such that the header ultimately becomes a list of addresses, each of the form user@proxy. Each proxy places its address at the head of the list, such that at the end of the signaling path, the first entry in the list is that of the last proxy used.

A 200 response to the INVITE will include the Record-Route: header as received by the ultimate destination of the INVITE. Hence, this response contains an ordered list of all the proxies used in the INVITE. The content of the header is propagated back through the network, along with the 200 response. When the 200 response is received at the client that originated the INVITE, the information contained in the Record-Route: header is used in subsequent requests (such as ACK or BYE) related to the same call. The client takes the information in the Record-Route: header and places it in a Route: header in reverse order, so that the Route: header contains a list of proxies in the direction from calling client to called server. At the end of the list, the client adds the content of any Contact: header received from the called server. Finally, it removes the first proxy entry from the list and sends the next message (e.g., ACK) to that proxy. At each proxy along the path from caller to callee, the first entry in the Route: header is removed before forwarding the message to that entity, be it a proxy or the ultimate called server. The net effect of these actions is that the first Route: header entry will always contain the address of the next hop along the path from caller to callee. Therefore, each proxy automatically knows where to forward the request.

Forking Proxy   A proxy can "fork" requests. This action would occur when a particular user is registered at several locations. An incoming INVITE for the user would be sent from the proxy to each of the registered locations. If one of the locations answered, then the proxy could issue a CANCEL to the other locations so that they do not continue ringing when the called user has answered elsewhere. In order to handle such forking, a proxy must be stateful.

Figure 5-13 gives an example of a forked request. For the sake of clarity, the Subject:, Contact:, Record-Route:, Route:, and entity header fields are

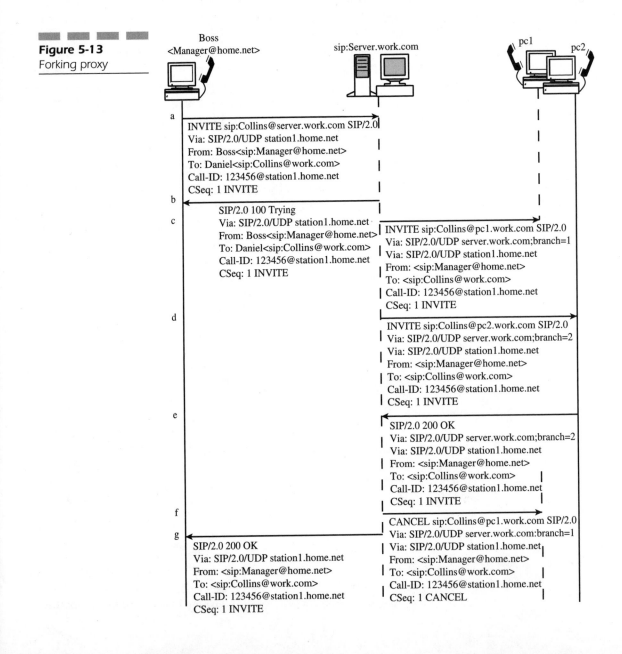

Figure 5-13
Forking proxy

Boss
<Manager@home.net>

sip:Server.work.com

pc1   pc2

a   INVITE sip:Collins@server.work.com SIP/2.0
    Via: SIP/2.0/UDP station1.home.net
    From: Boss<sip:Manager@home.net>
    To: Daniel<sip:Collins@work.com>
    Call-ID: 123456@station1.home.net
    CSeq: 1 INVITE

b   SIP/2.0 100 Trying
c   Via: SIP/2.0/UDP station1.home.net
    From: Boss<sip:Manager@home.net>
    To: Daniel<sip:Collins@work.com>
    Call-ID: 123456@station1.home.net
    CSeq: 1 INVITE

    INVITE sip:Collins@pc1.work.com SIP/2.0
    Via: SIP/2.0/UDP server.work.com;branch=1
    Via: SIP/2.0/UDP station1.home.net
    From: <sip:Manager@home.net>
    To: <sip:Collins@work.com>
    Call-ID: 123456@station1.home.net
    CSeq: 1 INVITE

d   INVITE sip:Collins@pc2.work.com SIP/2.0
    Via: SIP/2.0/UDP server.work.com;branch=2
    Via: SIP/2.0/UDP station1.home.net
    From: <sip:Manager@home.net>
    To: <sip:Collins@work.com>
    Call-ID: 123456@station1.home.net
    CSeq: 1 INVITE

e   SIP/2.0 200 OK
    Via: SIP/2.0/UDP server.work.com;branch=2
    Via: SIP/2.0/UDP station1.home.net
    From: <sip:Manager@home.net>
    To: <sip:Collins@work.com>
    Call-ID: 123456@station1.home.net
    CSeq: 1 INVITE

f   CANCEL sip:Collins@pc1.work.com SIP/2.0
g   Via: SIP/2.0/UDP server.work.com:branch=1
    Via: SIP/2.0/UDP station1.home.net
    From: <sip:Manager@home.net>
    To: <sip:Collins@work.com>
    Call-ID: 123456@station1.home.net
    CSeq: 1 CANCEL

    SIP/2.0 200 OK
    Via: SIP/2.0/UDP station1.home.net
    From: <sip:Manager@home.net>
    To: <sip:Collins@work.com>
    Call-ID: 123456@station1.home.net
    CSeq: 1 INVITE

omitted. In this scenario, Boss (at home) calls Collins (at work). Collins happens to be logged in at two different workstations, however— `pc1.work.com` and `pc2.work.com`. When the proxy receives the INVITE, it immediately returns a 100 (trying) response. RFC 2543 requires a forking proxy to return this provisional response immediately.

The proxy then forwards the INVITE request in parallel to the two locations. Not only does the proxy add a Via header to each outgoing INVITE, but it also adds a branch parameter (`;branch=xxx`). The purpose of the branch parameter is to distinguish between the various outgoing requests, and more importantly, the responses to those requests. Given that the response to a request will contain the same Via: header field information as contained in the request itself, the branch parameter enables clear distinction between the responses.

Collins answers the incoming call at `pc2.work.com`. Therefore, a 200 OK response is returned from `pc2.work.com`. Upon receiving the response, the proxy immediately issues a CANCEL to `pc1.work.com` so that any resources allocated by that machine can be released. pc1 responds to the CANCEL with a 200 (OK) response (not shown). The response from `pc2.work.com` is then relayed back to the originator of the INVITE request. The remainder of the call (ACK, BYE, etc.) will proceed either via the proxy or directly end-to-end, depending on whether the two endpoints each inserted a Contact: header field and whether the proxy made use of the Record-Route: and Route: header fields.

The Session Description Protocol (SDP)

The examples so far focused on various SIP requests, responses, and header fields. While we have made some mention of the message body and entity header fields, we have not described the message body in any detail. We did make mention, however, of the fact that the message body will contain information about the media to be exchanged, such as RTP payload type, addresses, and ports. Moreover, we have seen that the format of the description will most often be according to SDP. That protocol is specified in RFC 2327, and the next part of this chapter is devoted to describing this protocol in some detail.

The Structure of SDP

SDP simply provides a format for describing session information to potential session participants. Basically, a session consists of a number of media streams. Therefore, the description of a session involves the specification of a number of parameters related to each of the media streams. Information also exists that is common to the session as a whole, however. Therefore, we have session-level parameters and media-level parameters. Session-level parameters include information such as the name of the session, the originator of the session, and the time(s) that the session is to be active. Media-level information includes media type, port number, transport protocol, and media format. Figure 5-14 illustrates this general structure.

Because SDP simply provides session descriptions and does not provide a means for transporting or advertising the sessions to potential participants, we must use SDP in conjunction with other protocols (such as SIP). For example, SIP carries SDP information within the SIP message body.

Similar to SIP, SDP is a text-based protocol that utilizes the ISO 10646 character set in UTF-8 encoding (RFC 2044). This coding enables the usage of multiple languages and includes US-ASCII as a subset. While SDP field names use only US-ASCII, textual information can be passed in any language. For example, the field name that indicates the name of a session must be in US-ASCII, but the actual string of characters that comprise the

Figure 5-14
SDP session
description structure

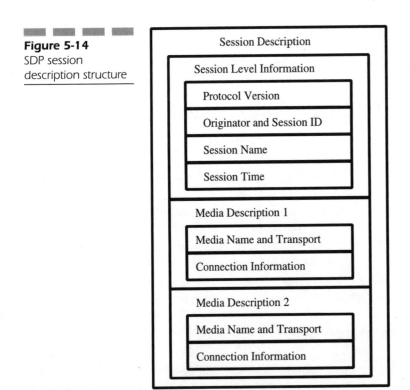

session name can be in any language. Although the use of ASCII coding in SDP as opposed to binary coding is a little bandwidth greedy, SDP is written is a compact form to counteract bandwidth inefficiency. Thus, where English words might have been used for field names, we use single characters instead. For example, v=version, s=session name, and b=bandwidth information.

SDP Syntax

SDP conveys session information by using a number of lines of text. Each line uses the format `field=value`, where `field` is exactly one character (case-sensitive) and `value` is dependent on the field in question. In some cases, `value` might consist of a number of separate pieces of information separated by spaces. No spaces are permitted between `field` and the = (equals) sign or between the = sign and `value`, however.

Session-level fields must be included first, followed by media-level fields. The boundary between session-level data and media-level data occurs at the first occurrence of the media description field (m=). Each subsequent occurrence of the media-description field marks the beginning of data related to another media stream in the session.

Mandatory Fields SDP includes mandatory fields, which must be included in any session description, and optional fields (which can be omitted). The following fields are mandatory:

- *v= (protocol version)* This field also marks the start of a session description and the end of any previous session description.
- *o= (session origin or creator and session identifier)*
- *s= (session name)* This field is a text string that could be displayed to potential session participants.
- *t= (time of the session)* This field provides the start time and stop time for the session.
- *m= (media)* This field indicates the media type, the transport port to which the data should be sent, the transport protocol (e.g., RTP), and the media format (typically an RTP payload format). The appearance of this field also marks the boundary between session-level information and media-level information or between different media descriptions.

Optional Fields Of the optional fields in SDP, some apply only at the session level, and some apply at the media level. Some optional fields can be applied at both the session level and media level. In such cases, the value applied at the media level overrides the session-level value for that media instance. Thus, if a given field has a value X for the session and a value Y for media type 1, then value X applies to all media except for media type 1, where value Y applies. The following fields are optional:

- *i= (session information)*   This field is a text description of the session and is meant to provide greater detail than the session name. This field can be specified at both the session level and at the media level.

- *u= (URI of description)*   This field is a URI (e.g., a Web address) where we can obtain further session information. For example, a session agenda might be posted on a Web page, in which case the URI of the Web page would be specified. Only one URI can be specified per session.

- *e= (e-mail address)*   This field is an e-mail address for the individual who is responsible for the session. Several instances of this field might exist, and these fields would be used in the case that several individuals could be contacted for more information. This field applies only at the session level.

- *p= (phone number)*   This field is a phone number for the individual who is responsible for the session. As for the e-mail address field, there might be several instances of this field. This field applies only at the session level.

- *c= (connection information)*   This field provides connection data, including connection type, network type, and connection address. This field can be applied either at the session level or at the media level.

- *b= (bandwidth information)*   This optional field specifies the bandwidth required in kilobits per second. This field can be specified at either the session level or the media level (or both) and indicates the bandwidth required either by the whole conference or by a given media type. When used at session level, this field can be considered analogous to a traditional dial-in conference, where a maximum number of lines is available.

- *r= (repeat times)*   In the case of a regularly scheduled session, this field specifies how often and how many times a session is to be repeated.

- *z= (timezone adjustments)*   This field is used with regularly scheduled sessions. In such cases, the schedule of sessions might span a change from standard time to Daylight Savings Time (or vice-versa). Because such changes occur at different times in different parts of the world, confusion might exist as to the actual start time of a specific session, particularly if it is to occur in spring or fall. This field enables the session creator to specify up front when such timing changes will occur.

■ *k= (encryption key)* This field provides an encryption key or specifies a mechanism by which a key can be obtained for the purposes of encrypting and decrypting the media. The field can be applied either at the session level, at the media level, or both.

■ *a= (attributes)* This field is used to describe additional attributes to be applied to the session or to individual media.

Of the optional fields, the connection information field would perhaps be better described as conditional. Because this field defines the network address to which data should be sent, it must be included somewhere in the session description. In fact, this field must appear either at the session level or at each media level. If this field appears at both the session level and the media level, then the media-level value applies to that particular media type.

Ordering of Fields Because certain fields can appear both at the session level and at the media level, the ordering of fields is important to avoid ambiguity. SDP demands that the following order be used:

Session Level

■ Protocol version (v)

■ Origin (o)

■ Session name (s)

■ Session information (i) (optional)

■ URI (u) (optional)

■ E-mail address (e) (optional)

■ Phone Number (p) (optional)

■ Connection Information (c) (optional)

■ Bandwidth information (b)

■ Time description (t)

■ Repeat information (r) (optional)

■ Time zone adjustments (z) (optional)

■ Encryption key (k) (optional)

■ Attributes (a) (optional)

Media Level

- Media description (m)
- Media information (i) (optional)
- Connection information (c) (optional if specified at the session level)
- Bandwidth information (b) (optional)
- Encryption key (k) (optional)
- Attributes (a) (optional)

Subfields   Of the various fields specified in SDP, a number are structured in several subfields. In such cases, the value of the field is a combination of several values separated by spaces. The SDP specification shows this structure as follows:

```
Field = <value of subfield 1> <value of subfield 2> <value of
subfield 3>.
```

A number of the more important such fields are described here. Refer to the SDP specification for a complete definition of all fields.

Origin   Origin (o) has six subfields: username, session ID, version, network type, address type, and address.

The username is the originator's login identity at the host machine or is "-" if no login identity applies.

The session ID is a unique ID for the session, most likely created by the host machine of the originator. In order to ensure that the ID is unique, RFC 2327 recommends the ID to make use of a *Network Time Protocol* (NTP) timestamp.

The version is a version number for this particular session. When a session is advertised, the version number is important for the purpose of distinguishing a revised version of the session from an earlier version.

The network type is a text string indicating the type of network. The string IN refers to Internet.

The address type is the type of address in the network. SDP defines IP4 and IP6 to denote IP version 4 and IP version 6, respectively.

The address is the network address of the machine where the session was created. This address can be a fully-qualified domain name or can be the actual IP address. This address can be the actual IP address only if the IP address is unique, which might not be the case if the address is local, such as an IP address on a LAN behind a proxy.

Connection Data Connection Data (c) has three subfields: network type, address type, and connection address. Although some of these have the same names as the subfields defined for the origin, they have different meanings. They refer to the network and address at which media data are to be received, as opposed to the network and address at which the session was created.

Network type indicates the type of network in question. Currently, only the value IN is defined.

Address type will have the value IP4, meaning IP version 4 (because this value is the only value specified in the SDP specification to date).

The Connection address is the address to which data should be sent. While the address might be a dotted-decimal IP address, a fully qualified domain name is generally a better choice, because it enables more flexibility and less ambiguity.

Media Information Media Information (m) has four subfields: media type, port, transport, and formats.

The media type can be audio, video, application, data, or control. For voice, the media type will be audio.

The port subfield indicates the port number to which media should be sent. The port number depends on the type of connection in question and the transport protocol. For VoIP, however, the media are normally carried by using RTP over UDP. Thus, the port number will be an even value between 1024 and 65535.

The format subfield lists the various types of media format that can be supported. In the case where a given user can support voice that is coded in one of several ways, each supported format is listed with the preferred format listed first. Normally, the format will be an RTP payload format with a corresponding payload type. In this case, you only have to specify that the media is an RTP audio/video profile and specify the payload type.

If we have a system that wishes to receive voice on port number 45678 and can only handle speech coded according to G.711 mu-law, then RTP payload type 0 applies, and the media information is as follows:

```
m=audio 45678 RTP/AVP 0
```

If we have a system that wishes to receive voice on port 45678 and can handle speech that is coded according to any of G.728 (payload type 15), GSM (payload type 3), or G.711 mu-law (payload type 0), and if the system prefers to use G.728, then the media information is as follows:

```
m=audio 45678 RTP/AVP 15 3 0
```

Attributes   SDP includes a field called attributes (a), which enables additional information to be included. Attributes can be specified at the session level, media level, or both. Furthermore, multiple attributes can be specified for the session as a whole and for a given media type. Therefore, numerous attribute fields might be present in a single session description, with the meaning and significance of the field dependent upon its position within the session description. If an attribute is listed prior to the first media information field, then the attribute is a session-level attribute. If the attribute is listed after a given media information field (m), then it applies to that media type.

Attributes can have two forms. The first form is a property attribute, where the attribute specifies that a session or media type has a specific characteristic. The second form is a value attribute, which is used to specify that a session or media type has a particular characteristic of a particular value. SDP describes a number of suggested attributes.

Examples of property attributes described in SDP are `sendonly` and `recvonly`. The first of these specifies that the sender of the session description wishes to send data but does not wish to receive data. In this case, the port number has no significance and should be set to zero. The second specifies that the sender of the session description wishes only to receive data and not send data.

An example of a value attribute would be `orient`, which is used in a shared white-board session and indicates whether the white-board has a portrait or landscape orientation. These examples would be included in SDP by using the following syntax:

```
a=sendonly
a=recvonly
a=orient:landscape.
```

The `rtpmap` attribute provides one important use of attributes in a VoIP application. This attribute can be applied to a media stream and is particularly useful when the media format is not a static RTP payload type. `rtpmap` has the following format:

```
a=rtpmap:<payload type> <encoding name>/<clock rate>[/<encoding
parameters>].
```

The payload type corresponds to the RTP payload type. The encoding name is a text string. RTP profiles that specify the use of dynamic payload types must define the set of valid encoding names and/or a means to register encoding names if that profile is to be used with SDP. The clock rate is the rate in Hz. The optional encoding parameters can be used to indicate the number of audio channels.

Strictly speaking, the rtpmap attribute is only necessary for dynamic payload types. For example, standard G.711 speech is a static RTP payload type and would be fully described by a media field such as the following:

```
m=audio 45678 RTP/AVP 0.
```

A dynamic RTP payload type needs greater information for the remote end to fully understand the media coding, however. An example of such a dynamic payload type is 16-bit linear-encoded stereo audio sampled at 16KHz. If we wish to use the dynamic RTP/AVP payload type 98 for such a stream, then the media description must be followed by an rtpmap attribute as follows:

```
m=video 45678 RTP/AVP 98
a=rtpmap 98 L16/16000/2
```

If the media description indicates multiple payload types, then there might be a separate rtpmap attribute specified for each. Although the rtpmap attribute is not necessary for static RTP payload types, SIP recommends the attribute to always be included when SDP is used for a SIP message body, even in the case of static payload types.

Usage of SDP with SIP

SIP and SDP make a wonderful partnership for the transmission of session information. While SIP provides the messaging mechanisms for the establishment of multimedia sessions, SDP provides a structured language for describing those sessions. The message body in SIP, identified by the entity headers, provides a neat slot where SDP can be used. We will now discuss this usage.

If we take the example outlined in Figure 5-9 and assume that Collins can support voice that is coded according to G.726, G.723, or G.728, and we assume that Boss can only support voice that is coded according to G.728, then the messages would appear as shown in Figure 5-15. Note the usage of the rtpmap attribute for each of the media formats that can be supported. Note also that, for the avoidance of clutter, many of the SIP headers that should be included, such as Via:, From:, To:, Call-ID:, and Subject: have been omitted in the diagram. Moreover, the 180 (Ringing) response has also been omitted. The user should assume that these are included according to Figure 5-9.

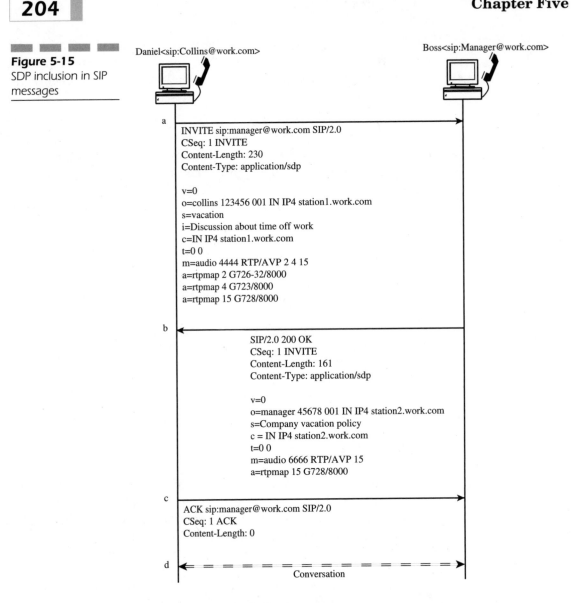

Figure 5-15
SDP inclusion in SIP
messages

Daniel<sip:Collins@work.com>

Boss<sip:Manager@work.com>

a

INVITE sip:manager@work.com SIP/2.0
CSeq: 1 INVITE
Content-Length: 230
Content-Type: application/sdp

v=0
o=collins 123456 001 IN IP4 station1.work.com
s=vacation
i=Discussion about time off work
c=IN IP4 station1.work.com
t=0 0
m=audio 4444 RTP/AVP 2 4 15
a=rtpmap 2 G726-32/8000
a=rtpmap 4 G723/8000
a=rtpmap 15 G728/8000

b

SIP/2.0 200 OK
CSeq: 1 INVITE
Content-Length: 161
Content-Type: application/sdp

v=0
o=manager 45678 001 IN IP4 station2.work.com
s=Company vacation policy
c = IN IP4 station2.work.com
t=0 0
m=audio 6666 RTP/AVP 15
a=rtpmap 15 G728/8000

c

ACK sip:manager@work.com SIP/2.0
CSeq: 1 ACK
Content-Length: 0

d

Conversation

In the scenario shown in Figure 5-15, both parties can support voice according to the same coding scheme. Therefore, media exchange (i.e., conversation) can take place.

Negotiation of Media

Likely, cases will exist where one party to a call will be unable to support the media formats specified by the other party. If, for example, in Figure 5-15

Boss could only support voice that is coded according to G.711 mu-law (RTP payload type 0), then there would be a mismatch between the capabilities of the called party and those of the calling party. In such a situation, the response to the INVITE should contain a 488 (Not Acceptable Here) or a 606 (Not Acceptable) status code. In addition, the response should contain a Warning: header field with the warning code of 304 (media type not available) or 305 (incompatible media format). In that case, the caller could choose to issue a new INVITE request.

This information applies to situations where the caller and called party cannot agree on any media stream. An INVITE can contain a session description with multiple media streams, however (e.g., an audio stream and a video stream). In such an event, the possibility exists that the called party can support one stream and cannot support another stream. For example, the called party might support audio but not video. In such an event, the called party should return a media description for the stream it can support. A media description for the unsupported stream should also be returned, but with a port number of zero.

Imagine again if Boss could only support G.711. Also imagine that Boss had incorrectly returned a 200 OK response with a session description specifying G.711. In such a case, Collins should not issue a revised session description in the ACK message, even if Collins could support G.711. The ACK message is not meant as an opportunity for the caller to change his mind about the media to be used. The ACK should contain a session description only in the event that the original INVITE did not contain a complete session description. Therefore, in the scenario just described, Collins should issue a new INVITE (after terminating the original call).

The fact that VoIP enables the use of a range of voice-coding techniques means that there can be a mismatch between the capabilities of different SIP users. When there is some doubt about a party's capability to handle a particular media type, the OPTIONS method can be used to provide a useful mechanism for finding out in advance. This procedure avoids the partial establishment of a session that is doomed to fail.

OPTIONS Method   A potential caller can use the OPTIONS method to determine the capabilities of a potential called party. The recipient of the OPTIONS request should respond with the capabilities supported.

Figure 5-16 shows an example of the use of the OPTIONS request. In this case, Collins (at work) queries the capabilities of Boss (at work). Boss responds with an indication that he can accept speech coded according to G.711 mu-law and GSM (RTP payload formats 0 and 3, respectively).

Figure 5-16
Usage of the
OPTIONS method

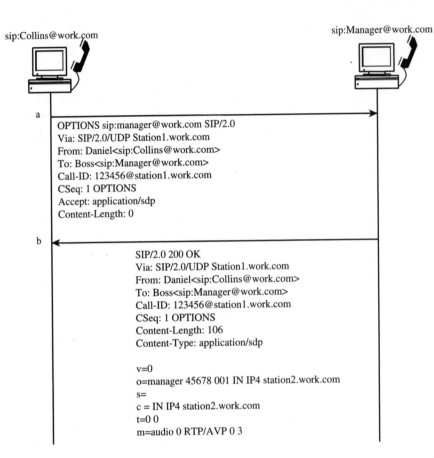

sip:Collins@work.com

sip:Manager@work.com

a

OPTIONS sip:manager@work.com SIP/2.0
Via: SIP/2.0/UDP Station1.work.com
From: Daniel<sip:Collins@work.com>
To: Boss<sip:Manager@work.com>
Call-ID: 123456@station1.work.com
CSeq: 1 OPTIONS
Accept: application/sdp
Content-Length: 0

b

SIP/2.0 200 OK
Via: SIP/2.0/UDP Station1.work.com
From: Daniel<sip:Collins@work.com>
To: Boss<sip:Manager@work.com>
Call-ID: 123456@station1.work.com
CSeq: 1 OPTIONS
Content-Length: 106
Content-Type: application/sdp

v=0
o=manager 45678 001 IN IP4 station2.work.com
s=
c = IN IP4 station2.work.com
t=0 0
m=audio 0 RTP/AVP 0 3

NOTE: *Note the use of the Accept header field. The issuer of the request uses this field to indicate what types of media are acceptable in the response. If the header field is omitted in the request, then a default value of application / sdp is assumed.*

Usage of SIP for Features and Services

We have already described a number of straightforward SIP messaging scenarios in this chapter. SIP can be used to implement a range of more advanced services, however. In fact, we have already seen that SIP inherently supports personal mobility through the use of registration. We have seen how a forking proxy can enable a one-number service, and we have also seen how the Retry-After: header can be used to support call-completion services. In many ways, these features are just the tip of the iceberg in terms of what SIP offers.

SIP messages can carry MIME (Multi-Purpose Internet Mail Extension) content as well as an SDP description. Therefore, a response to an INVITE could include a piece of text, an HTML document, an image, etc. Equally, because a SIP address is a URL, it can easily be included in Web content for click-to-call applications. Basically, SIP is well suited for integration with existing IP-based applications and can leverage those applications in the creation of new services. SIP not only supports new and exciting services, however, but it can also be used to implement the existing supplementary *Custom Local Area Signaling Service* (CLASS) services that exist in traditional telephony today—including call waiting, call forwarding, multi-party calling, call screening, etc.

In many cases, the signaling for a SIP call will be routed through a proxy. This process is useful in many ways because it enables the proxy to invoke various types of advanced-feature logic. The feature logic might reside at the proxy or might be located in a separate feature server or database. In fact, the proxy might also have access to many other functions, such as a policy server, an authentication server, etc. (as depicted in Figure 5-17). The

Figure 5-17
SIP service
architecture

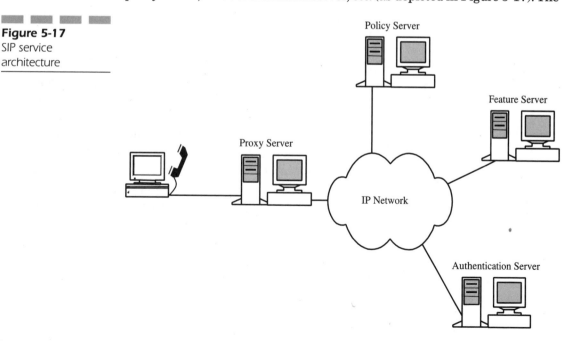

Policy server could hold call routing information and/or *Quality of Service* (QOS)-related information, while the feature server would hold subscriber-specific feature data, such as screening lists, forwarding information, etc. The proxy can possibly use the services of an *Intelligent Network* (IN) *Service Control Point* (SCP), with which communication might be conducted over the *IN Application Part* (INAP), as described in Chapter 7 ("VoIP and SS7"). While a distributed architecture has several advantages, it is also possible, of course, for the various functions depicted in Figure 5-17 to be integrated into the proxy in one monolithic implementation.

The following section describes a couple of examples of calling features that use SIP. These examples are simple and are certainly not meant to convey a complete picture of the features that SIP provides.

Call Forwarding

Figure 5-18 depicts a "call forwarding on busy" scenario. Note that, in order to avoid unnecessary clutter, the message body and many of the message headers have been omitted from the figure. These items would of course be included in a real signaling implementation.

User 1 calls User 2 via a proxy. User 2 wishes for calls to be forwarded to User 3 whenever User 2 is busy. Because User 2 happens to be busy, a 480 (Busy here) response is returned. Upon receiving the response, the proxy forwards the call to User 3. Note that the INVITE request from the proxy to User 3 contains the same To: field as in the INVITE to User 2, which enables User 3 to recognize that this call is a forwarded call originally sent to User 2. The 200 (OK) response from User 3 contains a Contact: header, indicating User 3 (which enables subsequent messages for this call to pass directly from User 1 to User 3). The proxy could have chosen to insert a Record-route: header in the INVITE, however. This action would ensure that future messages and responses are sent through the proxy.

A "call forwarding on no answer" scenario would be quite similar to the call forwarding on busy scenario. The difference is that the INVITE from the proxy to User 2 would time out. The proxy would then send a CANCEL to User 2 and then forward the call by sending an INVITE to User 3.

The proxy would completely take care of a "call forwarding unconditional" service. The proxy would not bother to send an INVITE to User 2 at all; rather, it would simply send an INVITE to User 3. The To: field of the INVITE to User B2 would still indicate user B1, as is the case for all call-forwarding scenarios.

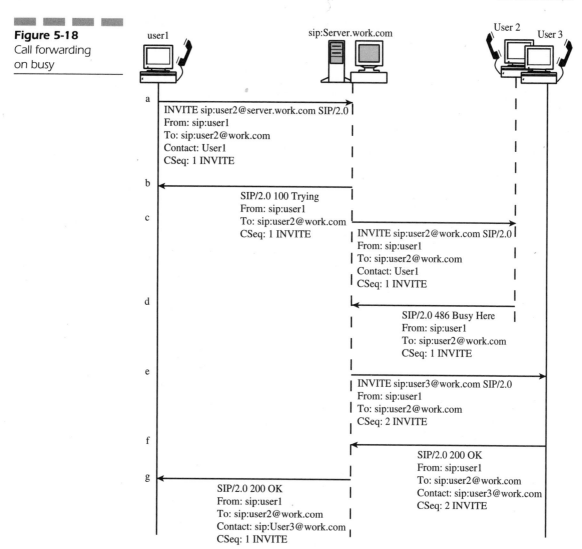

Figure 5-18
Call forwarding on busy

Consultation Hold

Figure 5-19 depicts a scenario in which User B calls User A, and after speaking for a while, User B places User A on hold and places a call to User C. Once the call to User C is finished, User B reopens communication with user A. Again, to avoid clutter, many of the SIP headers are omitted.

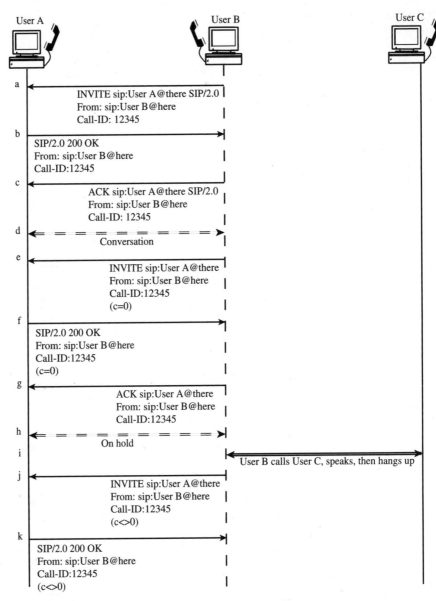

Figure 5-19
Consultation hold

Note the manner in which a call is placed on hold. A SIP INVITE is issued with the same Call-ID and other parameters as the original INVITE, but the SDP connection (c) field is set to an address of all zeros (message e in the figure). The 200 OK response to this INVITE also contains a session description, where the address for the media stream is all zeros. This 200

(OK) response receives an ACK, and at this point, the call is effectively on hold because no media streams are being transmitted.

In the example, User B places a call to User C in a normal manner. After that call is finished, User B reopens communication with User A by issuing another INVITE, again with the same Call-ID as the original call (message j in the figure). The message body in this INVITE no longer has a connection field of all zeros. Instead, the connection field uses the same address as in the original INVITE. Equally, the 200 (OK) response to the INVITE (message k) has a non-zero address in the SDP connection field. The 200 response is acknowledged (not shown), and communication between User A and User B is open again.

SIP Extensions and Enhancements

Since RFC 2543 was released in March 1999 as an Internet standards track protocol, SIP has attracted enormous interest. Many companies, including those from the traditional telecommunications environment and others such as cable TV providers and *Internet Service Providers* (ISPs), have recognized that SIP offers great potential. In the process, however, several individuals and companies have found that SIP would be better if some enhancements or extensions were added. A large number of extensions to SIP have been proposed and are being handled in the SIP working group of the IETF. At this writing, the enhanced version of SIP that is to include these various extensions is known as SIP-T. Because this name is not the first name that the enhanced version has been given, it might not be the last, either.

Just because a particular extension is proposed, however, there is no certainty that it will be included in an updated version of SIP. In fact, there have been extensions proposed in the past that have been rejected for one reason or another. Most of the extensions described here are useful, however, and align well with SIP as it is defined in RFC 2543. Some of them are particularly interesting and deserve mention. Moreover, some of the SIP extensions that have been proposed build upon other extensions. Thus, if a given extension is to be included in the protocol, it is likely that some others will, too. One thing is certain: RFC 2543 will be enhanced considerably to make it even more powerful before it finally becomes an Internet standard. Some, if not all, of the enhancements described here will be included. In fact, as of this writing, a new Internet draft has been prepared, which is an updated version of the original SIP protocol. This draft includes some of the enhancements described here, such as the 183 Session-Progress response, plus a number of clarifications and corrections to the original specification. We expect that the draft will be updated over time and will likely include even more of the enhancements described here. Note, however, that for the moment, the document is a draft and is subject to change or withdrawal.

The SIP 183 Session-Progress Message

This extension proposes the addition of a new response (status code 183), which enables the called user agent server to indicate to the calling user agent client that a one-way media path should be opened from called party to calling party. This process enables in-band call progress information to be provided to the calling client. Such information might include tones or announcements. The 183 (session progress) will include a session description to indicate the type of media to be returned to the caller.

In traditional telephony networks, circumstances exist in which in-band information needs to be passed from the called end to the calling end, prior to an answer from the called party. Such circumstances include announcements such as "Please hold while we attempt to connect your call" or other tones and announcements that are specific to the called network or system.

For those who are familiar with ISUP, this requirement is handled in traditional telephony networks with SS7 through the use of the *Address Complete Message* (ACM). The ACM is used to open a one-way audio path so that announcements or tones can be relayed back to the caller. In VoIP networks, the same basic need to open a one-way media path from the called end to the calling end will apply. Furthermore, if a SIP network resides between two PSTN networks and is connected to them via PSTN gateways, then those gateways will need to receive and send ACM messages. Hence, SIP needs to have an equivalent to the ACM message. The new 183 response is designed to fulfill this need.

Note that the SS7 ACM message, when used to open a one-way path, enables the ringing tone to be relayed from the called end to the calling end. In other words, when a person makes a phone call and hears the ringing tone, that tone is being produced at the far end. For this reason, calls to different countries produce a different ringing tone than domestic calls. We are not proposing, however, for the 183 (session-progress) response to be used to carry normal ring tones back to a calling party in a standard SIP-user to SIP-user call. The requirement to provide an alerting notification in such a scenario should be handled by the issuance of the existing 180 (ringing) response, causing the calling end to generate an alerting notification locally. In fact, in a SIP-to-SIP call, the alerting notification might not be a tone at all. The alert could come in the form of a textual indication, an icon on a screen, or any other appropriate notification.

The 183 response should only be used in cases where a temporary one-way media path is needed from the called end to the calling end until the call is answered. This response would be used for interoperating with the PSTN or under less-common circumstances, such as the need to provide special tones or announcements or to play music. The temporary media stream indicated in the 183 response should be a subset of that indicated in the original INVITE so that there will be some certainty that the caller will be able to handle the temporary media stream. The temporary media stream will be terminated as soon as the called user answers, at which point a 200 OK response will be returned, including a new session description.

The usage of the 183 Session Progress response in the context of interworking between SIP and the PSTN is described in greater detail a little later in this chapter.

The SIP INFO Method

A proposal was made to extend SIP by the addition of a new SIP method called the INFO method. The purpose of the INFO method is to enable the transfer of information generated as a result of events in the middle of a call. Examples of its use could include the following:

- Transfer of *Digital Tone Multi-Frequency* (DTMF) digits
- Transfer of account-balance information
- Transfer of mid-call signaling information generated in another network (e.g., PSTN) and passed to the IP network via a gateway

Through the use of the INFO method, application-layer information could be transferred in the middle of a call, thereby affording user agent servers, user agent clients, or proxies the opportunity to act upon the information.

The proposed extension is meant to be a flexible tool and as such does not define the information that might or might not be sent by using the INFO method. We envision, however, that the information to be carried in an INFO message would be included in a message body. The content and structure of the message body would depend on the information to be transferred. For any application that wishes to use the INFO method, it is expected that the specification of the application will also include a specification of the message body to be used. This technique is similar to how SDP specifies the message body to be used by SIP for multimedia calls. SIP itself does not specify media properties; rather, it simply provides a mechanism for signaling related to media exchange.

Clearly, the definition of the INFO method does not, by itself, introduce any new feature or service. Rather, the method is meant to be a generic means for transporting information related to mid-call services and other features that are as yet unknown. Whenever a new service or feature is designed to avail of the INFO method, appropriate specification work will need to be done to define how the service will work and how it will use the INFO method. Nonetheless, the INFO method offers a powerful and flexible tool within SIP for the support of new services. This example illustrates again the type of flexibility that is built into SIP so that new services can be accommodated, even if we have not yet defined those services.

The SIP-Supported Header

An Internet draft titled "The SIP Supported Header" was prepared to specify a new header to be included in SIP messages. Although this header is listed in Table 5-2, it is not specified in RFC 2543. This header has been included in a draft revision to the specification, however, and is one example of an extension that is planned for a new release of the specification.

RFC 2543 enables a client to indicate to a server that the server must support certain features or capabilities. This action can be performed through the use of the Require header. No mechanism exists, however, for a server to know about a client's capabilities. Thus, if a server wishes to issue a response to a client and wants to utilize some extension, it needs a way to know that the client supports that extension. Preferably, we need a way for a client to indicate to a server what extensions it supports, so that the server could then make use of this information if it so chooses. This goal is what the new header "Supported" enables.

The Supported header has the following syntax in *Augmented Backus-Naur Form* (ABNF) format:

```
Supported = ( "Supported" | "k" ) ":" 0#option-tag
```

Thus, as is to be expected, the header includes a tag for the client to indicate exactly which extensions it supports. Note that this particular SIP extension does not specify which types of capabilities or extensions can be offered. Merely, this extension is a mechanism to indicate when a particular extension or capability is supported. Therefore, the possible values for the option tag are not specified and could be any number of values. The only requirement is for the client and server to both understand what a particular value of the option tag means in order for the extension or capability to be invoked. For example, if the client supports an extension called "Felix," then the option tag would be the string "Felix." Provided that the client and server both have the same understanding of what "Felix" means, they can both take advantage of this information.

The Supported header is a general header and as such can be included in both requests and responses. This header can be included in BYE, CANCEL, INVITE, OPTIONS, and REGISTER requests. The header should not be included in the ACK request.

The Supported header must be included in a response to an OPTIONS request. After all, the whole point of the OPTIONS request is for a client to determine the capabilities of a server. If a server supports some extensions, then it must use the Supported header to indicate the extensions that it supports.

This extension to SIP also specifies that the Require header can be included in responses. The idea is that a response can include some extensions. In order for a client to process the response, it needs to support the extension(s) in question. By including the Require header in the response, the server lets the client know what support is needed within the client in order to properly process the response.

This extension to SIP also proposes a new response status code: 421 (Extension Required). A server returns this response when it requires a particular extension in order to process a request and when the request itself did not indicate that the client supports the extension. Imagine, for example, that a client sends an INVITE to a server. Let's suppose that the server exists only to provide some new and exotic service, and in order to provide that service, it must use some SIP extension. If the client does not indicate that it can support the necessary extension, then the server cannot invoke the service because it does not know that the client can support the service. When returning the 421 response, the server must include a Require header indicating what exactly it is that the client needs to support.

Figure 5-20 shows an example where a client issues an INVITE to a server. The client supports the extension "Rover" and indicates that support within the INVITE request. The server determines that the extension "Felix" must be supported in order for the request to be properly processed, however. Unfortunately, the client has not indicated that it is "Felix" capable. The

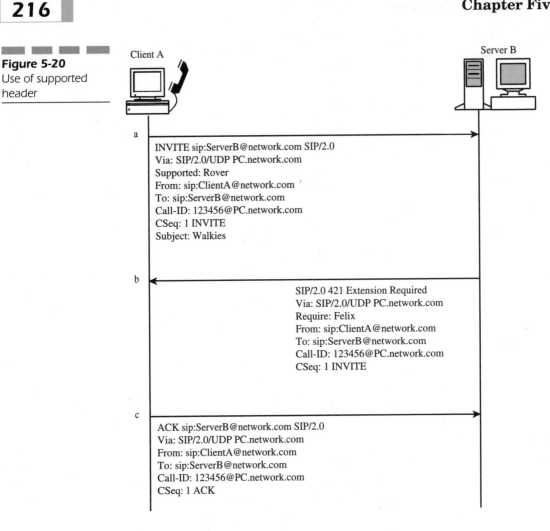

Figure 5-20
Use of supported
header

server returns a 421 (Extension Required) response, to which the client replies with ACK. Note that the ACK does not include a Supported header. In the figure, the session description is omitted.

Reliability of Provisional Responses

A SIP session is instigated with an INVITE message. One or more provisional responses might exist, such as 100 (trying) or 180 (ringing) prior to a final response to the INVITE. RFC 2543 enables messages to be sent over UDP, which is unreliable. Thus, the possibility exists that messages sent from one end might never reach the other end. SIP addresses some of these issues by specifying rules for retransmission of a request or response. RFC

2543 specifies that provisional responses are not answered with an ACK, however. Therefore, some provisional responses might be lost forever.

Imagine, for example, an INVITE sent from A to B. B responds with the provisional response, such as 180 (ringing), but A does not receive this response. The called user then picks up the phone, and B issues a final response of 200 (OK), which is received at A prior to expiry of the retransmission timer for the INVITE. A is happy that a final response has been received and that the call is established. There might be a problem, however, particularly in the case of interoperating with other networks. In the case of interworking, the 180 (ringing) response or the 183 (session-progress) response can be mapped to the Q.931 Alerting message or to the ISUP ACM. External network elements might be expecting such messages/responses to drive state machines, or certain applications might be relying upon guaranteed delivery to ensure that a particular feature or service operates correctly. A simple example would be a call to an unassigned number. The SS7 ACM message is used to open a one-way path from the called end to the calling end in order to relay an announcement such as, "The number you have called is not in service." If the ACM is mapped to a 180 or 183 response and that response is not delivered to the calling end, then the caller is left in the dark and does not understand why the call did not connect. Therefore, a SIP extension has been specified to provide reliability for SIP provisional responses.

The draft document describing this SIP extension is titled "Reliability of Provisional Responses in SIP." The extension involves the addition of two new headers: RSeq (response sequence) and RAck (response ACK). RAck is a request header field, and RSeq is a response header field. The extension also defines an option tag to be used with the "Supported" SIP header. This option tag is `option.ietf.sip. 100rel`[5]. The tag can also be included in the "Unsupported" header to indicate that a server does not understand the option. The extension also defines a new SIP method: *Provisional Response ACK* (PRACK).

The extension works according to the following example. A client that supports the reliable delivery of provisional responses sends an INVITE and includes the Supported header with the tag `option.ietf.sip.100rel`. When a server receives the INVITE and determines that a provisional response should be delivered reliably, it sends the response and includes the Rseq header. The server then waits for the client to send a PRACK message to confirm that the response has been received. If a PRACK message is received, then it is certain that the provisional response was received. Upon reception of the PRACK, the server returns 200 (OK) to indicate that

---

[5]Note that the draft revision to RFC 2543 states that extensions developed in the IETF no longer use the `org.ietf` prefix. In that case, the tag should just be `sip.100rel`.

the PRACK has been received. If the server does not receive the PRACK message, then it will retransmit the response 500ms later. If it still does not receive a PRACK, then the server retransmits the response after 1000ms more. The duration between retransmissions keeps doubling up to a maximum of 32 seconds. The server keeps sending the response every 32 seconds until a total of 96 seconds has elapsed, at which time the server assumes that a network failure has occurred.

The initial provisional response contains the RSeq header, which is defined to carry a 32-bit integer value between one and 4,294,967,295. The initial value of RSeq that is chosen the first time that a given response is returned is a random number between one and 2,147,483,647. Each time the response is retransmitted, the value of RSeq increments by one.

The PRACK sent by the client contains the RAck header. The RAck header contains two components. The first component is the value of RSeq copied from the provisional response for which the PRACK is being sent. The second is a CSeq number and method token that is copied from the response for which the PRACK is being issued.

The process is shown in Figure 5-21. In this example, the calling client uses the Supported: header to indicate in the INVITE that it supports acknowledgment of provisional responses. The first 180 (Ringing) response to the INVITE is lost. After a timeout, the server retransmits the response. Note that both instances of the response include the RSeq: header, but the value has been incremented in the second instance of the response.

Upon receipt of the retransmitted response, the calling client acknowledges the response through the use of the PRACK method. This method contains the RAck: header field. The content of the field is the value received in the RSeq: header of the response, plus the content of the CSeq: header in the response. Finally, the server confirms receipt of the PRACK by issuing a 200 (OK) response. Note that this 200 (OK) response relates to the PRACK and not to the initial INVITE.

Although reliability can be applied to provisional responses, it should not be applied to the 100 (trying) response, and the RSeq: header should not be included in such response. The 100 (trying) response is sent on a hop-by-hop basis, whereas the reliability mechanism described in this extension is end to end and does not apply to the 100 (trying) response.

Integration of SIP Signaling and Resource Management

Ensuring that sufficient resources are available to handle a media stream is an important aspect of providing a high-quality service and is therefore a necessity for a carrier-grade network. For example, when a called party

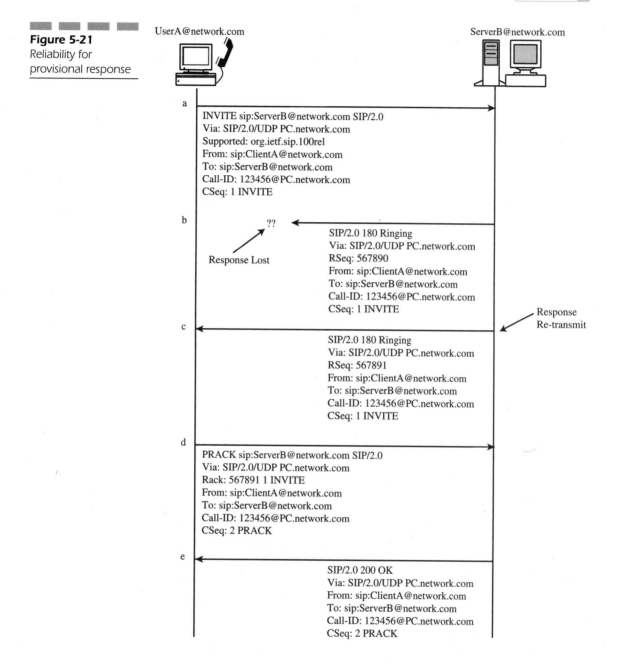

Figure 5-21
Reliability for
provisional response

UserA@network.com

ServerB@network.com

a

```
INVITE sip:ServerB@network.com SIP/2.0
Via: SIP/2.0/UDP PC.network.com
Supported: org.ietf.sip.100rel
From: sip:ClientA@network.com
To: sip:ServerB@network.com
Call-ID: 123456@PC.network.com
CSeq: 1 INVITE
```

b

??

Response Lost

```
SIP/2.0 180 Ringing
Via: SIP/2.0/UDP PC.network.com
RSeq: 567890
From: sip:ClientA@network.com
To: sip:ServerB@network.com
Call-ID: 123456@PC.network.com
CSeq: 1 INVITE
```

Response
Re-transmit

c

```
SIP/2.0 180 Ringing
Via: SIP/2.0/UDP PC.network.com
RSeq: 567891
From: sip:ClientA@network.com
To: sip:ServerB@network.com
Call-ID: 123456@PC.network.com
CSeq: 1 INVITE
```

d

```
PRACK sip:ServerB@network.com SIP/2.0
Via: SIP/2.0/UDP PC.network.com
Rack: 567891 1 INVITE
From: sip:ClientA@network.com
To: sip:ServerB@network.com
Call-ID: 123456@PC.network.com
CSeq: 2 PRACK
```

e

```
SIP/2.0 200 OK
Via: SIP/2.0/UDP PC.network.com
From: sip:ClientA@network.com
To: sip:ServerB@network.com
Call-ID: 123456@PC.network.com
CSeq: 2 PRACK
```

answers the phone and attempts to speak, it is imperative that the necessary resources are available to carry that speech back to the caller. Equally, resources must be available in the direction from caller to callee. While customers might expect to experience occasional congestion, that experience should only happen while attempting to establish the call. Once the far-end phone rings and is answered, a customer expects that a conversation can take place. Anything else is considered defective operation.

In our discussion so far, we have focused on the signaling related to call setup and tear-down. The bandwidth required for signaling is miniscule, however, compared to the bandwidth required for handling the media to be exchanged. Furthermore, the signaling might take a different path from the media. Therefore, we cannot assume that the successful transfer of signaling messages will automatically lead to the successful transfer of media. If carrier-grade operation is desired, the signaling must accommodate some mechanism to ensure that resources are available on the network for transport of the required media. While Chapter 8 describes some techniques for resource reservation and/or management, we should assume here that resource-reservation mechanisms can be available. Therefore, the task is to include capabilities within the signaling system so that these resource-management mechanisms can be utilized. The purpose of this extension is to do exactly that.

An Internet draft was prepared in 1999 describing a means for reserving network resources in advance of alerting the called user. This process involved the use of a new SIP header to be used in the INVITE, indicating to the called end that resources reservation is needed and that the user should not yet be alerted. After successful resource reservation, a second INVITE would be issued to set up the call. This approach, however, had some incompatibilities with standard SIP. When the new header was not supported by the called end, a call could be established without sufficient resources, because SIP states that new unrecognized headers should be ignored. Consequently, the extension was rejected. The basic need was recognized, however, and several groups got together to define a new Internet draft to address the issue in a slightly different way. That draft is called "Integration of Resource Management and SIP for IP Telephony." Like all Internet drafts, this new draft is subject to change.

The new approach proposes an extension to SIP, whereby a new method, called PRECONDITION-MET, is added. This new method is used to indicate that all prerequisites for call establishment have been met. The net result of the new extension is that the far-end phone will not ring until all preconditions have been met. Those preconditions will include reservation of resources for the call and/or the establishment of secure communication, but other requirements might also be applied. Thus, it is a more flexible approach to the problem as it is a mechanism that could be used to meet various requirements.

The draft also specifies extensions to SDP. In fact, the description of the prerequisites are included within SDP. This approach is reasonable, because SDP specifies the media to be exchanged. Given that the QOS requirements are related to the media, it is logical that they should be included as part of the session description, rather than within SIP itself. Moreover, if a given session is to include multiple media streams, then different QOS requirements might be applied to the different streams. For example, a session that includes both audio and video might have stronger QOS requirements for the audio in the event that it is considered more important. If different pre-conditions are to be applied to different media streams, then there is no choice but to specify those conditions within the message body, rather than within SIP.

Of course, because SIP is used to carry the session description, it is not possible to completely divorce the SIP signaling from the media to be exchanged (hence the need for the new SIP method). SIP should not care what the preconditions happen to be, however. By specifying the preconditions at the media level, we can define any number of preconditions in SDP in the future without having to revise SIP every time.

The basic methodology is depicted in Figure 5-22. An INVITE carries a session description that describes the media to be exchanged and preconditions to be met before the media exchange. At the far end, the session description is analyzed. If the preconditions are understood and it is within the capabilities of the called user agent to meet those preconditions, then a 183 Session-Progress response is returned. This response contains a message body indicating the preconditions to be met in the reverse direction. This provisional response must be sent reliably, because further action depends upon the receipt of the response at the calling end.

The 183 response might indicate whether confirmation of the resource reservation is needed. If it is needed, then the calling client, after receipt of the 183 response and after the resource reservation has occurred, will send the new PRECONDITION-MET message to the called end. This message also contains a session description with attributes indicating whether the resource reservation has succeeded or failed. Assuming that the resource reservation has succeeded, the user can be alerted, and the call can continue as normal.

NOTE: *Note that Figure 5-22 shows the PRECONDITION-MET message being sent end-to-end, rather than via any intervening proxies. This process is not a requirement. The message could equally be sent via an intervening proxy if the proxy utilized the Record-Route: header in conjunction with the initial INVITE. The same statement applies to the final ACK.*

Figure 5-22
Integration of
resource reservation
and SIP signaling

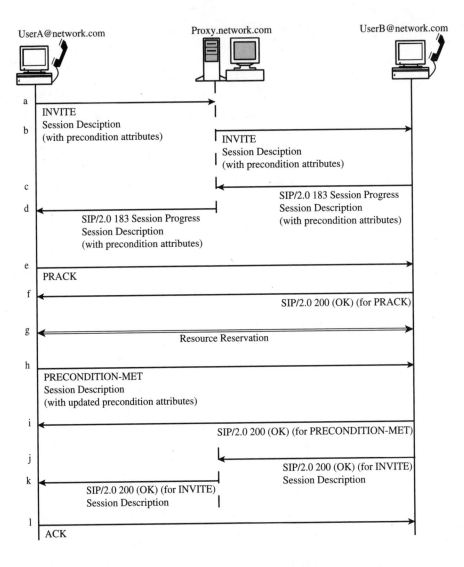

The preconditions to be met are specified in SDP through the addition of two new attributes called qos and security. The definition of the qos attribute is as follows:

```
qos-attribute = "a=qos:" strength-tag SP direction-tag
                              [SP confirmation-tag]
strength-tag = ( "mandatory" | "optional" | "success" | "failure" )
direction-tag = ( "send" | "recv" | "sendrecv" )
confirmation-tag = "confirm"
```

Mandatory strength means that the receiver of the session description must not proceed with the session unless resources have been met. Optional strength means that the receiver of the session description should reserve resources and proceed with the call, but the resource reservation is not an absolute requirement. The `success` strength is used with the PRECONDITION-MET method to indicate that resource reservation has been successful. Similarly, `failure` indicates that an attempt to reserve resources was not successful. The direction indicates whether resources should be reserved in just one direction or in both directions. `Confirm` indicates that receiver of the session description must send a PRECONDITION-MET message to the far end in order to indicate success or failure in the resource reservation. In other words, in Figure 5-22, the PRECONDITION-MET message could have been sent from either or both parties, depending on the content of the session descriptions.

The `security` attribute has the following ABNF syntax:

```
security-attribute = "a=secure:" strength-tag SP direction-tag
                                 [SP confirmation-tag]
```

The definitions of `strength`, `direction`, and `confirmation` for `security` are the same as for the `qos` attribute.

This SIP extension also specifies a new response status code, 580 (Precondition-Failure). This response is final, indicating that a session could not be established due to the failure to meet the required preconditions. This message might be sent as a response to the initial INVITE if the called system cannot support the preconditions at all, or it might be sent later if a resource reservation failed, for example.

Note that in the event that resources cannot be reserved, the decision not to complete a call might be made only as a last resort—even if the reservation is mandatory. In the case where sufficient resources cannot be made available, the option always exists to select a lower-bandwidth codec to reduce the resource requirements. This avenue should be explored before refusing to establish a call. You can only explore this option, however, if each end supports a lower bandwidth codec than its first choice. Moreover, a lower-bandwidth codec often means lower speech quality, something that needs to be seriously considered before assuming that a lower-bandwidth codec is really an option.

Interworking

Obviously, SIP-based networks will never be the only types of networks in existence. To begin, circuit-switched networks will continue to be with us for a long time. Therefore, an obvious need exists for SIP-based networks to be able to interwork with the circuit-switched networks of the PSTN. Furthermore, although many view SIP as the future of IP telephony, there is already an embedded base of H.323 systems—and more H.323 systems are being deployed. Therefore, there is also a need for SIP-based networks to interwork with H.323-based networks. RFC 2543 itself does not specifically address how such internetworking should be achieved, however.

PSTN Interworking

For interworking with the PSTN, gateways will obviously be required to provide the conversion from circuit-switched media to packet, and vice-versa. In fact, we already saw that H.323 makes this aspect a central part of H.323 architecture. Not only must there be media interworking, but there must also be signaling interworking. After all, SIP is a signaling protocol. If calls are to be established between a SIP-based network and the PSTN, then the SIP network must be able to communicate with the PSTN according to the signaling protocol used in the PSTN. In most cases, this protocol is SS7. For the establishment, maintenance, and tear-down of speech calls, the applicable SS7 protocol is ISUP.

The usage of gateways and the control of those gateways in a SIP-based network are discussed in some detail in Chapter 6, "Media Gateway Control and the Softswitch Architecture." Furthermore, interworking with the PSTN through the use of signaling gateways is discussed in detail in Chapter 7, "VoIP and SS7." For the moment, however, it is sufficient to assume that there is a gateway entity in the network for interworking between SIP and the PSTN. Let's call that entity a *network gateway* (NGW). In such a case, let's consider a call from Manager (at work) to Collins (at home), where Collins has a standard residential phone connected to a PSTN switch. Such a scenario would appear as in Figure 5-23. Note that most of the fields within the SIP messages and responses are omitted to avoid clutter.

Recall that SIP enables a SIP URL to have the form of a telephone number. The INVITE includes such a URL in the To: field, which is mapped to the calling-party number of the IAM. The From: field of the INVITE could also have a URL in the form of a telephone for the calling party. This feature would enable the called party to still take advantage of the PSTN Caller-ID service.

The PSTN switch responds with an ACM, which is mapped to the SIP 183 (session-progress) response. This response contains a session description and enables in-band information to be returned from the called switch to the SIP caller. Once the call is answered, an ISUP ANM is returned,

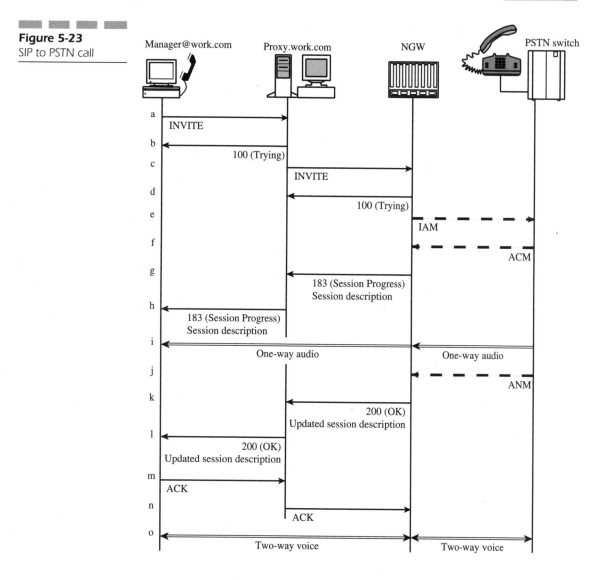

Figure 5-23
SIP to PSTN call

which is mapped to the 200 (OK) response. This response will have a final session description, because the session description in the 183 response is considered temporary.

If the call were to be in the opposite direction—from the PSTN to a SIP user—then it would appear as in Figure 5-24. In this case, the 180 Ringing response is used instead of the 183 Session-Progress response. The reason for this difference is that a SIP entity will not normally include in-band information prior to acceptance of the call. The NGW uses the 180 Ringing response to trigger the ringing tone towards the calling party locally. As is the case in the previous example, not every field is shown for the SIP messages and responses in Figure 5-24. Furthermore, for the avoidance of clutter, we do not show the usage of reliable signaling for SIP provisional responses in either example. We can assume that reliability of these responses might be applied so that PRACK methods and corresponding 200 OK responses might occur.

While these two interworking examples might seem quite straightforward, seamless interworking between two different protocols is not often quite that easy—because there is rarely a one-to-one mapping of messages and parameters between one protocol and the other. Sometimes the

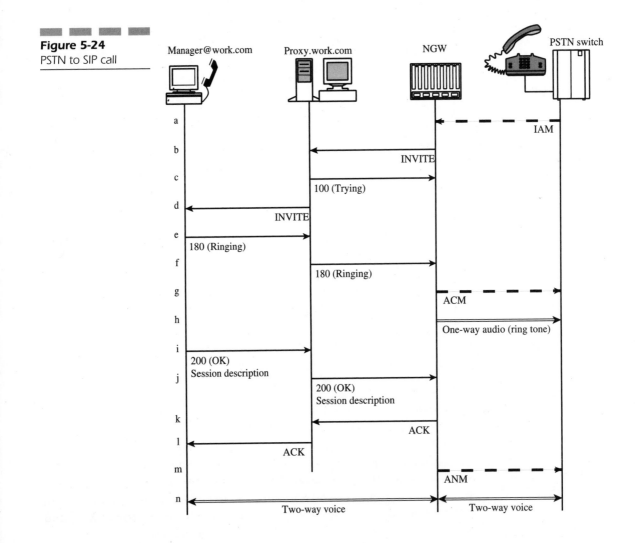

Figure 5-24
PSTN to SIP call

mapping from protocol A to B means that what is sent in protocol B is only an approximation of what was received in protocol A. If there is dual interworking; i.e., from protocol A to protocol B and back to protocol A, what comes out at one end might not be the same as what was inputted at the other end. Given that there are VoIP networks that provide a transit service from the PSTN at one end to the PSTN at the other end, this lack of transparency is a concern. To address this issue, an Internet draft has been prepared and titled "MIME media types for ISUP and QSIG objects." The purpose is to enable ISUP or QSIG messages to be carried within a SIP message body in hex format. In such a case, the SIP message body could be a multi-part message body, including both an SDP session description as one part and an ISUP or QSIG message as another.

The whole issue of interworking with SS7 is fundamental to the success of VoIP in the real world. This concern is major, not just to enable SIP-based networks to interwork seamlessly with the PSTN for regular phone calls but also to enable SIP-based systems to take advantage of services in the PSTN that rely on SS7 signaling. (These services include IN services, such as toll-free calling.) Therefore, Chapter 7 of this book is devoted to this topic and addresses the transit issue just described, as well as many other aspects.

Interworking with H.323

Interworking with H.323 is also a major consideration, particularly because H.323 is already widely deployed. To address the issue, an Internet draft has been prepared entitled "Interworking between SIP/SDP and H.323." This draft might, at some time in the future, become an informational RFC. Nevertheless, this draft addresses an important issue: how to provide end-to-end VoIP service between two different architectures.

The proposed approach for interworking between a SIP network and an H.323 network involves the use of a gateway, as depicted in Figure 5-25. Depending on the function to be performed, the gateway might appear to the SIP network as a user agent client or user agent server. To the H.323 network, the gateway will appear as an H.323 endpoint. On the SIP side, the gateway might also include a SIP registrar function and/or proxy function—and, on the H.323 side, it could contain a gatekeeper function.

Figure 5-26 depicts an example scenario of a call from a SIP caller to an H.323 terminal. The H.323 Fast-Connect procedure is used, and we assume that a gatekeeper is not involved in the call. If a gatekeeper were to be used, then the RAS signaling would occur on the H.323 side of the gateway and

Figure 5-25
SIP-H.323
interworking
gateway

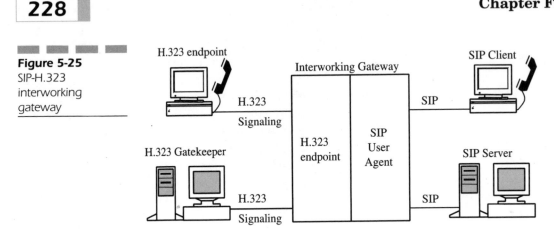

would not be seen by the SIP network in any case. The SDP information in the SIP INVITE is mapped to the logical-channel information included in the faststart element of the Q.931 SETUP message from the gateway to the H.323 terminal. In the example, a faststart element is conveniently returned from the called terminal in the Connect message, rather than in the Alerting message. In this case, the logical-channel information within the Connect message can be mapped directly to the session description within the SIP 200 (OK) response. If the faststart element had been included in the Alerting message or some other messages (such as Call Proceeding), then the gateway would simply have had to store the information until it was ready to send it to the SIP client in the 200 (OK) response. Note that the text included with the messages in the figure is for explanatory purposes only and does not match the actual syntax of the messages as they should appear in a real implementation.

If we assume that the called H.323 terminal network does not support the Fast-Connect procedure, then the sequence becomes more long winded, as shown in Figure 5-27. In this case, the gateway does support Fast Connect, but the called terminal does not. Thus, the Setup message includes the faststart element, but the messages returned from the called H.323 terminal do not. Therefore, upon receipt of the Connect message, the gateway must perform H.245 logical channel signaling to/from the called terminal before returning the 200 (OK) to the calling SIP client. The H.245 messaging includes capability exchange and the establishment of logical channels. For clarity, interim responses from the called terminal (Call Progress, Alerting, etc.) are not shown in the figure.

A call in the opposite direction, from H.323 to SIP, is relatively straightforward if the H.323 terminal and the gateway both support the H.323

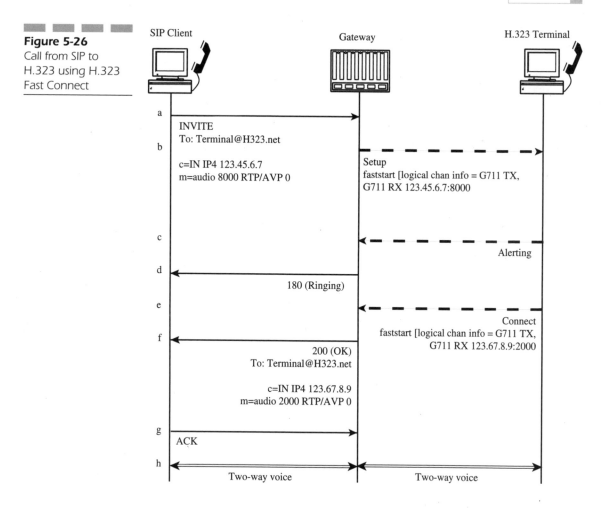

Figure 5-26

Call from SIP to H.323 using H.323 Fast Connect

Fast-Connect procedure. Such a scenario is depicted in Figure 5-28. In this case, the session information is included in the H.225/Q.931 SETUP message and this can be mapped to the SDP information within the SIP INVITE. Equally, the session description accompanying the 200 OK response from the called SIP user agent can be mapped to the information within the faststart element of the Connect message back to the H.323 terminal.

If, on the other hand, the calling H.323 terminal does not support the Fast-Connect procedure, then the interworking becomes more complex for two reasons. First, the fact that Fast Connect is not used means that separate H.245 logical channel signaling is required on the H.323 side of the gateway. Second, there is no logical channel information in the SETUP

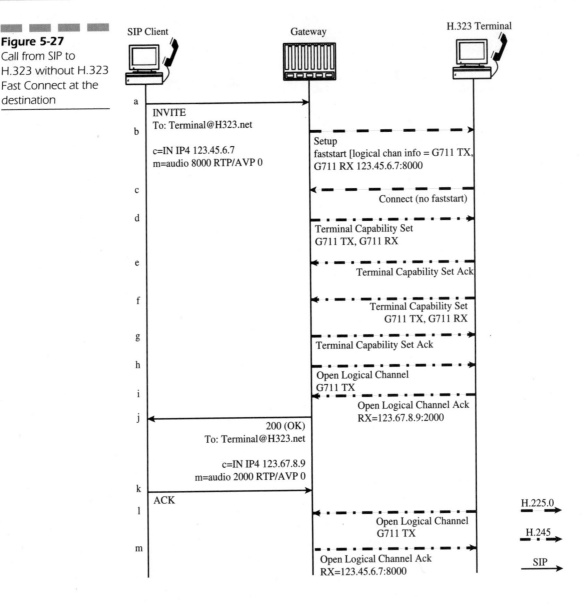

Figure 5-27
Call from SIP to H.323 without H.323 Fast Connect at the destination

message. Therefore, there is nothing to map to a session description in the SIP INVITE. In that event, the INVITE contains a default session description. The IP address in the connection (c=) line of the session description should contain a default value, such as the IP address of the gateway, while the media (m=) line should indicate a port number of 0, as shown in Figure 5-29. Only in the SIP ACK message is a real session description sent to the

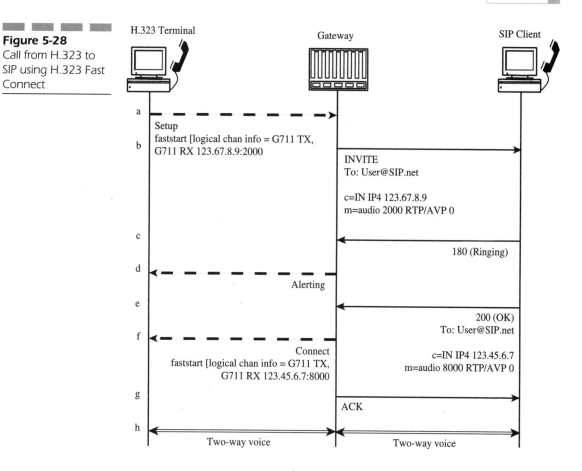

Figure 5-28
Call from H.323 to
SIP using H.323 Fast
Connect

called SIP user agent. The gateway only receives the information as part of the H.245 logical channel signaling. We can provide a new session description within ACK. In general, the ACK should not be used to carry a different session description to that in the INVITE. The ACK can contain a session description in the case that the initial INVITE is incomplete, however.

You can find numerous example scenarios for SIP/H.323 internetworking in the Internet draft mentioned previously. You can retrieve copies of the draft from the IETF's Web site.

Summary

Although quite new, SIP has already established itself as the wave of the future for signaling in VoIP networks. Its strengths include the fact that it

Figure 5-29
Call from H.323 to
SIP without H.323
Fast Connect

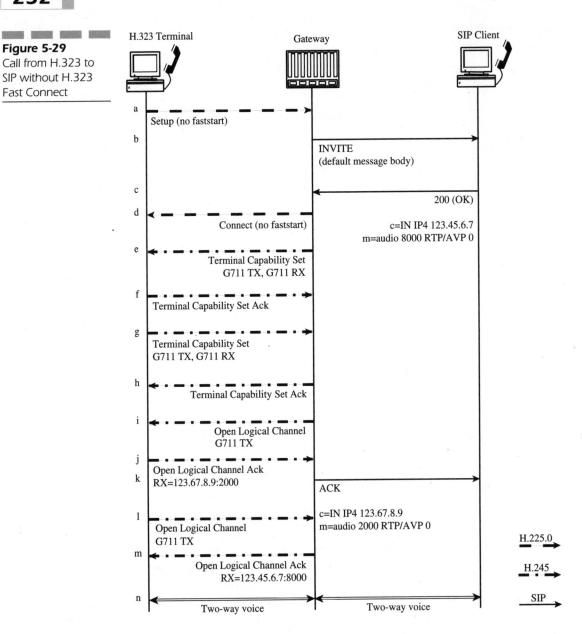

is far simpler than H.323 and is therefore easier to implement. Its simplicity belies its flexibility and power, however. SIP can support all of the features that are already supported by traditional telephony networks and more. As we will see in the next chapter, SIP also fits well with those protocols that are used for media gateway control—and as such, it forms part of the overall architecture known as softswitch.

Media Gateway Control and the Softswitch Architecture

Introduction

Previous chapters described some of the advantages that *Voice over IP* (VoIP) can offer over traditional circuit-switched telephony, such as lower cost of network implementation, integration of voice and data applications, new features, and reduced bandwidth for voice calls. Ideally, all telephony could be carried over IP so that these advantages could be made available globally. Unfortunately, that day will not arrive for quite a while. Replacing all existing circuit-switched networks with VoIP solutions is not feasible and perhaps not even desirable. Among other reasons, the cost would simply be exorbitant. Rather, one can expect a gradual evolution from circuit-switched networks to IP-based networks such that both types of networks will exist side-by-side for a long time.

Given that VoIP networks and traditional circuit-switched networks will have to coexist for many years, they must interoperate as seamlessly as possible. In other words, users of existing circuit-switched systems should be able to place calls to VoIP users, and VoIP users should be able to place calls to subscribers of circuit-switched networks. In each case, the user should not have to do anything radically different from simply calling another user on the same type of network. The challenge, therefore, is to develop solutions that make such seamless interworking possible.

Separation of Media and Call Control

Previous chapters discussed the interworking between VoIP networks and circuit-switched networks through the use of gateways. The primary function of gateways is to make the VoIP network appear to the circuit-switched network as a native circuit-switched system (and vice-versa). In other words, VoIP networks must accept signaling and media from circuit-switched networks in native format and convert both to the format that the IP network uses.

We also saw in both Chapters 4 and 5 that the signaling involved in a VoIP call might take a different path from the media path itself. The media might go directly from end to end, while the signaling might pass through one or more intermediate entities, such as gatekeepers or proxies. In fact, the logical separation of signaling and media is not a new concept; this idea has existed for many years in the *Signaling System 7* (SS7) architecture.

Putting these two considerations together means that a network gateway has two related but separate functions: signaling conversion and media conversion. The signaling aspect is coupled to whatever call-control functions are operating in the network. In fact, signaling can be considered the language that call-control entities use to communicate. The media conversion and transport can be considered a slave function, invoked and manipulated to meet the needs dictated by call control and signaling. In terms of network architecture, this situation leads to a configuration of the kind depicted in Figure 6-1. This figure shows a scenario in which a VoIP network sits between two external networks, as might be the case for a VoIP network that provides long-distance service. Figure 6-1 illustrates the separation of call control and signaling from the media path. Moreover, this figure illustrates the master-slave relationship between call control and media, even within the network gateway itself. Even if the gateway is just a single box that handles both signaling and media, there will be some internal control system between the call-control functions and media-handling functions.

Given that such a logical separation exists, there is no compelling reason why the separation could not also be made physical. In fact, such a physical separation has some significant advantages. First, this separation enables media conversion to take place as close as possible to the traffic source or sink within the circuit-switched network while centralizing the call-handling functions. Second, a smaller number of gateway controllers or call agents located more centrally can control multiple gateways placed in various locations. This situation leads to another benefit: new features could be rolled out more quickly, because they would need to be implemented only in the centralized call-handling nodes rather than at every node in the network. This approach leads us to the type of architecture shown in Figure 6-2.

Of course, the physical separation that could bring these benefits requires a well-standardized control protocol between the *Media Gateway Controller* (MGC) or call agent and media gateway. Fortunately, two such

Figure 6-1

VoIP network gateways in tandem between external networks

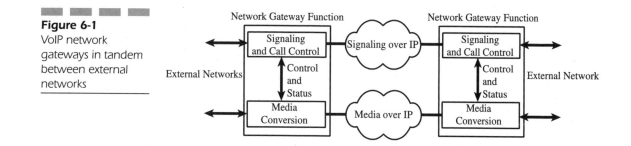

Figure 6-2
Distributed system
architecture

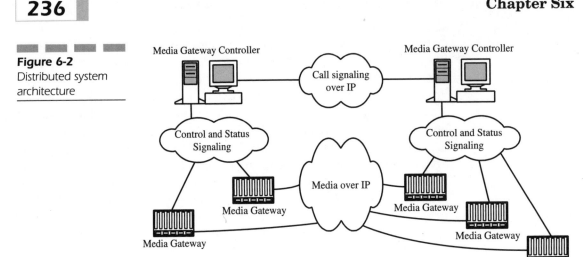

standardized protocols have been developed: the *Media Gateway Control Protocol* (MGCP) and MEGACO/H.248. Using one or another of these protocols for media gateway control and using *Session Initiation Protocol* (SIP) as the call-signaling protocol, the architecture shown in Figure 6-2 is a reality and is known as softswitch.

Softswitch Architecture

The softswitch architecture involves the separation of the media path and media-conversion functions from the call control and signaling functions, as depicted in Figure 6-2. The entities that handle call control are known as call agents or media gateway controllers, while the entities that perform the media conversion are known as media gateways. The name softswitch is used because many of the switching functions traditionally handled by large monolithic systems in the circuit-switched world are instead emulated by software systems. In fact, in some quarters, the term softswitch is used to refer to a call agent or MGC, as opposed to the overall architecture.

The softswitch architecture has many supporters within the VoIP industry. In fact, many of the leaders in the industry have come together in an organization known as the International Softswitch Consortium. This organization promotes the softswitch concept and related technologies, enables interested parties to share ideas and to cooperate in the ongoing development and deployment of the technology, and facilitates interwork testing between different implementations.

One of the reasons why the softswitch approach is so popular is the fact that it is a distributed architecture. A network operator can source different network components from different vendors so that the best in the class is chosen in each area. For equipment vendors, they can focus their efforts on one area without having to develop or acquire expertise in all areas. For example, a company can focus on development of a call agent without having expertise in voice-coding techniques and related *Digital Signal Processors* (DSPs) used within media gateways. Equally, a company that does have the necessary voice-coding expertise can develop a media gateway without having the call-management expertise needed to develop a call agent. Another distinct advantage is the rapid rollout of new features described previously.

In the softswitch architecture, SIP is typically used as the signaling protocol between the call agents so that a call agent appears as a SIP user agent in the SIP network. Therefore, when interworking between SIP and the *Public Switched Telephone Network* (PSTN), the call agent handles the functional interoperation between SIP and PSTN signaling, such as *ISDN User Part* (ISUP), while the gateways take care of the media-conversion functions, as shown in Figure 6-3.

As mentioned, two protocols exist for media gateway control. MGCP has been developed within the *Internet Engineering Task Force* (IETF), and MEGACO/H.248 is being developed as a cooperative effort between the IETF and the *International Telecommunications Union* (ITU)-T Study

Figure 6-3
Softswitch/PSTN
internetworking

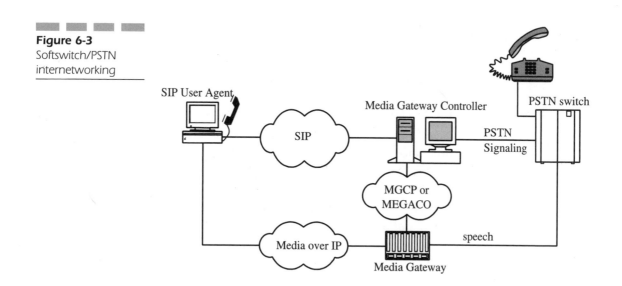

Group 16. This reason is why MEGACO/H.248 has such a strange name. Within the IETF, the protocol is known as MEGACO (for media gateway control), and within the ITU, it is known as H.248.

Requirements for Media Gateway Control

Before delving deeply into the details of MGCP or MEGACO, we must first consider the functions that the protocols should support and what requirements exist for interoperation with other protocols. This knowledge will clarify the reasons behind the protocols in the first place.

RFC 2805 ("Media Gateway Control Protocol Architecture and Requirements") is an informational RFC that describes the functions to be supported by a distributed gateway system, as depicted in Figure 6-3. Although the focus of this chapter is interworking between IP-based networks and circuit-switched networks, this document addresses generic requirements to be met for interworking with any type of network—not just circuit-switched networks. At a high level, the requirements for the control protocol are that it should:

- Enable the creation, modification, and deletion of media streams across a media gateway, including the capability to negotiate the media formats to be used

- Enable the specification of the transformations to be applied to media streams as they pass through a media gateway

- Enable the MGC to request and the MG to report the occurrence of specified events within the media stream (e.g., DTMF tones or call-associated signaling) and enable the MGC to specify the actions that the MG should automatically take when such events are detected

- Enable the MGC to request the media gateway to apply tones or announcements, either by direct instruction or upon the occurrence of certain events (such as applying a dial tone to an analog line when an off-hook event is detected)

- Enable the establishment of media streams according to certain *Quality of Service* (QOS) requirements and to change QOS during a call

- Enable the reporting of QOS statistics from a media gateway to a media gateway controller. Because the media gateway controller is not involved in the media transfer, it has no visibility of quality considerations (such as delay, jitter, and packet loss) and will need to be informed for network-management purposes.

- Enable the reporting of billing/accounting information, such as call start and stop times, the volume of content carried by media flow (e.g., the number of packets/bytes), and other statistics

- Support the management of associations between a media gateway and a media gateway controller, such as the capability for a given gateway to be controlled by a different media gateway controller in the case of failure of a primary media gateway controller

- Support a flexible and scalable architecture in which an MGC can control different media gateways with different capacities and different interfaces

- Facilitate the independent upgrade of MGs and MGCs, including the use of a version indicator in message exchanges

 # Protocols for Media Gateway Control

Several protocols have been written for media gateway control, and of these, two have been widely embraced. The first is MGCP (RFC 2705), an interesting document in that it is purely an informational RFC (rather than a standards-track protocol), although it includes a protocol specification. The reason why it is categorized as informational is because it is to be succeeded by MEGACO/H.248. Work on MEGACO started much later than MGCP, however. Even while MGCP was still an Internet draft, many systems developers included it within their product development rather than waiting for MEGACO. Therefore, the decision was made to release MGCP as an informational RFC.

As of this writing, the specification of the second protocol, MEGACO/H.248, is almost complete. The specification represents a first for both the traditional telephony world and the IP world. In the traditional telephony world, the ITU is considered the ultimate standards body, while the IETF is the corresponding standards body in the IP world. Both organizations, recognizing the need for a comprehensive control protocol, agreed in 1999 to cooperate on the development of a common standard: MEGACO/H.248. Unlike MGCP, it is planned that MEGACO/H.248 will become a full-fledged standard. In fact, the development of MEGACO is the main reason why we do not expect MGCP to be released as a standard.

At this writing, ongoing work is being conducted on specification of the MEGACO protocol. That work has already come a long way, however, and a great deal of the specification is already defined. We expect that the *Internet Engineering Steering Group* (IESG) will upgrade MEGACO to an RFC in summer 2000 and that the ITU will release MEGACO at the same time. Meanwhile, specification work on MGCP has stopped, although systems developers continue to implement MGCP within their systems. Given that MGCP is the control protocol being implemented in VoIP systems today and that MEGACO is not that far away, both should be described here.

MGCP

MGCP is a master-slave protocol in which controllers, known as call agents, control the operation of media gateways. The call agent takes care of call-control intelligence and related call signaling, while the media gateway takes instruction from a call agent and basically does what the call agent commands. The call-agent commands generally relate to the establishment of connections and tear-down of connections from one side of the gateway to another. In most cases, the call agent tells the media gateway to make a connection from a line or trunk on the circuit-switched side of the gateway to an RTP port on the IP side of the gateway, as depicted in Figure 6-4.

Figure 6-4
Basic approach to MGCP connection establishment

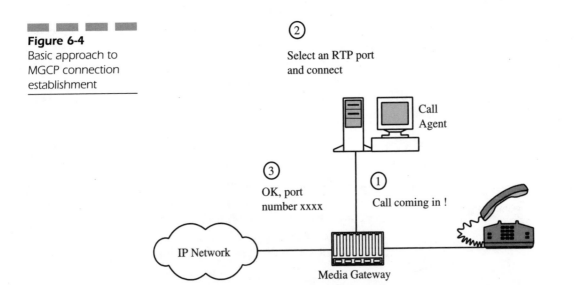

MGCP only address the communication between a call agent and a media gateway and does not address communication from one call agent to another. Because instances will arise in which media is to be exchanged between gateways that are controlled by different call agents, a need will exist for signaling communication between call agents. MGCP assumes that such a signaling protocol exists but does not specify what this protocol is. In reality, the protocol is likely to be the *Session Initiation Protocol* (SIP).

The MGCP Model

MGCP describes a model containing endpoints and connections. Endpoints are sources or sinks of media and include such elements as trunk interfaces or *Plain Old Telephone Service* (POTS) line interfaces. Endpoints exist within the media gateways, and depending on the type of endpoint, it might or might not have one or more external channel or line interfaces. For example, an analog-line endpoint would be connected to a physical analog line, which would in turn connect to an analog telephone. On the other hand, endpoints also exist that have no external connections. One example is an announcement endpoint that would reside internally in the gateway. Other entities in the IP network would connect to the endpoint for the purpose of retrieving an announcement.

Connections are the allocation of IP resources to an endpoint so that the endpoint can be accessed from the IP network. For example, when a call is to be established across a gateway from a circuit-switched line to an IP interface, an ad-hoc relationship is established between the circuit-switched line and an RTP port on the IP side. This relationship is called a connection, and a single endpoint can have several connections. Figure 6-5 provides a general representation of endpoints and connections.

MGCP Endpoints

MGCP describes a number of different endpoints, as follows:

- *DS0 channel* A digital channel operating at 64kbps. This channel is typically multiplexed within a larger transmission facility, such as a

Figure 6-5

General representation of an endpoint in MGCP

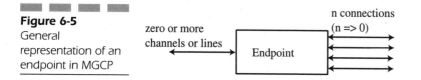

DS1 (1.544Mbps) or E1 (2.048Mbps). In most cases, the DS0 will carry voice that is coded according to either G.711 mu-law or G.711 A-law. In some cases, the DS0 will carry signaling instead of voice, such as an ISDN D-channel, in which case it is necessary for the signaling information received to be passed from the media gateway to the call agent for processing.

■ *Analog line*   Typically, an endpoint that interfaces to a standard telephone line and provides service to a traditional telephone. In most cases, the media to be received on the analog line will be an analog voice stream, although it could also be audio-encoded data from a modem—in which case the gateway is required to extract the data and forward it as IP packets.

■ *Announcement server access point*   An endpoint providing access to a single announcement. Normally, a connection to this type of endpoint will be one way, because the requirement is for media to be sent *from* the announcement rather than *to* it. We assume that the announcement server is within the IP network and that IP-based connections are to be made *to* the endpoint. Thus, the endpoint does not have any external circuit-switched channels or lines associated with it.

■ *Interactive Voice Response* (IVR) *access point*   An endpoint providing access to an IVR system, where prompts or announcements can be played and responses from the listener can be acted upon

■ *Conference bridge access point*   An endpoint where media streams from multiple callers can be mixed and provided again to some or all of the callers

■ *Packet relay*   A specific form of conference bridge supporting only two connections. This type of endpoint will use packet transport on both connections. A typical example would be a firewall between an open and a protected network, where the media stream must pass through such a relay point rather than directly from end-to-end.

■ *Wiretap access point*   A point from which a connection is made to another endpoint for the purposes of listening to the media transmitted or received at that endpoint. Connections to a wiretap access point will be one way only.

■ *Asynchronous Transfer Mode* (ATM) *trunk-side interface*   An endpoint corresponding to the termination of an ATM trunk, where such a trunk might be an ATM virtual circuit

Each endpoint is identified with an endpoint identifier. This identifier consists of the domain name of the gateway to which it belongs and a local name within the gateway. The local name within the gateway obviously depends on the gateway in question but will generally have a hierarchical form, such as X/Y or X/Y/Z. Y is an entity within X, and Z is an entity within Y. An example would be a gateway that supports multiple channelized T3 interfaces. If one wished to identify a single DS0 channel on the gateway, then one would have to identify the particular T3 (X), the particular DS1 within the T3 (Y), and the particular DS0 (Z) within that DS1. If we wished to identify DS0 number 7 within DS1 number 12 on DS3 number 4, then the identifier could be something like `trunkm4/12/7@gateway.somenetwork.net`.

We can issue a wild card to certain components of the identifier by using either a dollar sign ($), which indicates any, or an asterisk (*), which indicates all. Thus, if we wanted to refer to any DS0 within the fifth DS1 within the first T3 in our example, then we could do so through the term `trunk-1/5/$@gateway.somenetwork.net`. This type of wild card is useful when a call agent wishes to create a connection on an endpoint in a gateway and does not really care which endpoint is used. Through the use of the $ wild card, the choice of particular endpoint can be left to the gateway itself. Through the use of the * wild card, the call agent can instruct the gateway to perform some action related to a number of endpoints, such as a situation in which the call agent requested statistical information related to all endpoints on a gateway.

MGCP Calls and Connections

A connection is the relationship established between a given endpoint and an RTP/IP session. Consider, for example, a DS0 endpoint. If the DS0 is idle, then there is no connection associated with the endpoint—and no resources are allocated to the endpoint on the IP side of the gateway. When the DS0 is to be used to carry voice traffic, however, then IP resources are allocated to the endpoint—and a link is created from the DS0 on the line side of the gateway to an IP session on the IP side of the gateway. In MGCP terminology, this link is called a connection. If two endpoints are involved in a call, then two connections exist.

A call is a grouping of connections such that the endpoints associated with those connections can send and/or receive media to/from each other. Figure 6-6 shows a scenario in which a call is established between two analog phones, such that they communicate across an IP network by using two gateways. On each gateway, a connection is created between an endpoint on the line side of the gateway and IP resources (e.g., IP address and RTP port

Figure 6-6

Relationship between
calls and connections

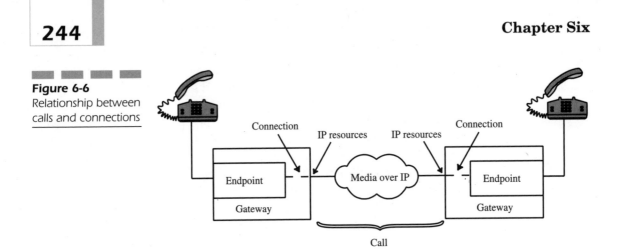

number) on the IP side of the gateway. Once a connection is created on each
of the two endpoints, the call can be considered the linking of these two connections—such that media can be passed between the endpoints.

When a gateway creates a connection for an endpoint and allocates IP
resources, those resources are described in terms of an SDP session description. The establishment of a call involves the endpoints sharing their
respective session descriptions with each other so that they can exchange
media streams. The primary function of MGCP is to enable those connections to be created and to enable the session descriptions to be exchanged.
MGCP does so through the use of commands and responses between call
agents and media gateways.

Overview of MGCP Commands

MGCP defines nine commands, some of which are sent from the call agent
to the gateway (and some from the gateway to the call agent). MGCP commands are composed of a command line, a number of parameter lines, and
optionally, a session description. The command line and parameters are
text, using the US-ASCII character set. The session description uses SDP,
also a text-based protocol.

The command line and the various parameters are separated from each
other by a *carriage return and line-feed* (CRLF) character. The session
description is separated from the command line and the various parameter
lines by a single empty line, similar to SIP.

The command line is composed of the "requested verb" (i.e., the name of
the command), a transaction identifier, the endpoint for which the command applies (or from which the command has been issued), and the pro-

tocol version (MGCP 1.0). These four items are separated from each other by a *space* (SP) character. Thus, the command appears as follows:

```
CommandVerb SP TransactionID SP EndpointID SP MGCP 1.0
```

MGCP defines the nine commands in the following list. For each command, the requested verb is coded as a four-letter upper-case ASCII string.

- *EndpointConfiguration* (EPCF) Issued by a call agent to a gateway in order to inform the gateway about bearer/coding characteristics on the line side of one or more endpoints. The only characteristic currently specified is whether the encoding on the line side is A-law or mu-law.

- *CreateConnection* (CRCX) Issued by a call agent to a gateway; used to create a connection at an endpoint

- *ModifyConnection* (MDCX) Issued by a call agent to a gateway; used to change some of the characteristics of an existing connection. One of the primary functions of this command is to provide information to the endpoint regarding the far end of the call.

- *DeleteConnection* (DLCX) Issued by a call agent to a gateway in order to instruct the gateway to terminate a connection. The command can also be issued by a gateway to a call agent (for example, where the gateway has detected that line-side connectivity has been lost).

- *NotificationRequest* (RQNT) Issued by a call agent to a gateway, requesting that the gateway notify the call agent when certain events occur. The gateway could be instructed, for example, to detect an off-hook signal from an analog line.

- *Notify* (NTFY) Issued by a gateway to a call agent in order to report the occurrence of certain events

- *AuditEndpoint* (AUEP) Issued by a call agent to a gateway to query the status of a given endpoint and/or other data regarding the endpoint. The gateway will respond with the requested information.

- *AuditConnection* (AUCX) Issued by a call agent to a gateway in order to query information regarding a specific connection

- *RestartInProgress* (RSIP) Issued by a gateway to a call agent in order to inform the call agent when one or more endpoints are taken in or out of service

Command Encapsulation MGCP supports the concept of encapsulation, where one command can be included within another. For example, when instructing a gateway to create a connection (CRCX command), we can also instruct the gateway to notify the gateway of certain events. Thus, we can find a NotificationRequest encapsulated within a CreateConnection command. This feature is particularly useful in cases where certain actions need to be taken or when events need to be detected only in conjunction with other activities. For example, a call agent might require a media gateway to detect and report DTMF tones, but only when a call is in progress. In such a case, a NotificationRequest command instructing the gateway to detect and report DTMF tones would be encapsulated within a Create-Connection command, so that the tone detection would only occur in the context of the connection created.

MGCP permits only one level of encapsulation. In other words, one command cannot be encapsulated within a second command (which is, in turn, encapsulated within a third command). The possibility exists, however, to send several commands at the same time (e.g., within the same UDP packet).

MGCP Parameters Each parameter is identified by a code that consists of one or two upper-case characters. A parameter code is followed by a single colon, one space character, and the parameter value. In some cases, the parameter value might be a single numeric value. In other cases, it might be a single hexadecimal string. In still other cases, it might be a comma-separated list of values or might comprise a list of subfields. The same set of parameters is available for use in commands and responses. MGCP defines the following parameters, where the parameter code is indicated within parentheses:

- *ResponseAck* (K) A list of one or more transaction ID ranges

- *BearerInformation* (B) Specifies the line-side encoding as A-law or mu-law. For example, B: e:mu means mu-law encoding.

- *CallId* (C) Unique identifier for a single call, comprised of hexadecimal digits

- *ConnectionId* (I) Unique identifier for a connection on an endpoint; also comprising hexadecimal digits

- *NotifiedEntity* (N) An address for a call agent to which notify commands should be sent from a gateway

- *RequestIdentifier* (X) Used in RQNT and NTFY commands to identify a particular request for information. The purpose of the parameter is to enable a call agent to correlate a given notification from a gateway with a particular notification request previously sent from the call agent to the gateway.

- *LocalConnectionOptions* (L)   A list of connection options, such as bandwidth, packetization period, silence suppression, gain control, echo cancellation, and encryption. These characteristics are connection-related characteristics that a call agent suggests to a gateway when a connection is to be established. In the syntax, the parameter is identified by L:. The individual components are separated by commas, and each is identified by one or two characters followed by a colon and followed by the specific value. For example, to turn echo cancellation off and to turn silence suppression on, the parameter would be L: e:off, s:on.

- *ConnectionMode* (M)   Options such as send only, receive only, and send-receive. These choices enable a given endpoint to be configured for one-way or two-way media transfer.

- *RequestedEvents* (R)   A list of events for which an endpoint must watch. Associated with each event, the endpoint can be instructed to perform one or more actions, such as notify the call agent immediately, collect digits, or apply a signal to the line. For example, an endpoint might be instructed to detect an off-hook even on an analog line, and upon detection of the event, to apply a dial tone to the line.

- *SignalRequests* (S)   Signals to be applied by an endpoint to the media stream, such as ring tone or dial tone

- *DigitMap* (D)   A digit map is a representation of a dialing plan and specifies rules related to which combinations of digits can be dialed. The purpose of the digit map is to enable the gateway to understand which types of digit strings a caller can send. When equipped with a digit map, the gateway has the capability to collect a set of digits and send them en-bloc to the call agent, rather than sending each digit individually. The digit map is sent in a *Notification Request* (RQNT) command from the call agent to the gateway.

- *ObservedEvents* (O)   A list of events detected by an endpoint. This parameter is used to report the events to the call agent. Generally, the events reported are a subset of those requested by the RequestedEvents in a command previously received from the call agent. The previously described RequestIdentifier is used to correlate events requested by the call agent with events reported by the gateway.

- *ConnectionParameters* (P)   Connection-related statistical information, such as average latency, average jitter, number of packets sent/received, number of lost packets, etc. These statistics are reported to a call agent at the end of a call, within the response to the DeleteConnection command. They also can be returned to the call agent as a response to the AuditConnection command.

- *ReasonCode* (E)   Used by a gateway when deleting a connection to indicate the cause of the deletion or after a restart of an endpoint to indicate the reason for the restart

- *SpecificEndpointID* (Z)   Used to indicate a single endpoint, particularly in a response to a command in which a call agent used a wild card. For example, when creating a connection using CRCX, the call agent might not care exactly which endpoint is used for the connection and might let the gateway make the choice. In such a case, the gateway would respond to the CRCX command and would use this parameter in that response to indicate the identity of the exact endpoint chosen.

- *SecondEndpointID* (Z2)   Used to indicate a second endpoint on the same gateway. This parameter would apply in the case where a call agent wishes to create a connection between two endpoints on the same gateway. This parameter can be used in the command to specify the second endpoint to be included in the connection. Its use in a command might involve the use of the "any" ($) wild card. In such a case, the parameter will be used in the response to the command to indicate the specific endpoint chosen as the second endpoint.

- *SecondConnectionId* (I2)   Connection identifier associated with the connection on a second endpoint. When a command such as CreateConnection specifies a second endpoint, then two connections are actually created. Each has its own identifier, and each can be manipulated through commands such as ModifyConnection.

- *RequestedInfo* (F)   Information to be provided by a gateway to a call agent in response to either of the two audit requests. The requested information can include items such as the current value of RequestedEvents, the current value of DigitMap, and the current value of NotifiedEntity. This parameter should not be confused with the RequestedEvents parameter.

- *QuarantineHandling* (Q)   Indicates how an endpoint should handle events that occur during the period in which the gateway is waiting for a response to a Notify command sent to a call agent (the quarantine period). The choices are that the endpoint should go ahead and process the events or simply discard them.

- *DetectEvents* (T)   A list of events that an endpoint should detect during the quarantine period. These events might include the detection of off-hook, on-hook, hook-flash, DTMF digits, etc.

- *RestartMethod* (RM)   Used to indicate whether a restart at an endpoint is graceful (existing connections are not terminated) or forced (existing connections are terminated); also used to indicate that an endpoint is about to be brought back into service after a specified delay

- *RestartDelay* (RD)   A number of seconds indicating when an endpoint will be brought back into service

- *EventStates* (ES)   Used in response to an audit command, this parameter is a comma-separated list of events associated with the current state of the endpoint and can indicate which events caused the endpoint to be in that particular state.

- *Capabilities* (A)   Coded in the same way as LocalConnectionOptions; used in response to an audit to indicate the capabilities supported by an endpoint

- *RemoteConnectionDescriptor* (RC)   An SDP session description related to the far end of a call; indicates an address and port to which media should be sent and describes the format of that media

- *LocalConnectionDescriptor* (LC)   An SDP session description for the current connection on an endpoint, indicating at which IP address and port number the endpoint is expecting to receive media and in what format. In a two-party call, the LocalConnectionDescriptor for the connection at one end corresponds to the RemoteConnectionDescriptor for the connection at the other end.

Not all parameters will appear in all commands. Rather, for each command, certain parameters are mandatory; certain parameters are optional; and certain parameters are forbidden. Table 6-1 shows the mapping of command parameters to commands, where M equals mandatory, O equals optional, and F equals forbidden.

Overview of MGCP Responses

For each MGCP command, a response is returned. Responses consist of a response line, optionally followed by a number of response parameters. The response line is made up of a return code, followed by the transaction identifier, and optionally followed by a commentary or reason phrase. Each of these is separated by a single *space* (SP) character, and the response line is terminated by a *carriage return line feed* (CRLF) and can be represented as follows:

```
Return code SP TransactionID SP Commentary CRLF
```

Table 6-1

Mapping of
command
parameters to
commands

Parameter Name	EPCF	CRCX	MDCX	DLCX	RQNT	NTFY	AUEP	AUCX	RSIP
ResponseAck	O	O	O	O	O	O	O	O	O
BearerInformation	M	O	O	O	O	F	F	F	F
CallId	F	M	M	O	F	F	F	F	F
ConnectionId	F	F	M	O	F	F	F	M	F
NotifiedEntity	F	O	O	O	O	O	F	F	F
RequestIdentifier	F	O	O	O	M	M	F	F	F
LocalConnection Options	F	O	O	F	F	F	F	F	F
ConnectionMode	F	M	M	F	F	F	F	F	F
RequestedEvents	F	O	O	O	O	F	F	F	F
SignalRequests	F	O	O	O	O	F	F	F	F
DigitMap	F	O	O	O	O	F	F	F	F
ObservedEvents	F	F	F	F	F	M	F	F	F
ConnectionParameters	F	F	F	O	F	F	F	F	F
ReasonCode	F	F	F	O	F	F	F	F	O
SpecificEndpointID	F	F	F	F	F	F	F	F	F
SecondEndpointID	F	O	F	F	F	F	F	F	F
SecondConnectionId	F	F	F	F	F	F	F	F	F
RequestedInfo	F	F	F	F	F	F	M	M	F
QuarantineHandling	F	O	O	O	O	F	F	F	F
DetectEvents	F	O	O	O	O	F	F	F	F
RestartMethod	F	F	F	F	F	F	F	F	M
RestartDelay	F	F	F	F	F	F	F	F	O
EventStates	F	F	F	F	F	F	F	F	F
Capabilities	F	F	F	F	F	F	F	F	F
RemoteConnection Descriptor	F	O	O	F	F	F	F	F	F
LocalConnection Descriptor	F	F	F	F	F	F	F	F	F

A return code is an integer value. Return codes are divided into four categories:

1XX (100 to 199) Provisional responses; a final response will follow later

2XX (200 to 299) Command successfully completed

4XX (400 to 499) Failure due to a transient error

5XX (500 to 599) Failure due to a permanent error

Table 6-2 gives a complete listing of the return codes defined in MGCP and their meanings.

We must correlate a response with the command that triggered the response. Therefore, the TransactionID appears in both commands and responses. The response to a command will use the same TransactionID as the command that generated the response.

The same parameters that are available for use in MGCP commands are also available for use in MGCP responses. The allowed usage in responses differs from the allowed usage in commands, however. For example, the LocalConnectionOptions parameter is an optional parameter in a CRCX command but is forbidden in a response to a CRCX command. Table 6-3 provides a mapping of response parameters to the commands that generate the response, where M equals mandatory, O equals optional, and F equals forbidden. We see from the table that responses generally contain fewer parameters than commands.

Command and Response Details

From the information in Tables 6-1 and 6-3, we can identify the structure of each command and the corresponding response. All that remains is to indicate where a given command can be encapsulated within another command. This information is included in the following command and response details, where optional parameters and optional encapsulated commands appear within square brackets. Note that the representation of the command and response formats here is slightly different than in certain parts of the MGCP specification. The MGCP specification defines both a protocol and an *Application Programming Interface* (API). The format here is based on the protocol and shows the complete command line with the EndpointId and MGCP version included. The format shows the complete response line with the ReturnCode, TransactionId, and Commentary all included. The information is presented in this way to enable a better visualization of the

Table 6-2

Return codes in
MGCP

Return Code	Description
100	The transaction is currently being executed. An actual completion message will follow later.
200	The requested transaction was completed normally.
250	The connection was deleted.
400	The transaction could not be executed due to a transient error.
401	The phone is already off hook.
402	The phone is already on hook.
403	The transaction did not execute because the endpoint does not have sufficient resources at this time.
404	Insufficient bandwidth exists at this time.
500	The transaction was not executed because the endpoint is unknown.
501	The transaction was not executed because the endpoint is not ready.
502	The transaction was not executed because the endpoint does not have sufficient resources.
510	The transaction was not executed because a protocol error was encountered.
511	The transaction was not executed because the command contained an unrecognized extension.
512	The transaction was not executed because the gateway is not equipped to detect one of the requested events.
513	The transaction was not executed because the gateway is not equipped to generate one of the requested signals.
514	The transaction was not executed because the gateway cannot send the requested announcement.
515	The transaction refers to an incorrect connection-id (might already have been deleted).
516	The transaction refers to an unknown call-id.
517	Unsupported or invalid mode
518	Unsupported or invalid package
519	The endpoint does not have a digit map.
520	The transaction could not be executed because the endpoint is restarting.

Return Code	Description
521	Endpoint redirected to another call agent
522	No such event or signal
523	Unknown action or illegal combination of actions
524	Internal inconsistency in LocalConnectionOptions
525	Unknown extension in LocalConnectionOptions
526	Insufficient bandwidth
527	Missing RemoteConnectionDescriptor
528	Incompatible protocol version
529	Internal hardware failure
530	CAS signaling protocol error
531	Failure of a grouping of trunks (e.g., facility failure)

commands and responses as actually transmitted by a call agent or gateway.

EndpointConfiguration Command

```
EPCF   TransactionId EndpointId MGCP 1.0
               BearerInformation
```

EndPointConfiguration Response

```
ReturnCode TransactionId Commentary
```

NotificationRequest Command

```
RQNT TransactionId EndpointId MGCP 1.0
            [NotifiedEntity]
            [RequestedEvents]
            RequestIdentifier
            [DigitMap]
            [SignalRequests]
            [QuarantineHandling]
            [DetectEvents]
            [encapsulated EndpointConfiguration]
```

Table 6-3

Mapping of
response
parameters to
MGCP commands

Parameter Name	EPCF	CRCX	MDCX	DLCX	RQNT	NTFY	AUEP	AUCX	RSIP
ResponseAck	F	F	F	F	F	F	F	F	F
BearerInformation	F	F	F	F	F	F	O	F	F
CallId	F	F	F	F	F	F	F	O	F
ConnectionId	F	O	F	F	F	F	F	F	F
NotifiedEntity	F	F	F	F	F	F	F	F	O
RequestIdentifier	F	F	F	F	F	F	O	F	F
LocalConnection Options	F	F	F	F	F	F	O	O	F
ConnectionMode	F	F	F	F	F	F	F	O	F
RequestedEvents	F	F	F	F	F	F	O	F	F
SignalRequests	F	F	F	F	F	F	O	F	F
DigitMap	F	F	F	F	F	F	O	F	F
ObservedEvents	F	F	F	F	F	F	O	F	F
ConnectionParameters	F	F	F	O	F	F	F	O	F
ReasonCode	F	F	F	F	F	F	O	F	F
SpecificEndpointID	F	O	F	F	F	F	F	F	F
SecondEndpointID	F	O	F	F	F	F	F	F	F
SecondConnectionId	F	O	F	F	F	F	F	F	F
RequestedInfo	F	F	F	F	F	F	F	F	F
QuarantineHandling	F	F	F	F	F	F	O	F	F
DetectEvents	F	F	F	F	F	F	O	F	F
RestartMethod	F	F	F	F	F	F	O	F	F
RestartDelay	F	F	F	F	F	F	O	F	F
EventStates	F	F	F	F	F	F	O	F	F
Capabilities	F	F	F	F	F	F	O	F	F
RemoteConnection Descriptor	F	F	F	F	F	F	F	O	F
LocalConnection Descriptor	F	M	O	F	F	F	F	O	F

NotificationRequest Response

```
ReturnCode TransactionId Commentary
```

Notify Command

```
NTFY TransactionId EndpointId MGCP 1.0
          [NotifiedEntity]
          RequestIdentifier
          ObservedEvents
```

Notify Response

```
ReturnCode TransactionId Commentary
```

CreateConnection Command

```
CRCX TransactionID EndpointID MGCP 1.0
          CallId
          [NotifiedEntity]
          [LocalConnectionOptions]
          Mode
          [RemoteConnectionDescriptor | SecondEndpointId]
          [Encapsulated NotificationRequest]
          [Encapsulated EndpointConfiguration]
```

CreateConnection Response

```
ReturnCode TransactionId Commentary
                    ConnectionId
                    [SpecificEndpointId]
                    [LocalConnectionDescriptor]
                    [SecondEndpointId]
                    [SecondConnectionId]
```

ModifyConnection Command

```
MDCX TransactionId EndpointId MGCP 1.0
          CallId
          ConnectionId
          [NotifiedEntity]
          [LocalConnectionOptions]
          [Mode]
          [RemoteConnectionDescriptor]
          [encapsulated NotificationRequest]
          [encapsulated EndpointConfiguration]
```

ModifyConnection Response

```
ReturnCode TransactionId Commentary
                    [LocalConnectionDescriptor]
```

DeleteConnection Command (from Call Agent to Gateway)

```
DLCX TransactionId EndpointId MGCP 1.0
          CallId
```

```
ConnectionId
[encapsulated NotificationRequest]
[encapsulated EndpointConfiguration]
```

DeleteConnection Response (from Gateway to Call Agent)

```
ReturnCode TransactionId Commentary
                    ConnectionParameters
```

DeleteConnection Command (from Gateway to Call Agent)

```
DLCX TransactionId EndpointId MGCP 1.0
         CallId
         ConnectionId
         ReasonCode
         Connectionparameters
```

DeleteConnection Response (from Call Agent to Gateway)

```
ReturnCode TransactionId Commentary
```

AuditEndpoint Command

```
AUEP TransactionId EndpointId MGCP 1.0
         [RequestedInfo]
```

AuditEndpoint Response

```
ReturnCode TransactionId Commentary
                    EndpointIdList | {
                    [RequestedEvents]
                    [DigitMap]
                    [SignalRequests]
                    [RequestIdentifier]
                    [NotifiedEntity]
                    [ConnectionIdentifiers]
                    [DetectEvents]
                    [ObservedEvents]
                    [EventStates]
                    [BearerInformation]
                    [RestartReason]
                    [RestartDelay]
                    [ReasonCode]
                    [Capabilities]}
```

AuditConnection Command

```
AUCX TransactionId EndpointId MGCP 1.0
         ConnectionId
         RequestedInfo
```

AuditConnection Response

```
ReturnCode TransactionId Commentary
                    [CallId]
                    [NotifiedEntity]
```

```
                              [LocalConnectionOptions]
                              [Mode]
                              [RemoteConnectionDescriptor]
                              [LocalConnectionDescriptor]
                              [ConnectionParameters]
```

RestartInProgress Command

```
        RSIP TransactionId EndpointId MGCP 1.0
                RestartMethod
                [Restartdelay]
                [ReasonCode]
```

RestartInProgress Response

```
        ReturnCode TransactionId Commentary
                                [NotifiedEntity]
```

Call Setup Using MGCP

Utilizing the foregoing descriptions of commands and responses, we can describe MGCP-based call setup in some detail. The following description of the process assumes that two endpoints are to be used on two different gateways, but that both gateways are connected to the same call agent.

Call setup using MGCP involves the creation of at least two connections, as shown in Figure 6-7. First, a call agent uses the CRCX com-

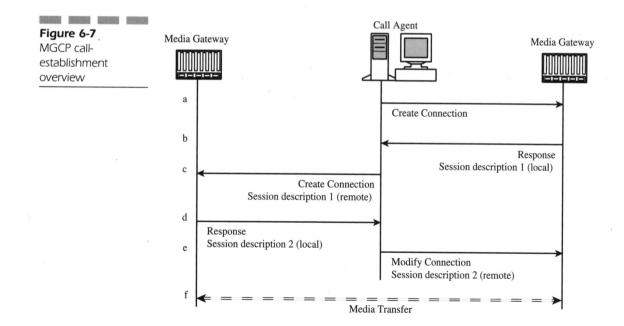

Figure 6-7
MGCP call-establishment overview

mand to instruct a gateway to create a connection on an endpoint. The call agent can specifically identify the endpoint to be used or can use a wild-card notation to enable the gateway to choose the endpoint. The gateway creates the connection and responds to the call agent with a session description (LocalConnectionDescriptor) that is applicable to the connection. The LocalConnectionDescriptor is an SDP description of the session as it applies to the connection just created. Therefore, this description contains information such as an IP address, port number, and media coding.

Upon receipt of the response from the first gateway, the call agent instructs the second gateway to create a connection on an endpoint. The call agent has more information available for inclusion in the CRCX command to the second gateway that it had for the CRCX command for the first gateway. Specifically, the call agent is armed with the knowledge of the session description applicable to the connection on the first endpoint, and it can pass this information to the second endpoint. Therefore, the call agent issues a CRCX command to the second gateway and includes the optional RemoteConnectionDescriptor parameter, which contains the session description obtained for the connection made on the first endpoint. Thus, the second endpoint in the call now has a session description related to the first endpoint and knows where to send media. For the moment, the first endpoint in the call is still in the dark.

The second endpoint sends a response to the CRCX command and includes the LocalConnectionDescriptor parameter, which is a session description related to the connection created on the second endpoint.

Third, the call agent issues a MDCX command to the first gateway. The command contains a RemoteConnectionDescriptor parameter, which corresponds to the session description received from the second gateway for the second endpoint. At this stage, both endpoints now hold a session description for the other and know exactly where to send the media stream and in what format. Media can now be exchanged.

Figure 6-8 shows a detailed example of call establishment between two gateways, each supporting DS0 endpoints and each managed by the same call agent, including the session descriptions involved.

a. The Call agent sends a CRCX command to MGA, specifying that a connection is to be created on endpoint 1 (EP1). The command also specifies the TransactionId (1111) and the CallId (1234567). No BearerInformation is sent in this case, because the endpoint is specifically identified—and we assume that the bearer-related information for the line side of the endpoint is already known. If it

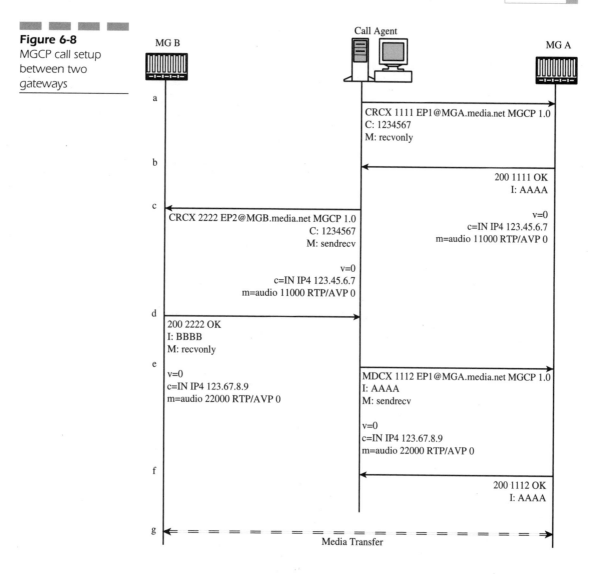

Figure 6-8
MGCP call setup between two gateways

Call Agent

MG B

MG A

a

CRCX 1111 EP1@MGA.media.net MGCP 1.0
C: 1234567
M: recvonly

b

200 1111 OK
I: AAAA

c

CRCX 2222 EP2@MGB.media.net MGCP 1.0
C: 1234567
M: sendrecv

v=0
c=IN IP4 123.45.6.7
m=audio 11000 RTP/AVP 0

v=0
c=IN IP4 123.45.6.7
m=audio 11000 RTP/AVP 0

d

200 2222 OK
I: BBBB
M: recvonly

e

v=0
c=IN IP4 123.67.8.9
m=audio 22000 RTP/AVP 0

MDCX 1112 EP1@MGA.media.net MGCP 1.0
I: AAAA
M: sendrecv

v=0
c=IN IP4 123.67.8.9
m=audio 22000 RTP/AVP 0

f

200 1112 OK
I: AAAA

g

Media Transfer

were not known or if the selection of the endpoint were wild carded, then the bearer information would specify either A-law or mu-law. The CRCX command specifies the mode to be receive-only for now. In other words, it can receive media from the IP network but cannot send media to the IP network. The reason for this choice is the fact that the endpoint does not yet have a session description for the remote end, and hence cannot know where to send a media stream. The remote end is about to be presented with a session description for the local end and could start to send media before the local endpoint receives a session description for the remote end. Therefore, the local endpoint should be prepared to receive media from the remote endpoint.

b. MG A responds to the CRCX command with the same TransactionID, a ConnectionId (AAAA), and a session description indicating voice coded according to G.711 mu-law (RTP payload format 0). Note that this voice coding is applicable to the IP side of the endpoint, as opposed to the line side of the endpoint. The coding could equally be some other coding choice, such as G.728 or G.729.

c. The call agent sends a CRCX command to the second gateway (MG B). Again, a specific endpoint (EP2) is indicated in the command. The TransactionId is 2222, and the same CallId (1234567) is used. The call agent includes a RemoteConnectionDescriptor, which is simply the session description received earlier from MG A. This time, the CRCX command specifies the mode as send-receive, because a session description for the remote end is available. In other words, the endpoint has an address and port to which it can send media.

d. MG B sends a positive response to the CRCX, including a ConnectionId (BBBB) and a session description corresponding to that connection. The session description in this case also happens to specify G.711 mu-law coded voice.

e. The call agent sends a Modify Connection (MDCX) command to the first gateway. The command specifies the ConnectionID and includes a RemoteConnectionDescriptor, which is the session description received from MG B. Given that a session description is available for the remote end, MG A now has an address and port to which it can send media. Therefore, the MDCX command specifies the mode as send-receive.

f. MG A sends a positive response to the MDCX command.

g. Media can now be exchanged between the two endpoints.

Call Setup Based on Detected Events

Our example in Figure 6-8 assumes that the call agent knows that a call is to be established between the two gateways. This process involves an assumption that the call agent has received external signaling regarding the call that needs to be established. Such signaling could, for example, have come from an external SS7 gateway terminating SS7 links and passing ISUP messages to the call agent. Alternatively, the call agent might have terminated one or more SS7 links directly.

A separate situation exists when the signaling related to a call is received at the media gateway itself. This situation would be the case, for example, if a call were to be received at a residential gateway equipped with analog-line endpoints. The signals and digits related to the call would be received directly at the analog-line endpoint and would need to be passed from the gateway to the call agent, where call routing decisions are made.

Figure 6-9 shows the establishment of a call between two gateways. One is a residential gateway terminating an analog line, and the other is a trunking gateway with DS0 endpoints.

a. The call agent begins by sending a NotificationRequest (RQNT) command to the residential gateway. This command uses the asterisk (*) wild card to indicate that the command is applicable to all endpoints on the gateway. The Digitmap parameter (D:) loads a digit map into the gateway. The digit map is effectively a dial plan, indicating which combinations of digits are acceptable to dial. In this case, the subscriber can dial 0 for the operator or any seven-digit number starting with a digit between 2 and 9. The next parameter is the RequestedEvents parameter (R:), which indicates the events that the gateway should detect and what to do when such events occur. In this case, the gateway is to detect an off-hook event (hd). An embedded request exists, E(. . .   ), specifying what should happen in the event that an off-hook event is detected. The gateway should detect an on-hook event (hu). The gateway should also detect any of the digits 0 to 9 and the characters * and #, and if these are detected, the gateway should treat them according to the digit map (D). Of course, upon detection of the off-hook event, the gateway should send a dial tone. The embedded request is equivalent to a second notification request within the first. Its purpose is to instruct an endpoint to detect certain events or to apply certain signals only if another event has occurred. For example, when a subscriber goes off hook, then the gateway should be ready to detect an on-hook event.

b. The gateway sends a positive acknowledgment to the RQNT command.

c. Some time later, a subscriber picks up the phone and EP1 detects the off-hook transition. EP1 is the endpoint connected to the subscriber's line. In accordance with the received RQNT command, the endpoint applies a dial tone and begins to collect digits. The analog-line

Figure 6-9
MGCP call setup
between residential
and trunking
gateways

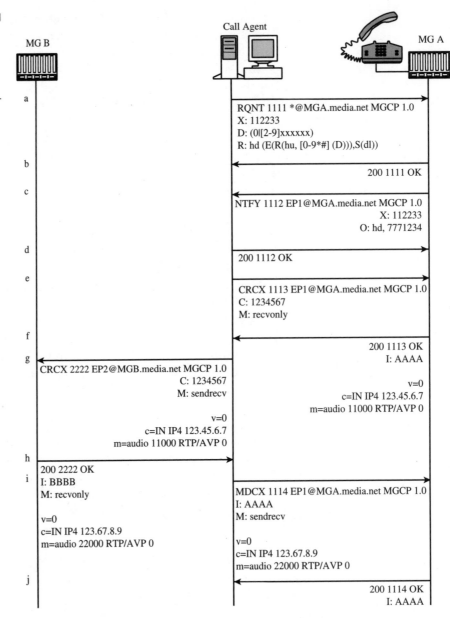

subscriber happens to dial 7771234. Because this number is acceptable according to the digit map, the endpoint sends a *Notify* (NTFY) command to the call agent, including the ObservedEvents (O) parameter. The two observed events are the subscriber going off hook and the collected dialed digits.

d. The call agent acknowledges the notification.

e. The call agent instructs the endpoint to create a connection, specifying the mode as receive-only.

f. The gateway creates the connection and responds to the CRCX command with a ConnectionID and a session description.

g–j. The call establishment proceeds in the same manner as described in Figure 6-8.

MGCP Events, Signals, and Packages

Figure 6-9 introduced a number of events to be detected, such as off-hook and on-hook, and at least one signal to be applied (dial tone). Depending on the function of the endpoint, there might be a range of events that can be detected and a range of signals that can be applied. For a residential line, the obvious events include off-hook, on-hook, and hook flash. The obvious signals include dial tone, stutter dial tone, call-waiting tone, and congestion tone (fast busy), among others. Similarly, a gateway that terminates MF trunks will need to detect events such as seizure signals, and MF signals and will need to send similar signals.

Because different types of endpoints will need to detect different events and apply different signals, MGCP groups events and signals into packages, where a package applies to a given type of endpoint. These packages have package names, and a given event or signal is identified by a package name and an event or signal name with the format `package name/event name`. For example, a gateway that terminates analog lines will implement the line package (L). The off-hook event has the symbol hd. Therefore, when referring to an off-hook event detected by an analog-line endpoint, the formal identification is L/hd. Incidentally, the package name is optional when it comes to identifying an event or signal, because each event or signal has a unique identifier. The use of the package name is useful with wild carding, however. If, for example, a gateway were required to detect every event in the line (L) package, then the events to be detected can be represented by L/*. This format is a lot easier than listing every event that might occur. You can also apply wild carding to the package name. If, for example, the event surprise happened to be included in more than one package, then the syntax */surprise would refer to the surprise event in every package.

Strictly speaking, MGCP does not differentiate between events and signals. As far as MGCP is concerned, signals are events. Therefore, for a gateway that terminates residential lines, it is legal in MGCP for a call agent to instruct a gateway to apply an on-hook signal or to detect a dial tone, rather than the other way around. Of course, such applications do not make sense, but the fact that events and signals are treated in the same way does make the definition of packages a little easier and reduces the complexity of the protocol syntax. The application programmer just needs to be a little careful. One would not want to detect on-hook after the occurrence of off-hook and end up applying on-hook instead. Strange things might happen, and the customer would certainly have difficulty making calls.

When the SignalRequests and RequestedEvents parameters are both specified in an RQNT command, the signals that are requested are synchronized with the events that are to be detected. For example, if the ringing signal is to be applied to a line and an off-hook event is to be detected, then the ringing signal will be stopped as soon as the off-hook event occurs.

A number of signals and events are defined in the MGCP specification and are grouped into various packages. Should there be a need for new events or packages to be defined in the future, then these should be documented (and the names of the events or packages should be registered with the Internet Assigned Numbers Authority, or IANA). Prior to registration with IANA, developers might wish to test new packages. Such experimental packages should have names beginning with the two characters x-. These characters indicate an experimental extension to the protocol, and packages starting with these characters do not need to be registered.

Interworking between MGCP and SIP

The previous two examples assume that two gateways are used but that the same call agent controls those two gateways. Equally likely is the situation where separate call agents control the two gateways. In such a case, a need exists for signaling between the two call agents. In the softswitch architecture, we use SIP.

Figure 6-10 is an adaptation of Figure 6-8, with the difference that two separate call agents control the two gateways. This example shows how MGCP and SIP fit well together, particularly because they both utilize SDP for describing sessions.Once the connection is created on the first endpoint (line b of Figure 6-10), the returned session description becomes the message body of a SIP INVITE that is sent to the other call agent (line c).

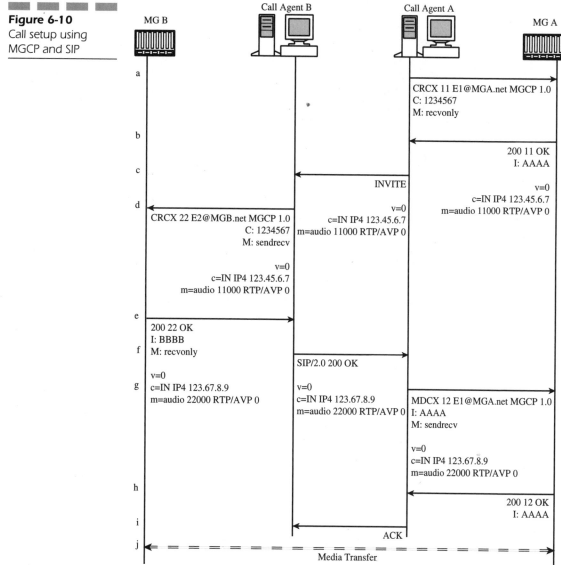

Figure 6-10
Call setup using
MGCP and SIP

NOTE: *Most of the SIP headers and other fields have been omitted, and to avoid clutter, the names of the endpoints and the TransactionIDs have been shortened compared to Figure 6-8.*

Once the second connection is created on the second gateway (line e), then the corresponding session description becomes the message body of the SIP 200 (OK) response (line f). In turn, the session description is mapped to the MDCX command that is sent to the first endpoint (line g). Finally, the positive acknowledgment to the MDCX command (line h) is mapped to the SIP ACK (line i), thereby completing the MGCP and SIP signaling sequences.

MGCP Call Release

Releasing a call with MGCP is normally done by issuing a *Delete Connection* (DLCX) command from a call agent to a gateway but can be instigated by the gateway in the event that the endpoint is no longer willing or capable to communicate. In the case of a residential analog line, the endpoint will not normally issue a DLCX command to the call agent when the subscriber hangs up. Instead, the gateway will simply notify the event to the call agent. The call agent then takes responsibility for deleting the connection and issues the DLCX command.

Figure 6-11 is a follow-on example from the scenario depicted in Figure 6-9. The gateway has been instructed by the call agent to report the occurrence of an on-hook event. Once such an event is detected, the gateway notifies the call agent by issuing an NTFY command (line a).

The call agent acknowledges the notification (line b) and then sends a DLCX command to the endpoint (line c) to remove the connection. Note that the DLCX command includes an encapsulated notification request with a request identifier (X:) and a requested event (R:) of off-hook. In other words, the call agent is instructing the gateway to clear the connection but to keep an eye on the line for the next time that the subscriber picks up the phone.

The gateway responds to the DLCX command (line d) with a positive response (250 = Connection has been deleted). The response includes connection parameters providing feedback statistics regarding the call. These statistics include the number of packets sent, the number of octets sent, the number of packets received, the number of octets received, and the number of packets lost. Note that MGCP defines additional connection data that can also be reported in response to a DLCX command.

The call agent issues a second DLCX command to delete the connection at the second gateway (line e), and the second gateway responds with similar connection parameters (line f). The call is now completely disconnected, and the IP resources are available for use by other sessions.

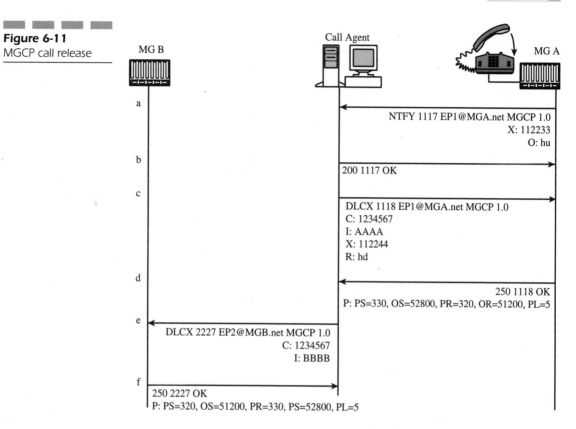

Figure 6-11
MGCP call release

MG B

Call Agent

MG A

a

NTFY 1117 EP1@MGA.net MGCP 1.0
X: 112233
O: hu

b

200 1117 OK

c

DLCX 1118 EP1@MGA.net MGCP 1.0
C: 1234567
I: AAAA
X: 112244
R: hd

d

250 1118 OK
P: PS=330, OS=52800, PR=320, OR=51200, PL=5

e

DLCX 2227 EP2@MGB.net MGCP 1.0
C: 1234567
I: BBBB

f

250 2227 OK
P: PS=320, OS=51200, PR=330, PS=52800, PL=5

MEGACO/H.248

Known as MEGACO within the IETF and H.248 within the ITU, this protocol has widespread backing within both the Internet and telecommunications communities. In fact, the cooperation between the two industry organizations reflects the convergence that is occurring between telecommunications as we have known it and the relatively new world of the Internet.

Given that the term MEGACO/H.248 is rather a mouthful, for the purpose of brevity, the remainder of the book uses only the term MEGACO.

MEGACO Architecture

MEGACO has a number of architectural similarities to MGCP. Similar to MGCP, it defines *Media Gateways* (MGs), which perform the conversion of media from the format required in one network to the format required in another. MEGACO also defines *Media Gateway Controllers* (MGCs), which control call establishment and tear-down within MGs and are akin to call agents as defined by MGCP. The similarities are hardly that surprising, given that they were both designed to address a similar set of requirements.

The MEGACO protocol involves a series of transactions between MGCs and MGs. Each transaction involves the sending of a TransactionRequest by the initiator of the transaction and the sending of a TransactionReply by the responder. A TransactionRequest consists of a number of commands, and the TransactionReply comprises a corresponding number of responses. For the most part, an MGC requests transactions, and the corresponding actions are executed within an MG. As in MGCP, however, a number of cases exist where an MG initiates the transaction request.

The MEGACO specification is written so that it provides both a text encoding and a binary encoding of the protocol. The text encoding is written according to *Augmented Backus-Naur Form* (ABNF), described in RFC 2234. The binary encoding is written according to *Abstract Syntax Notation One* (ASN.1), described in ITU-T Rec X.680, 1997. This fact reflects one difference between the IETF, which normally uses ABNF, and the ITU, where ASN.1 is the format of choice. MGCs are strongly recommended to support both types of encoding. MGs can support both types of encoding but are not required to do so. Within this book, the ABNF form is used when offering examples of MEGACO usage, because it is considered more human readable than ASN.1.

Terminations   MEGACO defines terminations, which are logical entities on an MG that act as sources or sinks of media streams. Certain terminations are physical. They have a semi-permanent existence and can be associated with external physical facilities or resources. These would include a termination connected to an analog line or a termination connected to a DS0 channel. They exist as long as they are provisioned within the MG and are akin to endpoints, as defined in MGCP.

Other terminations have a more transient existence and only exist for the duration of a call or media flow. These terminations are known as ephemeral terminations and represent media flows, such as an RTP session. These terminations are created as a result of a MEGACO Add command, which is somewhat similar in function to the CreateConnection command in MGCP. They are destroyed by means of the Subtract command, which is somewhat similar to the MGCP DeleteConnection command.

Terminations have specific properties, and the properties of a termination will vary according to the type of termination. Clearly, a termination connected to an analog line will have different characteristics to a termination connected to a TDM channel, such as a DS0. The properties associated with a termination are grouped into a set of descriptors. These descriptors are included in MEGACO commands, thereby enabling termination properties to be changed according to instruction from MGC to MG.

A TerminationID references a termination. The MG chooses this identifier. Similar to MGCP, MEGACO enables the use of the wild cards "all" (*) and "any" or "choose" ($). A special TerminationID exists called Root. This TerminationID is used to refer to the gateway as a whole, rather than to any specific terminations within the gateway.

Contexts A Context is an association between a number of terminations for the purposes of sharing media between those terminations. Terminations can be added to contexts, removed from contexts, or moved from one context to another. A termination can exist in only one context at any time, and terminations in a given gateway can only exchange media if they are in the same context.

A termination is added to a context through the use of the Add command. If the Add command does not specify a context to which the termination should be added, then a new context is created as a result of the execution of the Add command. This mechanism is the only mechanism for creating a new context. A termination is moved from one context to another through the use of the Move command and is removed from a context through the use of the Subtract command. If the execution of a Subtract command results in the removal of the last termination from a given context, then that context is deleted.

The relationship between terminations and contexts appears in Figure 6-12, where a gateway is depicted with four active contexts. In context C1, we see a simple two-way call across the MG. In Context C2, we see a three-way call across the MG. In contexts C3 and C4, we see a possible implementation of call waiting. In context C3, terminations T6 and T7 are involved in a call. Another call arrives from termination T8 for termination T7. If the user wished to accept this waiting call and place the existing call on hold, then we could move termination T7 from context C3 to context C4.

The existence of several terminations within the same context means that they have the potential to exchange media. The existence of terminations in the same context does not necessarily mean that they can all send data to each other and receive data from each other at any given time, however. The context itself has certain attributes, including the topology, which indicates the flow of media between terminations (which terminations can send media to others/receive media from others). Also, there is the priority attribute, which indicates the precedence applied to a context when an

■■■ ■■■ ■■■ ■■■

Figure 6-12
MEGACO contexts
and terminations,
showing the possible
Move command

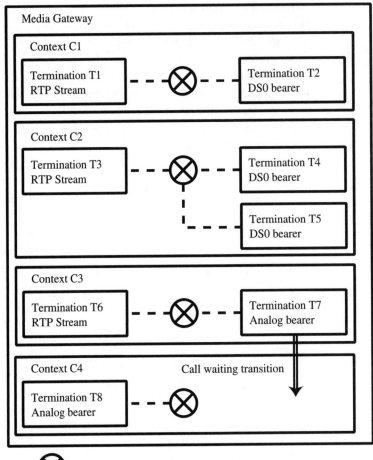

⊗ represents association between terminations

MGC must handle many contexts simultaneously, and an emergency attribute that is used to give preferential handling to emergency calls.

A ContextID, assigned by the MG, identifies a context and is unique within a single MG. As is the case for terminations, MECAGO enables wild carding when referring to contexts, such that the "all" (*) and "any" or "choose" ($) wild cards can be used. The "all" wild card can be used by an MGC to refer to every context on a gateway. The "choose" wild card ($) is used when an MGC requires the MG to create a new context.

A special context exists known as the null Context. This context contains all terminations that are not associated with any other termination (i.e., all terminations that do not exist in any other context). Idle terminations normally exist in the null context. The ContextID for the null context is simply "-."

Transactions MEGACO transactions involve the passing of commands and the responses to those commands. Commands are directed towards terminations within contexts. In other words, every command specifies a ContextID and one or more TerminationIDs to which the command applies. This situation is the case even for a command that requires some action by an idle termination that does not exist in any specific context. In such a case, the null context is applicable.

Multiple commands can be grouped in a transaction structure, where a set of commands related to one context can be followed by a set of commands related to another context. The grouped commands are sent together in a single TransactionRequest. This process can be represented as follows:

```
TransactionRequest (TransactionID {
        ContextID1 {Command, Command, ... Command},
        ContextID2 {Command, Command, ... Command},
        ContextID3 {Command, Command, ... Command}  } )
```

No requirement exists for a TransactionRequest to contain commands for more than one context or even to contain more than one command. A TransactionRequest containing just a single command for a single context is perfectly valid.

Upon the receipt of a TransactionRequest, the recipient executes the enclosed commands. The commands are executed sequentially in the order specified in the TransactionRequest. Upon completed execution of the commands, a TransactionReply is issued. This command has a similar structure to the TransactionRequest in that it contains a number of responses for a number of contexts. A TransactionReply can be represented as follows:

```
TransactionReply (TransactionID {
        ContextID1 {Response, Response, ... Response},
        ContextID2 {Response, Response, ... Response},
        ContextID3 {Response, Response, ... Response}  } )
```

As mentioned, commands within a transaction are executed in sequence. If a given command in the transaction happens to fail, then the subsequent commands in the transaction are not processed. One exception is when a transaction contains one or more optional commands. If an optional command fails, then processing of the remaining commands continues. Optional commands in a transaction are indicated by inserting the string O- immediately prior to the command name.

In the case where the recipient of a TransactionRequest might take some time to execute the request, the option exists of returning an interim reply so that the sender of the request does not assume that it has been lost. This interim reply is simply an indication that the TransactionRequest has been received and is being handled. The term applied to this response is TransactionPending. A TransactionID is a 32 bit integer. This reply simply returns the TransactionId received (and no parameters):

```
TransactionPending (TransactionID { } )
```

Messages Not only is it possible to combine multiple commands within a single transaction, but it is also possible to concatenate multiple transactions within a Message. In ABNF syntax, a Message begins with the word MEGACO followed by a slash, followed by the protocol version number (1), followed by a message ID (mId), followed by the message body. The mId can be the domain name or IP address (and optionally, the port) of the entity that is transmitting the message. As an alternative, the mId can be the MTP point code and network indicator of the entity that is issuing the message.

The following example combines the concept of Message, transactions, and commands and shows the text format of a message:

```
MEGACO/1 [111.111.222.222]:34567
Transaction = 12345 {
        Context = 1111 {
                Add = A5555,
                Add = A6666
        }
        Context = $ {
                Add = A7777
        }
}
```

In this example, an MGC issues the message from address 111.111.222.222 and port 34567. One transaction exists, which is identified by TransactionId 12345. Two commands exist that are related to context 1111. Both commands are Add commands, causing terminations A5555 and A6666 to be added to context 1111. One command is related to any context that the MG might choose. This command is an Add command, whereby termination A7777 is added to the context. The processing of this command would result in the MG creating a new context, for which the ContextID would be returned in the response.

Transactions within a message are treated independently. The order of the transactions within the message does not imply an order in which the recipient of the message should handle the transactions. Note that this situation is quite different from the handling of commands within a transaction, where the ordering of the commands is of significance.

Overview of MEGACO Commands

MEGACO defines eight commands that enable the control and manipulation of contexts and terminations. Most of the commands are sent from an

MGC to an MG. The exceptions are the Notify command, which is always sent from MG to MGC, and the ServiceChange command, which can be sent from either an MG or an MGC.

Add   The Add command adds a termination to a context. If the command does not specify a particular context to which the termination will be added, then a new context is created. If the command does not indicate a specific TerminationID but instead uses the choose ($) wild card, the MG will create a new ephemeral termination and add it to the context.

Modify   The Modify command is used to change the property values of a termination, to instruct the termination to issue one or more signals, or to instruct the termination to detect and report specific events.

Subtract   The Subtract command is used to remove a termination from a context. The response to the command is used to provide statistics related to the termination's participation in the context. These reported statistics depend upon the type of termination in question. For an RTP termination, the statistics might include items such as packets sent, packets received, jitter, etc. In this respect, the command is similar to the MGCP DLCX command. If the result of a Subtract command is the removal of the last termination from a context, then the context itself is deleted.

Move   The Move command is used to move a termination from one context to another. You should not use this command to move a termination from or to the null context, because these operations must be performed with the Add and Subtract commands, respectively. The capability to move a termination from one context to another provides a useful tool for accomplishing the call-waiting service.

AuditValue   The MGC uses the AuditValue command to retrieve current values for properties, events, and signals associated with one or more terminations.

AuditCapabilities   The MGC uses the AuditCapabilities command to retrieve the possible values of properties, signals, and events associated with one or more terminations. At first glance, this command might appear similar to the AuditValue command. The difference between them is that the AuditValue command is used to determine the current status of a termination, whereas the AuditCapabilities command is used to determine the possible statuses that a termination might assume. For example, AuditValue would indicate any signals that are currently being applied by a termination, where AuditCapabilities would indicate all of the possible signals that the termination could apply if required.

Notify   The Notify command is issued by an MG to inform the MGC of events that have occurred within the MG. The events to be reported will have previously been requested as part of a command from the MGC to the MG, such as a Modify command. The events reported will be accompanied by a RequestID parameter to enable the MGC to correlate reported events with previous requests.

ServiceChange   The MG uses the ServiceChange command to inform an MGC that a group of terminations will be taken out of service or will resume service. The command is also used in a situation where an MGC is handing over control of an MG to another MGC. In that case, the command is first issued from the controlling MGC to the MG in order to instigate the transfer of control. Subsequently, the MG issues the ServiceChange command to the new MGC as a means of establishing the new relationship.

Descriptors

MEGACO defines a number of descriptors that are available for use with commands and responses. These descriptors form the parameters of the command and/or response and provide additional information to qualify a given command or response. Depending on the command or response in question, a given descriptor will be mandatory, forbidden, or optional. In most cases, where a descriptor is not forbidden, it is optional. Relatively few cases exist where a descriptor is mandatory.

The general format of a descriptor is as follows:

```
Descriptorname=<someID>{parm=value, parm=value, …}
```

The parameters of a descriptor can be fully specified, under specified, or over specified. Fully specified means that the parameter is assigned a unique and unambiguous value that the command recipient must use in processing the command. Under specified involves the use of the "any" or "choose" wild card ($), enabling the command recipient to select any value that it can support. Over specified means that the sender of the command provides a list of acceptable parameter values in order of preference. The command recipient selects one of the offered choices for use. The response to the command will indicate which of the choices was selected.

Modem Descriptor   This descriptor specifies the modem type and associated parameters to be used in modem connections for audio, video, or data communications or text conversation. The following types of modem are included: V.18 (text telephony), V.22 (1200 bps), V.22bis (2400 bps), V.32 (9600 bps), V.32bis (14400 bps), V.34 (33600 bps), V.90 (56 kbps), V.91 (64 kbps), and synchronous ISDN.

By default, a termination does not have a modem descriptor. In other words, at startup, no modem-related properties will be specified for a termination. They will be applied later as a result of a command (e.g., Add or Modify) from an MGC to an MG.

Multiplex Descriptor   In multimedia communications, different media streams can be carried on different bearer channels. The Multiplex descriptor specifies the association between media streams and bearers. The following multiplex types are supported: H.221. H.223, H.226, and V.76.

Media Descriptor   The media descriptor describes the various media streams. This descriptor is hierarchical in that it contains an overall descriptor, known as a Termination State Descriptor, which is applicable to the termination in general. It also contains a number of Stream Descriptors that are applicable to each media stream. Furthermore, each Stream Descriptor also contains up to three subordinate descriptors, known respectively as the Local Control Descriptor, the Local Descriptor, and the Remote Descriptor. This type of hierarchy can be represented as follows:

Media Descriptor
       Termination State Descriptor
       Stream Descriptor
              Local Control Descriptor
              Local Descriptor
              Remote Descriptor

Termination State Descriptor   This descriptor contains the two properties ServiceStates and EventBufferControl plus other properties of a termination that are not specific to any media stream.

The ServiceStates property indicates whether the termination is available for use. Three values are applicable: "test," "out of service," and "in service." In service does not mean that the termination is currently involved in a call. Rather, it means that the termination is either actively involved in a call or is available for use in a call. In service is the default state of a termination.

The EventBufferControl property specifies whether events detected by the termination are to be buffered following detection or processed immediately. To begin, a termination will report events that the MGC has asked it to report, which the MGC will do through the issuance of a command that contains an EventsDescriptor. Whether the termination will actually report the events indicated in the EventsDescriptor depends on whether the EventBufferControl is set to off or lockstep. If set to off, then the termination will report detected events immediately. If the value is set to lockstep, then events will be buffered in a *first in, first out* (FIFO) buffer along with the time that the event occurred. The contents of the buffer will be examined upon receipt of a new EventsDescriptor, specifying events that the termination is to detect. Inclusion of the EventBufferControl within the TerminationStateDescriptor enables the MGC to turn on or off immediate notification of events without having to send a separate EventsDescriptor each time.

The Termination State Descriptor is optional within a Media descriptor.

Stream Descriptor   A stream is defined by its subordinate descriptors LocalControl Descriptor, Local Descriptor, and Remote Descriptor and is identified by a StreamID.

StreamID values are used between an MG and an MGC to indicate which media streams are interconnected. Within a given context, streams with the same StreamID are connected. A stream is created by specifying a new StreamID on a particular termination in a context. A stream is deleted by setting empty Local and Remote descriptors and by setting the values of ReserveGroup and ReserveValue in the subordinate LocalControl Descriptor to "false."

Local Control Descriptor   This descriptor contains properties that are specific to the media stream—in particular the Mode property, the ReserveGroup property, and the ReserveValue property.

Mode can take one of the values `sendonly`, `receiveonly`, `sendreceive`, `inactive`, or `loopback`. Send and receive directions are with respect to the exterior of the context. If mode is set to `sendonly`, then a termination can only send media to entities outside the context. The termination cannot pass media to other entities within the same context. If mode is set to `receiveonly`, then a termination can receive media from outside the context and pass it to other terminations in the context, but it cannot receive media from other terminations and pass it to destinations outside the context.

When an MGC wishes to add a termination to a context, it can specify a set of choices for the session (using local and remote descriptors) that the MGC would prefer the MG to use and specifies these in order of preference.

The MG is not obligated to select the first choice offered by the MGC. The MG might be obligated to reserve resources for the sessions indicated by the MGC, however. The properties ReserveValue and ReserveGroup indicate which resources should be reserved for the alternatives specified by the MGC in the Local Descriptor and Remote Descriptor.

The Local Descriptor and Remote Descriptor can specify several properties and/or property groups. For example, a given SDP description could indicate two groups of properties: one for G.711 A-law and one for G.729. If the value of ReserveGroup is true, then the MG should reserve resources for one of these property groups. The use of ReserveValue is similar, but this command is applied to reserve resources for a particular property value, as opposed to a group of properties.

Local Descriptor and Remote Descriptor When used with a text encoding of the protocol, the Local Descriptor and Remote Descriptor contain zero or more SDP session descriptions describing the session at the local end of a connection and the remote end of a connection, respectively.

The MEGACO usage of SDP includes some variations from the strict syntax of SDP, as specified in RFC 2327. In particular, the s=, t=, and o= lines are optional; the choose wild card ($) is permitted; and it is permissible to indicate several alternatives in places where a single parameter value should normally be used.

Each of the local and remote descriptors can contain several session descriptions, with the beginning of the session description indicated with the SDP v= line.

Events Descriptor This descriptor contains a RequestIdentifier and a list of events that the MG should detect and report, such as off-hook transition, fax tone, etc. The purpose of the RequestIdentifier is to correlate requests and consequent notifications. Normally, detected events will be reported immediately to the MGC. Events might be buffered, however, depending on the value of the EventControlBuffer property (which is a property specified in the TerminationState Descriptor).

In the case where events are buffered and subsequently need to be notified to the MGC, then the information related to those events is stored in the EventBufferDescriptor.

Signals Descriptor This descriptor contains a list of signals that a termination is to apply. Signals can be applied to just a single stream or to all streams in a termination. Typical signals will include, for example, tones applied to a subscriber line (such as ring tone or dial tone). Three types of signals exist:

- *On/off* The signal remains on until it is turned off.
- *Timeout* The signal persists until a defined period of time has elapsed or until it is turned off, whichever occurs first.
- *Brief* The signal is to be applied only for a limited time (such as would be the case for an MF signal on R1 MF trunks).

A SignalsDescriptor can include a sequential sequence of signals that are to be played one after another.

Audit Descriptor The Audit Descriptor specifies a list of information to be returned from an MG to an MGC. The Audit descriptor is simply a list of other descriptors that should be returned in the response. These descriptors can include the following: Modem, Multiplex, Events, Signals, ObservedEvents, DigitMap, Statistics, Packages, and EventBuffer.

Service Change Descriptor This descriptor is used only in association with the ServiceChange command and includes information such as the type of service change that has occurred or is about to occur, the reason for the service change, and a new address to be used after the service change.

The type of service change is defined by the parameter Service-ChangeMethod, which can take one of the defined values Graceful, Forced, Restart, Disconnected, Handoff, or Failover. Graceful indicates the removal of terminations from service after a specified delay, without interrupting existing connections. Forced indicates an abrupt removal from service, including the loss of existing connections. Restart indicates that restoration of service will begin after a specified delay. Disconnected applies to the entire MG and indicates that connection with the MGC has been restored after a period of lost contact. The MGC should issue an Audit command to verify that termination characteristics have not changed during the period of lost contact. Handoff is used from MGC to MG to indicate that the MGC is going out of service and that a new MGC is taking over. This method is also used from an MG to a new MGC when trying to establish contact after receiving a handoff from the previous MGC. Failover is sent from MG to MGC in the case where an MG has detected a fault and a backup MG is taking over.

The ServiceChangeDelay specifies a number of seconds indicating the delay that is applicable. This parameter would be used, for example, in conjunction with a Graceful service change.

As suggested by its name, the ServiceChangeReason parameter indicates the reason for the service change. Table 6-4 shows the reasons currently defined in MEGACO.

Table 6-4	Reason Code	Interpretation
Service change reasons	900	Service Restored
	901	Cold Boot
	902	Warm Boot
	903	MGC Directed Change
	904	Termination malfunctioning
	905	Termination taken out of service
	906	Loss of lower layer connectivity
	907	Transmission failure
	908	MG impending failure
	909	MGC impending failure
	910	Media capability failure
	911	Modem capability failure
	912	Mux capability failure
	913	Signal capability failure
	914	Event capability failure
	915	State loss

DigitMap Descriptor The DigitMap is a description of a dialing plan and is stored in the MG so that the MG can send received digits en-bloc to the MGC, rather than one-by-one. A digit map can be loaded into an MG through operation and maintenance actions or can be loaded from the MGC by MECAGO commands. In the case where it is loaded from the MGC, then the DigitMap Descriptor is used to convey the information.

According to the syntax of MEGACO, a digit map is a string or a list of strings, with each string comprising a number of symbols representing the digits 0 through 9 and the letters A through K. The letter x can be used as a wild card, designating any digit from 0 through 9. The dot (.) symbol is used to indicate zero or more repetitions of the immediately preceding digit or digit string. This symbol can be used to indicate an indeterminate number length in the dial plan. For example, in the United States, domestic calls use 10-digit numbers. International calls use numbers of variable length, however, so that a call to Germany might require more digits than a call to Iceland. To indicate calls to international numbers, the string `011x.` could be used. The `011` indicates an international call, and the `x.` indicates some number of digits—each in the range 0 to 9.

In addition, the string can contain three symbols indicating a start timer (`T`), a short timer (`S`), and a long timer (`L`). To understand the usage of these timers, consider what happens when someone picks up a standard residential telephone to make a call. First, the off-hook is detected, and the system (e.g., telephone exchange) applies a dial tone and gets ready to collect digits. The system will only wait for so long for the person to press a button on the phone. If the person does not press a button in time, the system will time out and take some other action, such as prompting the user. This time-out value would correspond to the start timer. Once the person starts to dial digits, then an inter-digit timer is applied. This timer will either be a short timer or a long timer, depending on the dial plan. If the person has dialed a certain number of digits and the system still needs more digits to be able to route the call, then a short timer will be applied while waiting for the next digit. On the other hand, if the person has dialed sufficient digits to route the call but further digits could still be received and would cause a different routing, then a long timer will be applied. The purpose of the long timer is for the system to be confident that the person has finished dialing. An example in the United States would be where dialing 0 causes routing to an operator from the local-exchange carrier, and dialing 00 causes routing to a long-distance operator. After the system has received the first 0, it must wait before routing the call to make sure that the person does not intend to dial a second 0 in the expectation of being connected to a long-distance operator.

Table 6-5 gives an example of a dial plan.

This dialing plan could be represented in a digit map as

(0|00|[1-7]xxx|8xxxxxxx|Fxxxxxxx|Exx|91xxxxxxxxxx|9011x.)

Note that the timers have been omitted from this digit map. This situation does not mean that timing does not occur at the MG; rather, it just means that the timers are applied by default, according to the usage rules just described. For example, each one of the patterns in this digit map could be preceded by the start timer symbol (`T`). The fact that the symbol is not listed in the digit map means that the MG acts as if it were placed at the start of each pattern in any case.

Statistics Descriptor This descriptor provides statistical information regarding the usage of a termination within a given context. The specific statistics to be returned depend upon the type of termination in question.

	Dialed Digits	Interpretation
Table 6-5	0	Local operator
Example dial plan	00	Long-distance operator
	Xxxx	Local extension number (start with 1-7)
	8xxxxxxx	Local number
	#xxxxxxx	Off-site extension
	*xx	Star services
	91xxxxxxxxxx	Long-distance number
	9011 + up to 15 digits	International number

Strictly speaking, this descriptor is always optional and can be returned as a result of several different commands. This descriptor should be returned when a termination is removed from a context through the Subtract command and should also be returned in response to the AuditValue command.

Observed Events Descriptor   This descriptor is a mandatory parameter in the Notify command, where it is used to inform the MGC of events that have been detected. This descriptor is optional in most other command responses, except for ServiceChange. When used in a response to the Audit-Value command, this descriptor is used to report events stored in the event buffer that have not yet been reported to the MGC. The descriptor contains a RequestIdentifier with a value matching that received in the Events-Descriptor that listed the events to be detected in the first place. This feature enables a correlation between the events requested and the events reported.

The descriptor enables the optional inclusion of a timestamp with each observed event indicating the time at which the event was detected. If the timestamp is to be used, it is structured in the form yyyymmddThhmmssss, which gives a precision of hundredths of a second. The T separates the year, month, and day from the hour, minute, and seconds. When used, the timestamp itself is separated from the specific event description by a colon (:).

Topology Descriptor   The Topology Descriptor is somewhat different from other descriptors in that it is relevant to a context only, rather than to a specific termination within a context. The purpose of the descriptor is to indicate how media streams should flow within a context (who can hear/see whom). By default, all terminations in a context can send and receive media to and from each other. If a different situation is desired, then the Topology descriptor is used.

The descriptor comprises a sequence of triples in the form (Termination 1, Termination 2, association). Such a triple indicates whether there is to be no media flow, one-way media flow, or two-way media flow between the two terminations. These three alternatives are indicated by isolate, oneway, and bothway, respectively. The order in which the terminations appear in the triple is important. For example, the triple (T1, T2, oneway) means the T1 can send media to T2, but T2 must not send media to T1.

In the case where there are more than two terminations in a context, then more than one triple is required. For example, if there are three terminations (T1, T3, and T3), then three triples are needed to describe the association between T1 and T2, the association between T1 and T3, and the association between T2 and T3. Figure 6-13 provides an example of several topologies between three terminations. We assume that the topologies are applied in sequence. The Topology descriptor can be a useful tool for

Figure 6-13
Context topologies

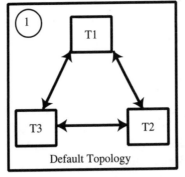

Default Topology

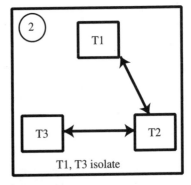

T1, T3 isolate

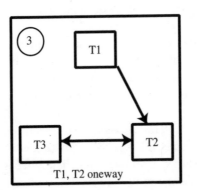

T1, T2 oneway

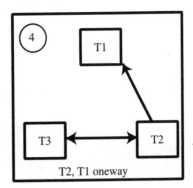

T2, T1 oneway

implementing services such as call waiting, where the controlling party needs to swap between two other parties. In a conference-call situation, this descriptor can enable one or more parties to the call to have a private discussion before returning to the whole group.

The Topology descriptor will not be found as a parameter to any of the MEGACO commands. The commands are directed at specific terminations within a context, and the topology is a characteristic of a context as a whole (rather than any specific termination). When the descriptor is to appear in a transaction, it appears before any commands for terminations in a particular context. The descriptor can only appear in a transaction for a context that already exists.

Packages

In the foregoing description of the various descriptors, reference is understandably made to specifics that depend on the type of termination in question. Clearly, different types of terminations exist, and the properties and capabilities of those terminations will vary according to type. For example, it is obvious that a termination connected to an analog telephone line will be quite different from a termination connected to a digital trunk. Consequently, it is not practical to define within the protocol itself all of the possible termination properties that could exist, all of the possible signals that could possibly be sent, all of the possible statistics that might apply, and all of the possible events that might be detected.

To address the variations between different types of terminations, MEGACO incorporates the concept of packages. Packages are groups of properties, signals, events, and statistics. Within the package, these items are defined and are given identifiers, and the parameters associated with them are specified. A given termination implements a set of packages, and the MGC can audit a termination to determine which packages it supports. Several basic packages are described in the MEGACO specification. Among these are the Tone Detection Package, the DTMF Generator Package, and the Analog Line Supervision Package, to name a few.

The MEGACO specification also provides guidance for the specification of new packages (which should be registered with the IANA).

Packages Descriptor In the foregoing discussion of descriptors, one descriptor is purposely not mentioned: the Packages descriptor. This descriptor was left out until the concept of packages could be introduced.

The Packages descriptor is used only in responses and provides the MGC with a list of the packages implemented in a termination.

MEGACO Command and Response Details

Having described the purpose of the various commands and descriptors, we now examine the details of how they are put together into transactions and messages. This task is performed here prior to specifying what descriptors are applicable to each command, so that the reader can have a good understanding of how commands and responses are syntactically structured. This knowledge might also help to better visualize the various command/descriptor combinations.

Syntax for Commands and Responses

Command Syntax As described previously, commands are applied to terminations within a specific context (or a $ context in some cases). Commands applicable to the terminations within the same context are grouped. Several such groups form a transaction, and several transactions form a message. In the text syntax of the protocol, groupings of commands are enclosed by the brackets { and }. Similarly, the contents of the different descriptors are enclosed in brackets, and commas separate individual descriptors. This structure leads to a syntax that is illustrated in the following example. The line numbers A to K are not part of the syntax but are used just for reference when describing the various components.

```
A.   MEGACO/1 [111.111.222.222]:3456
B.   Transaction = 12345 {
C.        Context = - {
D.            Modify = * {
E.                Events = 7777 {al/of},
F.                DigitMap = Map1 {
G.                    (0|00|[1-7]xxx|8xxxxxxx|Fxxxxxxx|Exx|91xxxxxxxxxx|9011x.) }
H.            }
I.        }
J.   }
```

This example shows a message using MEGACO version 1 from an entity at IP address 111.111.222.222. This message uses port number 3456 for MEGACO messages (line A). Only one transaction exists in this message, and it has a TransactionID of 12345 (line B). The transaction relates to only one context: the null context identified by the symbol - (line C). Only one command is specified—a Modify command—and it applies to all terminations in this context as indicated by the "all" (*) wild card (line D).

The command has two descriptors. The first is an Events descriptor with a RequestID value of 7777 (line E). Only one event is to be detected, and it belongs to the Analog Line Supervision package (al)—the off-hook (of) event (line E). Note that line E, by having both a right and a left bracket, totally describes this descriptor. Note also the comma, which indicates that another descriptor is to be expected.

The next descriptor is the DigitMap descriptor, which specifies a Digit-Map name of Map1 (line F). The details of the digit map are given on line G. This line also contains a right bracket, which terminates the DigitMap descriptor. The fact that there is not a subsequent comma indicates that no more descriptors are included for this command.

Line H simply contains a right bracket, terminating the Modify command. Line I simply contains a right bracket, terminating the set of commands applied to this context. Line J contains a right bracket, terminating the transaction.

Within the syntax, we know that this is a transaction request (as opposed to a transaction reply or a transaction pending), due to the use of the string `Transaction` on line B. A transaction reply would utilize the string `Reply`, and a transaction pending would utilize the string `Pending`.

Response Syntax   This example command leads to a fairly simple response, as follows:

```
A2.    MEGACO/1 [111.111.333.333]:5678
B2.    Reply = 12345 {
C2.         Context = - { Modify = *}
D2.    }
```

The same basic structure is applied to the response. In this case, the message is issued from the MG—and consequently, the IP address and port numbers are different (line A2). The TransactionID is the same as in the transaction request; however, the Reply token is used to indicate that this message is a reply (line B2). Line C2 specifies the context, the null context, as in the transaction request. Line C2 also lists the commands to which the response is issued (Modify) and the terminations for which the commands apply (all of them in this case). There are no descriptors in the response. Line D2 uses a right bracket to indicate the end of the reply.

Responses in the Case of Failure   The previous example showed a reply where the command was carried out successfully. There will be cases where one or more commands cannot be carried out, however. Situations might also arise where a complete transaction might fail. To accommodate such situations, MEGACO defines a special descriptor called an Error descriptor. This descriptor contains an error code, and optionally, an explanatory string.

Let's assume that the MG was completely incapable of executing the foregoing command due to an internal error. In that case, the MG would respond as follows:

```
MEGACO/1 [111.111.333.333]:5678
Reply = 12345 {
        Context = - { Modify = * {
                                Error = 500 { Internal Gateway Error}
        }
        }
}
```

By examining this response, we can see that the error being reported is applicable to the Modify command. Equally, the error descriptor might be applied higher up—say, at the level of a transaction if a complete transaction could not be executed. This situation could arise, for example, in the case of some internal error in the MG, a syntax error in the transaction request, or for any of a number of reasons. The MEGACO specification provides a complete list of error codes and associated explanatory strings.

Add, Modify, and Move Commands   The Add, Modify, and Move commands are practically identical in terms of the syntax used. The only syntactical difference between them is in the name of the command itself. They all include the descriptors Media, Modem, Mux, Events, Signals, DigitMap, and Audit—all of which are optional. To indicate that items are optional, in the ABNF syntax we enclose them in square brackets. Therefore, the syntax of the commands can be represented as follows, where the Command is Add, Move, or Modify:

```
Command ( TerminationID
          [, MediaDescriptor]
          [, ModemDescriptor]
          [, MuxDescriptor]
          [, EventsDescriptor]
          [, SignalsDescriptor]
          [, DigitMapDescriptor]
          [, AuditDescriptor]
        )
```

Note that this code is a representation of the command and is not exactly how the command will appear as it crosses the interface between the MGC and MG. The exact protocol structure as it crosses the interface will be as described in the Command Syntax section earlier in this chapter.

While the three commands are identical in terms of the descriptors that can be used, they obviously have different functions. This factor also has an impact on whether wild cards can be used and where. In the Add command, the ContextID can be "choose" ($), in which case a new context is created. Equally, the TerminationID can take the "choose" ($) value, in which case a new termination is created as a result of the execution of the Add command. The context cannot be the null context, however, because one does not add a termination to the null context. Rather, a termination exists in the null context when it has been subtracted from whatever other context in which it resided.

For the Modify command, the "choose" ($) option is not permitted if used either for the TerminationID or for the ContextID. We cannot make modifications to terminations that do not exist, and we cannot create a new context except with the Add command. We can, however, refer to the null context when issuing a modify command, because it is perfectly valid to modify a termination that exists in the null context.

For the Move command, the null context is not permitted, because the Move command is used only to move terminations from one active context to another. The Move command cannot be used to Move a termination to or from the null context. We can use the "all" (*) wild card as a TerminationID in the Move command, but we cannot use the "choose" ($) wild card.

Response A successful response to a command can also contain a number of descriptors. As is the case for the command itself, all of the allowed descriptors in a response to an Add, Modify, or Move command are optional. The response can be represented as follows. Again, the command can be either Add, Modify, or Move.

```
Command ( TerminationID
         [, MediaDescriptor]
         [, ModemDescriptor]
         [, MuxDescriptor]
         [, EventsDescriptor]
         [, SignalsDescriptor]
         [, DigitMapDescriptor]
         [, ObservedEventsDescriptor]
         [, EventBufferDescriptor]
         [, StatisticsDescriptor]
         [, PackagesDescriptor]
       )
```

At first glance, with the exception of some changes in the list of allowed descriptors, the response appears similar to the command. Within the syntax, however, the two are easily distinguished by the use of the token Transaction when sending a command and the use of the token Reply when responding.

Subtract Command The Subtract command is used to remove a termination from a context. If the termination removed by a Subtract command happens to be the last termination in a given context, then the context itself is deleted. The Subtract command includes only one optional descriptor: the Audit descriptor. The command is represented as follows:

```
Subtract ( TerminationID
                    [, AuditDescriptor]
            )
```

Response   Although the syntax of the Subtract command differs significantly from that of the Add, Modify, and Move commands, the syntax of the response to a Subtract command is the same as the syntax of a response to an Add, Move, or Modify command.

```
Subtract ( TerminationID
                    [, MediaDescriptor]
                    [, ModemDescriptor]
                    [, MuxDescriptor]
                    [, EventsDescriptor]
                    [, SignalsDescriptor]
                    [, DigitMapDescriptor]
                    [, ObservedEventsDescriptor]
                    [, EventBufferDescriptor]
                    [, StatisticsDescriptor]
                    [, PackagesDescriptor]
            )
```

While all of the descriptors in the response to a Subtract command are optional, the Statistics descriptor will be returned by default. In the case where the termination happened to represent an RTP session, this descriptor provides the MGC with information such as the number of packets sent, the number of packets received, the number of lost packets, average latency, etc.

The AuditValue Command   The AuditValue command includes just a single mandatory descriptor: the Audit descriptor. The command is used to collect current settings of attributes of a termination, such as properties and signals.

```
AuditValue ( TerminationID,
                    AuditDescriptor
            )
```

Response

```
AuditValue ( TerminationID
                    [, MediaDescriptor]
                    [, ModemDescriptor]
                    [, MuxDescriptor]
                    [, EventsDescriptor]
                    [, SignalsDescriptor]
                    [, DigitMapDescriptor]
                    [, ObservedEventsDescriptor]
                    [, EventBufferDescriptor]
                    [, StatisticsDescriptor]
                    [, PackagesDescriptor]
            )
```

As is to be expected, the response to the AuditValue command contains a significant number of optional descriptors, each of which can provide information regarding the current state of play at the termination.

The AuditCapabilities Command   With the exception of the command name, the syntax of the AuditCapabilities command is identical to the AuditValue command. Whereas the AuditValue command returns information regarding the current settings for properties, events, signals, etc., the AuditCapabilities command returns information about those properties, events, signals, etc. that a termination can support.

```
AuditCapabilities ( TerminationID,
                         AuditDescriptor
               )
```

Response

```
AuditCapabilities ( TerminationID
                       [, MediaDescriptor]
                       [, ModemDescriptor]
                       [, MuxDescriptor]
                       [, EventsDescriptor]
                       [, SignalsDescriptor]
                       [, DigitMapDescriptor]
                       [, ObservedEventsDescriptor]
                       [, EventBufferDescriptor]
                       [, StatisticsDescriptor]
                       [, PackagesDescriptor]
                 )
```

The Notify Command   The Notify command is issued only by an MG to an MGC. This command provides a list of events detected on a termination. As such, it contains just two descriptors: the mandatory Observed Events descriptor and the optional Error descriptor. Within the Observed Events descriptor is a RequestID, which matches the RequestID of an Events Descriptor previously sent from the MGC to the MG. This descriptor enables the MGC to correlate reported events with a particular request in order to detect and report such events.

```
Notify ( TerminationID
            ObservedEventsDescriptor
            [, ErrorDescriptor]
        )
```

Response   The only response that will be issued to a Notify command will occur in the case of an erroneous command, in which case the Error descriptor is used. This situation is the case for all failed commands.

The ServiceChange Command This command, which is used at restart, failover of an MG, or handoff from one MGC to another, carries only one mandatory descriptor: ServiceChange. The TerminationID might be specific, might be wild carded by using the "all" wild card (*), or might have the value Root to indicate the MG as a whole.

```
ServiceChange ( TerminationID
                             ServiceChangeDescriptor
             )
```

Response The response to the ServiceChange also includes the Service-Change descriptor, but the descriptor is optional in the response.

```
ServiceChange ( TerminationID
                            [ServiceChangeDescriptor]
             )
```

Call Setup Using MEGACO

In Figure 6-8, we examined the setup of a call between two gateways by using MGCP. Based on the foregoing descriptions of MEGACO commands and responses, we are in a position to describe how the same call establishment would be handled in the case where MEGACO is the control protocol (instead of MGCP). The corresponding call flow appears in Figure 6-14. The overall objective is to establish communication between termination T1 on MG A and termination T4 on MG B. MG A has an IP address of 311.311.1.1, while MG B has an IP address of 322.322.1.1. The MGC has an IP address of 333.333.1.1 and uses port number 3333 for MEGACO messages.

The figure includes only a representation of the MEGACO messages and commands used, and it does not include a complete syntax of the messages. The complete syntax is given as follows, along with descriptive text:

```
Line (a)
MEGACO/1 [333.333.1.1] : 3333
Transaction = 1 {
    Context = $ {
         Add = T1,
         Add = $ {
             Media {
                  Stream = 1 {
                      LocalControl {
                          Mode = receiveonly
                          }
                      Local {
```

```
v=0
c=IN IP4 $
m=audio $ RTP/AVP 0
v=0
c=IN IP4 $
m=audio $ RTP/AVP 15
                    }
                }
            }
        }
    }
}
```

The MGC issues a transaction request to MG A by using a TransactionID value of 1. The transaction is to be executed on a new context created by the MG, as indicated by the "choose" ($) wild card as the ContextID. Two Add commands are used: the first adds a termination T1 to the new context, and the second adds a new termination to the same context. This addition is indicated by the use of the $ as a TerminationID.

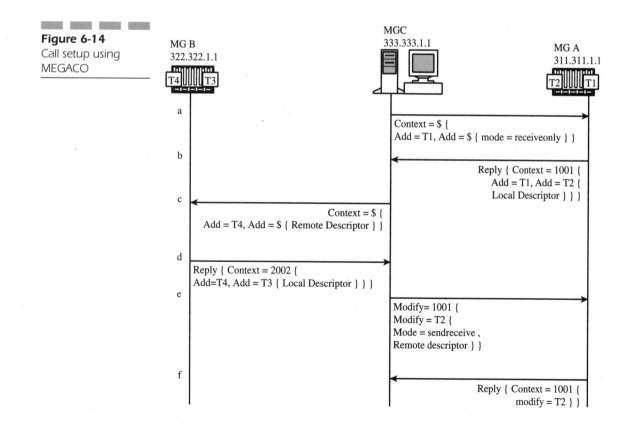

Figure 6-14
Call setup using
MEGACO

The media descriptor for the media associated with the new termination has a StreamID value of 1, which the MGC assigned. The LocalControl is set to `receiveonly`, which means that the termination can accept media from the remote end but not send any media.

The MGC makes two suggestions as to the media properties to be applied by the termination. We can see this action through the use of two session descriptions, with the v= line indicating the boundary between each description. The first is the use of RTP payload type 0 (G.711 mu-law). The second is the use of RTP payload type 15 (G.728). Note the use of the choose wild card within the SDP descriptions. The wild card is used for both the IP address and the port number to be used in each of the descriptions.

```
Line (b)
MEGACO/1 [311.311.1.1] : 1111
Reply = 1 {
     Context = 1001 {
          Add = T1,
          Add = T2 {
               Media {
                    Stream = 1 {
                         Local {
                              v=0
                              c=IN IP4 311.311.1.1
                              m=audio 1199 RTP/AVP 0
                         }
                    }
               }
          }
     }
}
```

MG A replies to the transaction by using the same TransactionID and provides the ContextID value of 1001 to indicate the newly created context. MG A also indicates that the addition of termination T1 is successful and indicates the TerminationID of the newly created termination, T2. Associated with this termination, MG A provides a media description which uses the same StreamID as specified by the MGC. MG A also provides a session description as the Local descriptor. In this case, the session description indicates voice coded according to G.711 mu-law (RTP payload type 0). Furthermore, the IP address and port for the media stream have been filled in by the MG.

```
Line (c)
MEGACO/1 [333.333.1.1] : 3333
Transaction = 2 {
     Context = $ {
          Add = T4,
          Add = $ {
               Media {
```

```
                                    Stream = 2 {
                                        LocalControl {
                                            Mode = sendreceive
                                            }
                                        Local {
                                            v=0
                                            c=IN IP4 $
                                            m=audio $ RTP/AVP 0
                                        },
                                        Remote {
                                            v=0
                                            c=IN IP4 311.311.1.1
                                            m=audio 1199 RTP/AVP 0
                                        }
                                    }
                                }
                            }
                        }
                    }
```

The MGC issues a message to MG B with the TransactionID value of 2. The transaction applies to a new context to be created on MG B, and two terminations are to be added to the new context. The first is termination T4, and the second is a termination that the MG needs to create, as indicated by the use of $ as the TerminationID.

For the new termination, the mode is set to sendreceive so that the endpoint can both send and receive media. The MGC makes a suggestion as to the local session description that should be applied, indicating RTP payload type 0, but leaves the choice of IP address and port number to MG B. The MGC specifies the remote descriptor in detail. This session description happens to be the one received from MG A for termination T2.

```
Line (d)
MEGACO/1 [322.322.1.1] : 2222
Reply = 2 {
    Context = 2002 {
            Add = T4,
            Add = T3 {
                Media {
                    Stream = 2 {
                        Local {
                            v=0
                            c=IN IP4 322.322.1.1
                            m=audio 2299 RTP/AVP 0
                        }
                    }
                }
            }
        }
    }
}
```

MG B replies to the transaction request, using the same value of TransactionID. MG B indicates the ContextID of the newly created context (2002) and also indicates that the terminations T4 and T3 have been successfully added to the context. T3 is a newly created termination. The reply also specifies media-related information for T3. Specifically, it includes a local descriptor in the form of an SDP description, indicating the IP address of 322.322.1.1, the port number of 2299, and a media format according to RTP payload type 0.

```
Line (e)
MEGACO/1 [333.333.1.1] : 3333
Transaction = 3 {
    Context = 1001 {
        Modify = T2 {
            Media {
                Stream = 1 {
                    LocalControl {
                        Mode = sendreceive
                    }
                    Remote {
                        v=0
                        c=IN IP4 322.322.1.1
                        m=audio 2299 RTP/AVP 0
                    }
                }
            }
        }
    }
}
```

The MGC issues a new message to MG A. The message has one transaction request with a TransactionID value of 3 and involves a Modify command for termination T2 in context 1001. The Modify command first changes the Local Control to `sendreceive` so that the termination can send media to the far end. Of course, the termination needs a session description for the far end so that it has an IP address and port number to which to send the media. This information is provided in the Remote descriptor, which contains the session description received by the MGC from MG B.

```
Line (f)
MEGACO/1 [311.311.1.1] : 1111
Reply = 3 {
    Context = 1001 {
        Modify = T2
    }
}
```

MG A sends a positive response to the MG to confirm that the modification has been performed. At this stage, termination T1 is in the same context as termination T2 and can share media. Terminations T2 and T3 have exchanged session descriptions and can share media. Terminations T3 and

T4 are in the same context and can share media. The net effect is that there is a two-way path between terminations T1 and T4, which means that the original objective has been achieved.

MEGACO Call Setup Based on Detected Events

As was the case with the MGCP-based example, the example in Figure 6-14 assumes that the MGC knows from some external signaling source that a call is to be established between the two media gateways. The possibility exists, of course, that the trigger for establishing the call could come from one of the media gateways. The outline for the messages in such a scenario is shown in Figure 6-15, which is effectively an adaptation of the scenario depicted in Figure 6-9 (but uses MEGACO instead of MGCP). Of course, Figure 6-15 does not show the full MEGACO context; rather, it just indicates the overall flow and the main commands involved. The full syntax is described as follows.

```
Line (a)
MEGACO/1 [333.333.1.1] : 3333
Transaction = 1 {
    Context = - {
        Modify = T1 {
            Events = 1111 { al/of } ,
            Digitmap = Plan 1 { (0|[2-9]xxxxxx) }
        }
    }
}
```

The MG modifies termination T1 in the null context. The modification requires the termination to notify the MGC upon the occurrence of an off-hook event. This event is defined in the analog line supervision package and is identified by `al/of`. The Modify command also loads a digit map with the name Plan 1.

```
Line (b)
MEGACO/1 [311.311.1.1] : 1111
Reply = 1 {
    Context = - { Modify = T1 }
}
```

The MG sends a positive response to the Modify command. The reply contains the same TransactionID as the transaction request.

```
Line (c)
MEGACO/1 [311.311.1.1] : 1111
Transaction = 2 {
```

```
Context = - {
      Notify = T1 {
            ObservedEvents = 1111 { 20000527T09460000:al/of }
      }
   }
}
```

At exactly 09:46 A.M. on May 27, 2000, the phone connected to termination T1 goes off hook. The MG sends a Notify command to the MGC, indicating the

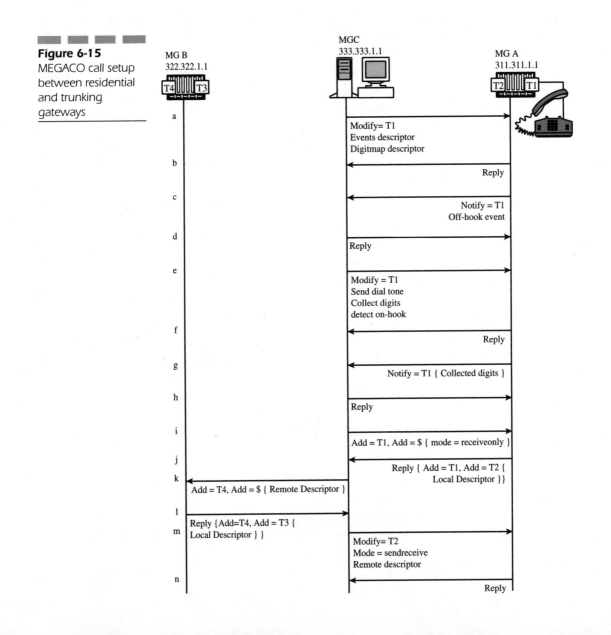

Figure 6-15
MEGACO call setup between residential and trunking gateways

time that this event occurred. Note that the same Request Identifier is used in the Notify command as was issued by the MGC in the Modify command. This feature enables the MGC to correlate detected events with the command that required the events to be detected in the first place.

```
Line (d)
MEGACO/1 [333.333.1.1] : 3333
Reply = 2 {
     Context = - { Notify = T1 }
}
```

The MGC acknowledges the Notify command from MG A.

```
Line (e)
MEGACO/1 [333.333.1.1] : 3333
Transaction = 3 {
     Context = - {
         Modify = T1 {
             Events = 1112 { al/on, dd/ce { DigitMap = Plan1 } } ,
             Signals = { cg/dt }
         }
     }
}
```

The MGC immediately sends a Modify command to termination T1. The command instructs the termination to listen for an on-hook event (al/on) and to also notify the MGC when a digit map completion occurs according to Plan 1. This instruction is indicated by the use of the DTMF detection package (dd) and the digit map completion event (ce). The MGC also instructs the termination to apply a dial tone (dt) signal. This signal is defined in the call progresstone-generation package (cg) and therefore is identified as cg/dt.

```
Line (f)
MEGACO/1 [311.311.1.1] : 1111
Reply = 3 {
     Context = - { Modify = T1 }
}
```

MG A sends a positive acknowledgment to the Modify command.

```
Line (g)
MEGACO/1 [311.311.1.1] : 1111
Transaction = 4 {
     Context = - {
         Notify = T1 {
             ObservedEvents = 1112 {
                 20000527T09460500:dd/ce {ds = "7771234" Meth=FM }
             }
         }
     }
}
```

MGA sends a Notify command to the MGC , indicating that a digit map completion event has occurred. The time of the event is indicated, which happens to be exactly five seconds after the phone went off the hook. The user of the analog phone dialed the digits 7771234, which is a full match (Meth = FM) with the Plan 1 digit map.

```
Line (h)
MEGACO/1 [333.333.1.1] : 3333
Reply = 4 {
     Context = - { Notify = T1 }
}
```

The MGC acknowledges the Notify command from MG A.

```
Line (i)
MEGACO/1 [333.333.1.1] : 3333
Transaction = 5 {
     Context = $ {
          Add = T1,
          Add = $ {
               Media {
                    Stream = 1 {
                         LocalControl {
                              Mode = receiveonly
                         }
                         Local {
                              v=0
                              c=IN IP4 $
                              m=audio $ RTP/AVP 0
                         }
                    }
               }
          }
     }
}
```

The MGC analyzes the dialed digits and determines that the call needs to be connected to termination T4. The first action that it performs is to instruct MG A to create a new context and to add termination T 1 and a new termination to that context. Both the new context and the new termination are indicated with the choose wild card ($).

The media descriptor for the media associated with the new termination has a StreamID value of 1, which the MGC assigns. The LocalControl is set to receiveonly, which means that the termination can accept media from the remote end but not send any media. The MGC also suggests for the new termination to operate by using the G.711 mu-law coding scheme. The choose wild card is used for both the IP address and the port number to be used.

```
Line (j)
MEGACO/1 [311.311.1.1] : 1111
```

```
Reply = 5 {
     Context = 1001 {
          Add = T1,
          Add = T2 {
               Media {
                    Stream = 1 {
                         Local {
                              v=0
                              c=IN IP4 311.311.1.1
                              m=audio 1199 RTP/AVP 0
                         }
                    }
               }
          }
     }
}
```

MG A replies to the two Add commands, indicating that both were successfully completed. The new context has a ContextID value of 1001, and the newly created termination has a TerminationID of T2. The media descriptor has a local descriptor that includes the session description for the local end. In this case, the session description indicates voice that is coded according to G.711 mu-law (RTP payload type 0). Furthermore, the MG has filled in the IP address and port for the media stream.

```
Line (k)
MEGACO/1 [333.333.1.1] : 3333
Transaction = 6 {
     Context = $ {
          Add = T4,
          Add = $ {
               Media {
                    Stream = 2 {
                         LocalControl {
                              Mode = sendreceive
                         }
                         Local {
                              v=0
                              c=IN IP4 $
                              m=audio $ RTP/AVP 0
                         },
                         Remote {
                              v=0
                              c=IN IP4 311.311.1.1
                              m=audio 1199 RTP/AVP 0
                         }
                    }
               }
          }
     }
}
```

The MGC issues a message to MG B. The transaction applies to a new context to be created on MG B, and two terminations are to be added to the

new context. The first is termination T4, and the second is a termination
that the MG needs to create (as indicated by the use of $ as the Termina-
tionID).

For the new termination, the mode is set to sendreceive so that the endpoint can both send and receive media. The MGC makes a suggestion as to
the local session description that should be applied, indicating RTP payload
type 0, but leaves the choice of IP address and port number to MG B. The
MGC specifies the remote descriptor in detail. This session description is
the one received from MG A for termination T2.

```
Line (1)
MEGACO/1 [322.322.1.1] : 2222
Reply = 6 {
     Context = 2002 {
           Add = T4,
           Add = T3 {
              Media {
                  Stream = 2 {
                       Local {
                              v=0
                              c=IN IP4 322.322.1.1
                              m=audio 2299 RTP/AVP 0
                          }
                       }
                  }
              }
         }
   }
```

MG B replies to the transaction request by using the same value of
TransactionID. MG B indicates the ContextID of the newly created context
(2002) and also indicates that the two terminations T4 and T3 were successfully added to the context. T3 is a newly created termination. The local
descriptor for Termination T3 indicates an IP address of 322.322.1.1, a port
number of 2299, and a media format according to RTP payload type 0.

```
Line (m)
MEGACO/1 [333.333.1.1] : 3333
Transaction = 7 {
    Context = 1001 {
         Modify = T2 {
             Media {
                 Stream = 1 {
                      LocalControl {
                           Mode = sendreceive
                           }
                      Remote {
                           v=0
                           c=IN IP4 322.322.1.1
                           m=audio 2299 RTP/AVP 0
                      }
```

```
                }
            }  '
        }
    }
}
```

The MGC issues a new message to MG A. This command modifies the termination T2 by enabling it to send and receive media and provides a remote descriptor, which happens to be the local descriptor of termination T3.

```
Line (n)
MEGACO/1 [311.311.1.1] : 1111
Reply = 7 {
    Context = 1001 {
            Modify = T2
    }
}
```

MG A sends a positive acknowledgment to the Modify command. Now, a communication path exists from termination T1 to termination T4 via terminations T2 and T3.

Interworking between MEGACO and SIP

Interworking between MEGACO and SIP works in much the same manner as interworking between MGCP and SIP. Once a Local descriptor is available at the gateway where the call is being originated, then this information can be passed in a SIP INVITE as a SIP message body. Upon acceptance of the call at the far end, the corresponding session description is carried back as a SIP message body within the SIP 200 (OK) response and is passed to the originating-side gateway. Once the gateway on the originating side has acknowledged receipt of the remote session description, then the MGC can send a SIP ACK to complete the SIP call setup.

No fundamental difference exists between MGCP and MEGACO in the manner in which SIP interworking is achieved. This statement is understandable, because MEGACO and MGCP deal only with MGC-MG control, and both handle call setup in quite similar ways.

CHAPTER **7**

VoIP and SS7

Introduction

For a long time, signaling in circuit-switched networks was such that the signaling related to a particular call followed the same path as the speech for that call. This approach is known as *Channel-Associated Signaling* (CAS), and this technique is still widely deployed today. Examples include the R1 *Multi-Frequency* (MF) signaling used in North America and the R2 *Multi-Frequency Compelled* (MFC) signaling used in many other countries. Although it is widely used in circuit-switched networks, CAS is considered old technology. The prevailing technology in newer circuit-switched networks is *Common Channel Signaling* (CCS).

CCS involves the use of a separate transmission path for call signaling, compared to the bearer path for the call itself (as shown in Figure 7-1). This separation enables the signaling to be handled in a different manner to the call itself. Specifically, other nodes on the network might analyze the signals and take action based on the content of the signals, without needing to be involved in the bearer path. Furthermore, CCS enables signaling messages to be exchanged in cases where no call needs to be established at all. Imagine,

Figure 7-1
CAS versus CCS

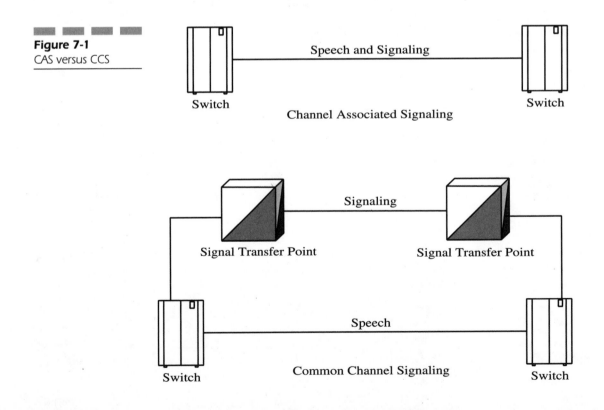

for example, dialing a star code to activate some feature. In such a scenario, the local switch might communicate with a service node to activate the feature on the network without actually establishing a call to the service node.

The standard for CCS today is *Signaling System 7* (SS7). Various versions of SS7 are deployed all over the world. In fact, SS7 could be considered the ultimate signaling standard in telecommunications. Anyone who wishes to implement a circuit-switched network and who does not implement SS7 is not really playing in the big leagues and cannot really call his or her network carrier grade.

Why is SS7 so important? SS7 enables a wide range of services to be provided to the end-user. These features include Caller ID, toll-free calling, call screening, number portability, and others. In fact, SS7 is the foundation for *Intelligent Network* (IN) services.

Given that SS7 is so prevalent and so important to telecommunications, it follows that any VoIP carrier that wishes to interwork with the *Public-Switched Telephone Network* (PSTN) must support SS7 in order to communicate with circuit-switched networks. If a VoIP network is to be considered carrier grade, then it must support SS7. The capability to speak the same language as the PSTN is not the only reason, however. Customers who use VoIP networks will expect the same types of features as are available in existing networks and will expect these services to work in the same way. In order to compete effectively, a new carrier has no alternative but to match (if not exceed) the features and functionalities already available. To do so, a carrier can utilize the capabilities of SS7 or try to create a totally new way of providing such services. Utilizing SS7 is the obvious choice. We do not mean that the carrier is limited to what SS7 can provide, however. In fact, IP provides the opportunity to offer a wide range of new services. Customers will still expect to see existing services such as toll-free calling, however, and a new carrier has little alternative but to use SS7 to provide many of those services.

This chapter provides a brief introduction to SS7, followed by a description of the technical solutions that can be deployed to support interworking between VoIP networks and SS7. The interworking that needs to be provided has two aspects. First, SS7 needs to be supported by applications in an IP environment, although SS7 was not originally designed for such an environment. Second, different network-configuration issues exist, depending on the architecture of the VoIP network itself. For example, the VoIP network might follow the H.323 model. Alternatively, the VoIP network might use a softswitch architecture based on the *Session Initiation Protocol* (SIP) and MCGP or MEGACO. To some degree, how the interworking is conducted is influenced by the network architectures in question.

Figure 7-2
SS7 protocol stack

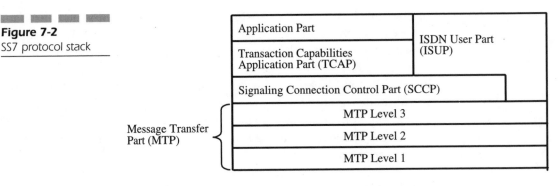

The SS7 Protocol Suite

SS7 is not a single protocol per se; rather, it is a suite of protocols operating in concert. In the same way that TCP or UDP use the services of IP in a stack arrangement, SS7 also defines a stack that has certain similarities with the *Open Systems Interconnection* (OSI) seven-layer model. Figure 7-2 shows the SS7 stack.

The Message Transfer Part

The three lower levels of the SS7 stack comprise the *Message Transfer Part* (MTP). This part of the protocol is responsible for getting a particular message from source to destination. Level 1 maps closely to layer 1 of the OSI stack. This layer handles the issues related to the signals on the physical links between one signaling node and another. These links are known as signaling links and operate at 56kbps or 64kbps.[1] At a node that supports SS7, the physical termination of a signaling link is known as a signaling terminal.

MTP level 2 deals with the transfer of messages on a given link from one node to another. Signaling messages from higher-layer protocols are passed down the stack to MTP2, which packages them for transmission on a given signaling link and takes care of getting the message across that link. As well as providing functions for passing messages on behalf of MTP users, MTP2 also has some messages of its own, such as the *Link Status Signal*

---

[1]This statement applies only to narrow-band SS7. A version of SS7 also exists, known as broadband SS7, which is used for signaling in *Asynchronous Transfer Mode* (ATM) networks.

Unit (LSSU) and *Fill-In Signal Unit* (FISU). The LSSU is used between two ends of a link to ensure alignment and correct link functioning, and FISUs are sent when there is nothing else to send.

MTP level 3 (MTP3) deals with the management of messaging in the signaling network as a whole. For a message that is destined for a particular signaling destination, there might be many different paths that the message could take from its source to its destination. These paths could include a direct signaling link if one exists, or a path via a number of different intermediate nodes (known as Signal Transfer Points, or STPs). MTP3 takes care of determining which outgoing link should be used for a particular message. Therefore, MTP3 includes functions for the mapping of signaling destinations to various signaling link sets. MTP3 manages load-sharing across different signaling links and also handles the rerouting of signaling in the case of link failure, congestion, or the failure of another node on the signaling network. In order to help manage the overall signaling network, MTP3 includes a complete signaling-network management protocol for ensuring proper operation of the SS7 network.

Furthermore, MTP3 provides a number of services to the protocol layer above it. These services involve the transfer of messages to and from the upper layer, indicating (for example) availability or unavailability or a particular destination and signaling network status, such as congestion. These services are provided through the primitives MTP-Transfer request, MTP-Transfer indication, MTP-Pause indication, MTP-Resume indication, and MTP-Status indication.

ISUP and SCCP

Above MTP, two main alternatives exist: the *ISDN User Part* (ISUP) and the *Signaling Connection Control Part* (SCCP).[2] Let's deal with ISUP first.

Strictly speaking, ISUP is a signaling protocol that provides services to ISDN applications. In reality, ISUP is most used as the protocol for setting up and tearing down phone calls between switches and is the most commonly used of the SS7 applications. Examples of ISUP messages include call setup messages, such as the *Initial Address Message* (IAM) that is used to initiate a call between two switches; the *Answer Message* (ANM) that is used to indicate that a call has been accepted by the called user; and

---

[2]Other alternatives exist, most notably the *Telephony User Part* (TUP). TUP can be considered a simpler version of ISUP. Numerous national variants of TUP exist.

the *Release message* (REL) that is used to initiate a call disconnect. ISUP is a connection-oriented protocol, which means that it relates to the establishment of connections between users. Therefore, ISUP messages between source and destination are related to a bearer path between the source and destination, although the path of the messages and the path of the bearer might be extremely different. In terms of mapping to the *Open Systems Interconnection* (OSI) model, ISUP maps to levels 4 through 7.

SCCP provides both connection-oriented and connectionless signaling, although it is most often used to provide connectionless signaling. Connectionless signaling refers to signaling that needs to be passed from one switch or network element to another without the need to establish a signaling connection. In other words, connectionless signaling means that information is simply sent from source to destination, without the prior establishment and subsequent release of a dedicated signaling relationship. For example, consider a mobile phone user who roams away from his home network. When the user arrives at the visited network, the visited network and home network must communicate so that the visited network can learn more about the subscriber (and so that the home network can know where the subscriber is). Such signaling is connectionless and it uses the services of SCCP.

SCCP also provides an enhanced addressing mechanism to enable signaling between entities—even when those entities do not know each other's signaling addresses (which are known as point codes). This addressing is known as Global Title Addressing. Basically, this system is a means through which some other address, such as a telephone number, can be mapped to a point code—either at the node that initiated the message or at some other node between the origin and destination of the message.

The *Transaction Capabilities Application Part* (TCAP) enables the management of transactions and procedures between remote applications, such that operations can be invoked and the results of those operations can be communicated between the applications. TCAP is defined for connectionless signaling only. TCAP provides services to any number of application parts. Common application parts include the *Intelligent Network Application Part* (INAP) and the *Mobile Application Part* (MAP).

SS7 Network Architecture

SS7 is purely a signaling protocol that enables services and features in the telephony network. The SS7 network should be considered a separate net-

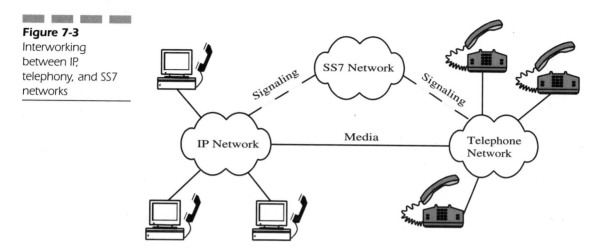

Figure 7-3
Interworking between IP, telephony, and SS7 networks

work in its own right for a number of reasons, however. First, SS7 involves the separation of signaling from the bearer. Second, SS7 involves a number of network elements that deal only with signaling itself and that are tied together in a network of their own. Consequently, when we think about SS7 in the context of interworking between circuit-switched networks and VoIP networks, then we should think in terms of three networks, as shown in Figure 7-3. The VoIP system and the circuit-switched network pass signaling via the SS7 network and pass media directly to and from one another.

Figure 7-4 depicts a typical SS7 network arrangement. The two switches shown do not communicate signaling to each other via direct paths. Instead, the signaling is carried via one or more STPs, typically arranged in a quad. This configuration serves several purposes. To start, this configuration avoids the need for direct signaling links between all endpoints (switches), meaning that a fully meshed signaling network is not required. Instead, a separate, more compact signaling network can be created. Second, the quad arrangement ensures great robustness. Each switch will be connected to two STPs, and there are multiple paths between different STPs. Thus, the failure of a single link will never result in a complete loss of signaling capability between different endpoints.

In fact, SS7 networks in general are extremely robust. The design of SS7 is such that reliable signaling networks can be built. The complete failure of a node or a number of signaling links does not mean catastrophe for the network. As far as the applications are concerned, once information is sent to the SS7 stack, it is as good as delivered to the ultimate destination. SS7 really is that reliable—and that fast.

Figure 7-4
An SS7 network

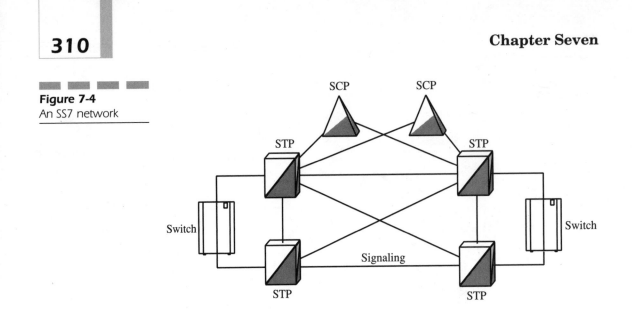

Signaling Points

Each node in an SS7 network is a *signaling point* (SP) and has one or more signaling addresses, each of which is known as a *Signaling Point Code* (SPC). SPs that are directly connected are said to be adjacent, and the links between such adjacent SPs are known as signaling links. Each of these is a 56kbps or 64kpbs link. Several signaling links might exist between two adjacent signaling points for capacity or security reasons, and the set of links is not surprisingly called a linkset. More formally, a linkset is a group of signaling links that directly connect two SPCs.

Signal Transfer Point

The function of an STP is to transfer messages from one SPC to another. Basically, this system works as depicted in Figure 7-5. Let's assume that Switch A, with SPC value 1, wishes to send a message known as a *Message Signal Unit* (MSU) to Switch B, which has an SPC value of 2. Let's also assume that no signaling link exists between them but that they are both connected to an STP that has SPC value 3. In that case, the routing tables at Switch A could be set up to send the MSU to the STP. The MSU contains the destination signaling address of Switch B; i.e., SPC value 2. When the message is received at the STP, the STP first checks to see whether the message is destined for it by checking the destination address. Given that the message is not, the STP then checks its routing tables to see where mes-

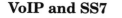

Figure 7-5
STP functions

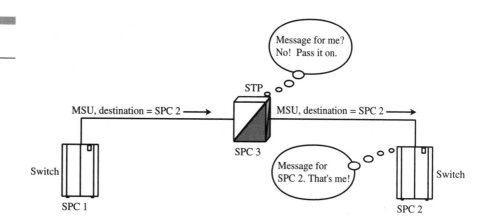

sages for SPC 2 should be sent. The STP determines that such MSUs should be sent on the link to Switch B. When the MSU arrives at Switch B, the same process is repeated—except that Switch B recognizes that the MSU is destined for it and sends the content of the message to the appropriate signaling application.

Service Control Point (SCP)

A *Service Control Point* (SCP) is a network entity that contains additional logic and that can be used to offer advanced services. To use the service logic of the SCP, a switch needs to contain functionality that will enable it to act upon instructions from the SCP. In such a case, the switch is known as a *Service Switching Point* (SSP). In the case where a particular service needs to be invoked, the SSP sends a message to the SCP asking for instructions. The SCP, based upon data and service logic that is available, will tell the SSP which actions need to be taken. A good example of such a service is a call to a toll-free 800 number. When a subscriber dials such a number, the SSP knows that it needs to query the SCP for instructions (which it does). The SCP contains translation information between the dialed 800 number and the real number to which the call should be sent. The SCP responds to the SSP with a routable number, which the SSP uses to route the call to the correct destination. Note that the signaling between the SSP and the SCP is connectionless signaling. Hence, this signaling uses the services of SCCP. In fact, this process involves an application that uses the services of TCAP, which in turn uses the services of SCCP.

Message Signal Units (MSUs)

As mentioned, the messages that are sent in the SS7 signaling network are known as MSUs. Each MSU has the format shown in Figure 7-6. The *Service Information Octet* (SIO) contains the Service Indicator, which indicates the upper-level protocol to which the message applies (e.g., SCCP or ISUP). The SIO also contains the sub-service field, which indicates the signaling numbering plan in use. We will discuss this topic in more detail shortly. The *Signaling Information Field* (SIF) contains the actual user information that is being sent.

If we look at the details of the SIF, then an MSU has one of the formats shown in Figure 7-7, depending on whether the ANSI version of SS7 or the ITU-T version of SS7 is being used. Of particular importance is the routing label, which contains two important fields: the *Destination Point Code* (DPC) and the *Originating Point Code* (OPC). These are the signaling

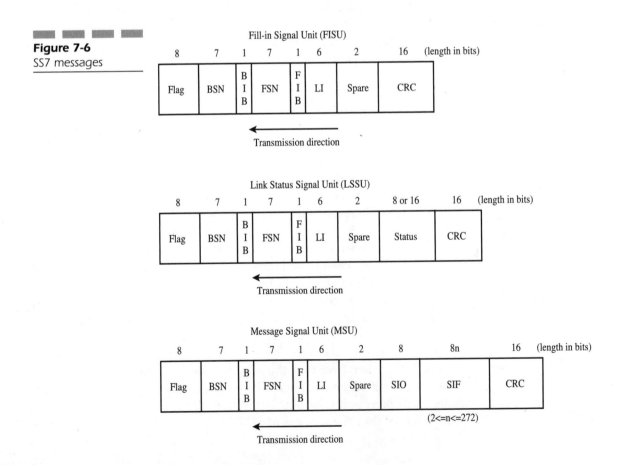

Figure 7-6
SS7 messages

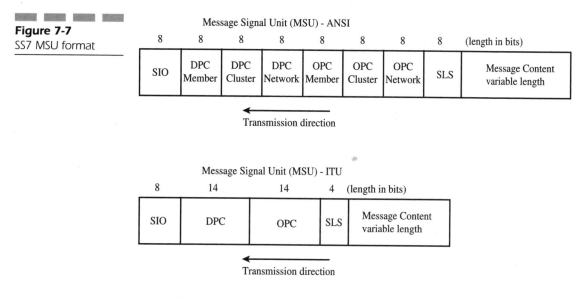

Figure 7-7
SS7 MSU format

Message Signal Unit (MSU) - ANSI

SIO = Service Information Octet
SLS = Signaling link selection
The ANSI SLS is 8 bits, but historically was 5 bits with three bits spare.

addresses of the destination of the message and the originator of the message, respectively. The *Signaling Link Selection* (SLS) field also exists, which indicates the particular signaling link to be used for carrying this message between the two nodes in question.

SS7 Addressing

As mentioned, signaling entities in an SS7 network are addressed by SPCs. A quick look at Figure 7-7 shows that unfortunately, different versions of SS7 exist with different point-code structures. For example, in North America, point codes are 24 bits long—whereas in most other countries, point codes are 14 bits long. A notable exception is China, where point codes are 24 bits long.

In North America, point codes have a hierarchical structure with the format network-cluster-member, each consisting of eight bits. The operator of a large network would be allocated a network code, enabling that operator to have about 65,000 different signaling point codes.

Given that there are different SS7 formats and addressing schemes, how do SS7 entities in different countries communicate? The answer is through

Figure 7-8
International
signaling

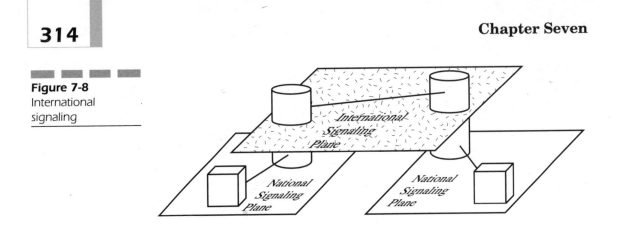

international signaling gateways. These gateways are signaling points that support the national SS7 variant at one side and the ITU-T international variant at the other side. This concept brings us back to the sub-service field mentioned previously. Recall that this field is used to indicate the signaling numbering plan in use. Four variants actually exist: National, National Spare, International, and International Spare. A network element that acts as an international gateway will have at least two point codes, one national point code, and one international point code. Messages to and from other nodes in the national network will be addressed with the national point code, and messages to or from other international nodes will use the international point code. Thus, a hierarchy exists in the SS7 network in which messages between countries must go from the national signaling plane to the international signaling plane and back down to the destination country, as shown in Figure 7-8.

The ISDN User Part (ISUP)

In regular telephony, ISUP is the most-used SS7 application. This protocol is used for the establishment and release of telephone calls. Figure 7-9 shows a typical call establishment and release.

The call begins with the IAM, which contains information about the called number, the calling number, the transmission requirement (typically 64kbps), the type of caller (ordinary, operator, pay phone, etc.), and other information such as whether a satellite link has been included in the call so far.

Upon receipt of the IAM at the destination switch, an *Address Complete Message* (ACM) is returned. This ACM indicates that the call is through-connected to the destination. This message causes a one-way audio path to

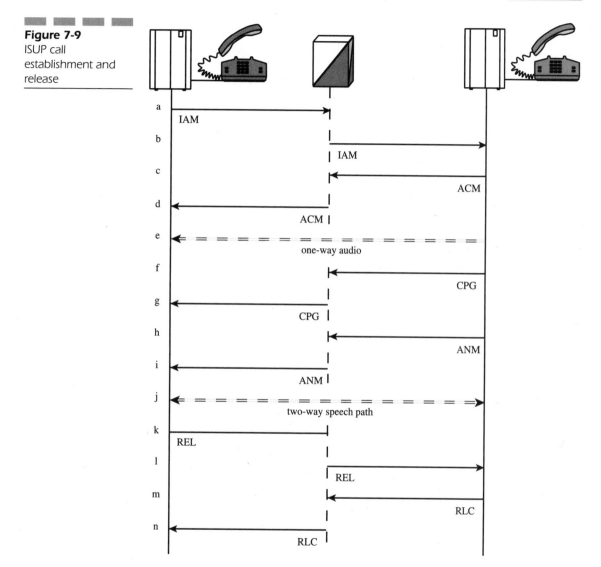

Figure 7-9
ISUP call
establishment and
release

be opened from destination switch to originating switch so that the caller
can hear the ring-back tone. Note that the tone is generated at the destination end. Strictly speaking, the ACM is an optional message. Although it is
returned in most calls, that is not always the case. Situations exist where an
ACM is not returned. If it is not returned, then the caller does not hear a
ring-back tone at all, and it appears as though the call is answered immediately. This situation happens quite frequently for calls to toll-free numbers, particularly if the call is answered automatically.

After the ACM, there might be one or more *Call Progress* (CPG) messages. The CPG is an optional message and is used to provide information to the calling switch regarding the handling of the call.

Once the called party answers, an *Answer Message* (ANM) is returned. This ANM typically has two purposes. The first is to open the transmission path in both directions so that the parties can converse. The second is to instigate charging for the call, because charging for most calls begins when someone answers the call.

After conversation, one of the parties hangs up. This action causes a *Release* (REL) message to be sent to the other end. The end that receives the REL responds with a *Release Complete* (RLC) message. At this point, the call is complete.

Note that the signaling passes through one or more STPs, whereas the speech takes a more direct transmission path from A to B. This process leads to one important question. Given that there might be many simultaneous calls between switch A and switch B, how does the ISUP signaling differentiate between calls? In particular, how is a given ISUP message correlated with a given speech path between the switches? The answer is the *Circuit Identification Code* (CIC), which is contained within each ISUP message. For an ISUP message between two switches, the CIC indicates the specific trunk between the switches to which the message applies. In fact, a given circuit between two switches is completely identified by the combination of OPC, DPC, and CIC. We can see the CIC in the ISUP message formats shown in Figure 7-10. The figure also serves to further highlight the differences between different SS7 variants.

Performance Requirements for SS7

Given that VoIP networks need to interwork with the PSTN, we clearly see that VoIP networks need to speak SS7 (at least, to the outside world). Not only should VoIP networks support the messages of SS7, but they must also support the performance requirements specified for SS7. For example, for MTP, Bellcore[3] specification GR-246-Core states that a given route set should not be out of service for more than 10 minutes per year. Other requirements are that no more than 1×10^{-7} messages should be lost, and no more than 1×10^{-10} messages should be delivered out of sequence. In ISUP, numerous timing requirements must also be met. For example, a two-

---

[3]Bellcore is now known as Telcordia. This organization has written many technical standards, most of which have been adopted by North American network operators.

Figure 7-10
ISUP message format

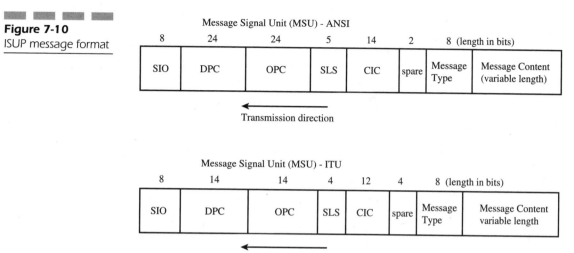

Message Signal Unit (MSU) - ANSI

8	24	24	5	14	2	8	(length in bits)
SIO	DPC	OPC	SLS	CIC	spare	Message Type	Message Content (variable length)

← Transmission direction

Message Signal Unit (MSU) - ITU

8	14	14	4	12	4	8	(length in bits)
SIO	DPC	OPC	SLS	CIC	spare	Message Type	Message Content variable length

← Transmission direction

SIO = Service Information Octet
SLS = Signaling link selection

second timer exists that is initiated when a Continuity Check Request is sent, requiring a continuity check tone to be returned within that time.

Imagine a VoIP network that operates as a long-distance network, as depicted in Figure 7-11. This network receives an IAM from a network at one side of the country, processes the information, and determines that the call should be sent to a network at the other side of the country. The network sends an IAM to the destination network, which responds with an ACM. The VoIP network must process the ACM and send it to the originating network. This process is conducted exactly the same as in a traditional long-distance network. If we think about the amount of internal signaling that goes on in an H.323 network or even in a softswitch network, however, we realize that a lot of work is done between the passing of external SS7 messages. Furthermore, if we think about the post-dial delay of today's telephone networks (a second or two), we realize just how fast this work needs to be done. The point is that a VoIP network that uses SS7 must meet the stringent requirements that are already met by today's circuit-switched networks. The issue, then, is how to make sure that VoIP networks can emulate the signaling performance of SS7. Fortunately, many groups have been working on this problem. In particular, the *Signaling Transport* (Sigtran) group of the IETF has made great progress in this area. The following sections of this chapter are devoted to describing the architectures and solutions developed within Sigtran.

Figure 7-11
A long-distance VoIP
network

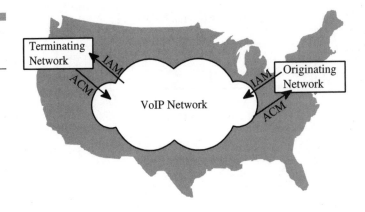

Sigtran

Sigtran is a working group within the IETF that is dedicated to addressing the issues regarding the transport of signaling within IP networks. In particular, the group addresses the issues related to signaling performance within IP networks and interworking with other networks, such as the PSTN.

In order to better understand the issues to be addressed, let us again consider the scenario depicted in Figure 7-11. Let's assume that the VoIP network has a softswitch architecture. Therefore, the VoIP network consists of a number of media gateways, a number of signaling gateways, and a number of media gateway controllers/call agents. The purpose of the signaling gateways is to connect directly with signaling points or STPs in the external SS7 network. The function of the media gateways and MGCs is described in Chapter 6, "Media Gateway Control and the Softswitch Architecture." Let's assume for now that the call agents communicate with each other by using SIP and that they communicate with media gateways by using MGCP.

Given that the MGC is the call-control entity within the network, it follows that ISUP signaling received at a signaling gateway must be passed to the MGC so that the MGC can appropriately control call establishment and call release. If we assume a straightforward call from a trunking gateway at one side to a trunking gateway at the other side, then the scenario would be similar to Figure 7-12.

The carriage of ISUP signaling information from *signaling gateway* (SG) to MGC or from MGC to SG is indicated in Figure 7-12 with the letters IP. For example, an IAM carried in the IP network is denoted IP IAM. Note that the content of the message does not change, because the MGC is assumed to contain a standard ISUP application. The only difference is that

Figure 7-12
SIP/MGCP/ISUP
interworking

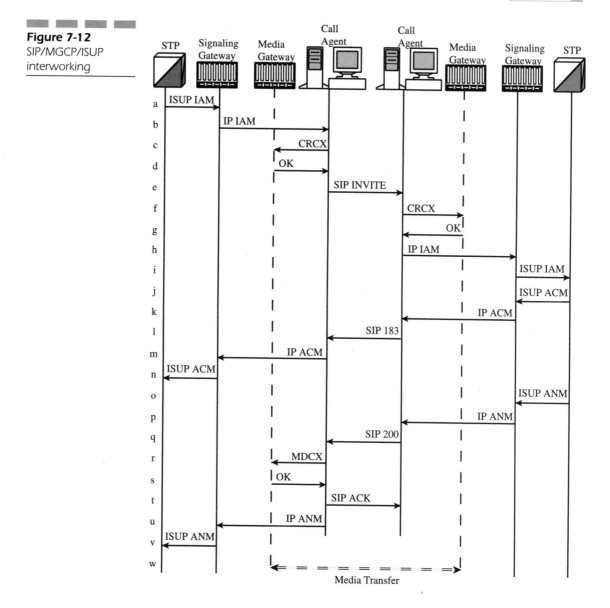

the ISUP messages are carried to and from the MGC over IP, rather than by standard MTP. Thus, all that the SG does is to take the ISUP message from the SS7 network and deliver it to the appropriate entity on the IP network. This task might sound simple—just a simple translation from point code to IP address, right? Unfortunately, things are rarely that simple.

Certainly, point code to IP address translation is one of the issues we must address. The issue of which transport (layer 4) protocol to use for carrying the messages also exists, however. Furthermore, there is the issue of using a standard SS7 application (such as ISUP) in an IP environment. How can we deploy such an application that expects certain services from lower layers such as MTP, when those lower layers do not exist on the IP network?

For the transport protocol, the ISUP messages must be carried in the IP network with the same speed and reliability as in the SS7 network. In some implementations, proprietary mechanisms are used for getting information from a signaling gateway to a call agent. This process requires the MGC and SG to be sourced from the same vendor, however, thereby reducing the network operator's vendor-selection choices. In some other implementations, the ISUP message is simply packaged as a UDP or TCP packet and is sent from the signaling gateway to the call agent. Problems with this approach might exist, however. First, there are strict performance requirements that should be met with regard to message loss and message sequencing. This situation leads us away from UDP, because it is inherently unreliable. We now have TCP as an alternative. Strict timing requirements need to be met, however. While TCP might ensure the accurate, in-sequence delivery of messages, this method is not particularly fast. This technique also suffers from head-of-line blocking. Therefore, TCP is not a great choice either. Something else is needed.

In order to address these issues, Sigtran has prepared a number of documents describing the issues and potential solutions. One of these is an Internet draft titled "VoIP Signaling Performance Requirements and Expectations." As the title suggests, this draft describes the performance objectives to be met by a signaling transport architecture. This architecture uses ITU and Telcordia (formerly Bellcore) requirements as input, thereby using the performance already met by standard telephony as a baseline. The draft also includes some analysis of *Post-Dial Delay* (PDD) in VoIP networks and concludes that the performance of VoIP systems can be made comparable to the PSTN. Of course, the main purpose of the document is to serve as an input to the Sigtran specification work. Also prepared is the Informational RFC 2719.

Sigtran Architecture

The Sigtran architecture is defined in RFC 2719, an informational RFC titled "Framework Architecture for Signaling Transport." The architecture uses the signaling components shown in Figure 7-13. Signaling over stan-

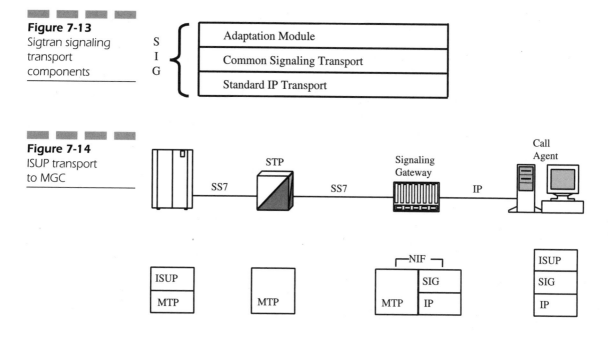

Figure 7-13
Sigtran signaling
transport
components

Figure 7-14
ISUP transport
to MGC

NIF = Nodal Interworking Function

dard IP uses a common transport protocol that ensures reliable signaling delivery and an adaptation layer supporting specific primitives, as required by a particular signaling application. In other words, the common signaling transport makes sure that messages are delivered error free and in sequence, regardless of the vagaries of the underlying IP network. The adaptation layer provides an interface to the SS7 applications so that those applications do not realize that the underlying transport is IP (rather than a standard SS7 network).

If we apply this architecture to a situation where a *signaling gateway* (SG) is connected to the SS7 network on one side and to an MGC on the other, then the transport of ISUP signaling would appear as in Figure 7-14. Within the SG, the *Nodal-Interworking Function* (NIF) is the function responsible for interworking between the SS7 network on one side and the IP network on the other.

Figure 7-15 shows the Sigtran architecture in more detail. The *Stream-Control Transmission Protocol* (SCTP) corresponds to the common signaling transport shown in Figure 7-13. SCTP ensures error-free, in-sequence delivery of user messages, supports fast delivery of messages, and avoids

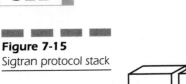

Figure 7-15
Sigtran protocol stack

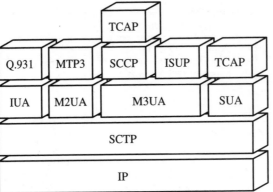

head-of-line blocking. Therefore, this protocol is more suitable than TCP. Furthermore, it supports network-level fault tolerance, something that is critical for carrier-grade network performance. SCTP is described in more detail a little later. Before delving into the details of SCTP, however, let's take a quick look at the other components of the Sigtran architecture. Several such components exist, all of which are briefly described here. In the following sections, we apply more focus to those components that support ISUP interworking, because ISUP is the most commonly used SS7 application.

Above SCTP, a number of user-adaptation modules exist:

■ *SS7 MTP2-User Adaptation Layer* (M2UA)   Provides adaptation between MTP3 and SCTP. M2UA provides an interface between MTP3 and SCTP such that standard MTP3 can be used in the IP network without the MTP3 application software realizing that messages are being transported over SCTP and IP instead of MTP2. For example, a standard MTP3 application implemented at the MGC could exchange MTP3 signaling-network management messages with the external SS7 network. In the same manner that MTP2 provides services to MTP3 in the SS7 network, M2UA provides services to MTP3 in the IP network. M2UA has a registered port number of 2904.

The usage of M2UA is depicted in Figure 7-16. In this scenario, two SGs provide an interface to the outside SS7 network. Both are connected to a call agent. On the call-agent side of the SGs, we have M2UA over SCTP over IP, whereas on the SS7 side, we have the standard SS7 MTP. At the call agent, we have a standard MTP3 operating over M2UA and IP. In a regular SS7 network, MTP3 utilizes the services of MTP2. In the scenario depicted in Figure 7-16, however, the MTP3 at the call agent utilizes the services of the MTP2 located at the SG without realizing that it is not local. The function of M2UA is to provide transparent access

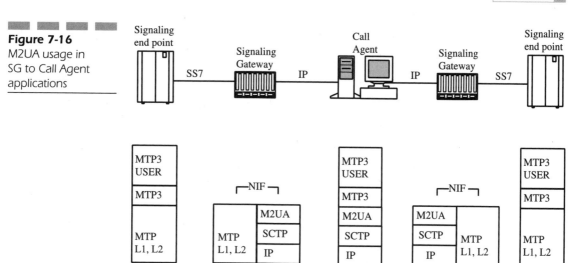

Figure 7-16
M2UA usage in
SG to Call Agent
applications

NIF = Nodal Interworking Function

from the standard MTP3 at the call agent to the standard MTP2 at the SG. In the example of Figure 7-16, the MTP3 application at the call agent can receive MTP3 signaling network-management messages, such as *Transfer Allowed* (TFA), *Transfer Prohibited* (TFP), etc. MTP3 can use that information to determine how to route messages from a higher-level MTP3-user, such as ISUP.

■ *SS7 MTP3-User Adaptation Layer* (M3UA)   Provides an interface between SCTP and those applications that typically use the services of MTP3, such as ISUP and SCCP. M3UA and SCTP enable seamless peer-to-peer communication between MTP3 user applications on the IP network and identical applications on the SS7 network. The application in the IP network does not realize that SCTP over IP transport is used instead of typical SS7. In the same manner that MTP3 provides services to applications such as ISUP on the SS7 network, M3UA offers equivalent services to applications on the IP network. M3UA has a registered port number of 2905.

We must recognize that M3UA is simply an adaptation layer between the upper protocols and SCTP. Therefore, although it provides the same primitives to the upper layer as MTP3 offers in standard SS7, it is not an IP-flavored MTP3. For example, M3UA does not implement standard MTP3 signaling network-management messages such as TFA, TFP, etc. This point is important and will be emphasized shortly.

Consider a call agent that needs to run an application such as ISUP. The agent can perform this task in two ways. One, it can run ISUP over MTP3 over M2UA over SCTP, as shown in Figure 7-16. Alternatively, it can implement ISUP over M3UA over SCTP, as shown in Figure 7-17. The difference between the two is a matter of where the MTP3 function really resides. In the scenario depicted in Figure 7-17, the real MTP3 exists at the SGs. M3UA simply enables the ISUP application at the MGC to remotely access the MTP3 function at the SG, without the ISUP application realizing that the MTP3 function is not local.

The call agent might have its own point code, separate from that of the SG. In that case, the SG function emulates an STP and appears as an STP to the outside SS7 network. The outside SS7 network views the call agent as a typical SS7 signaling endpoint to which access is achieved via one or more SG STPs.

Given that the M3UA provides services to ISUP, and given the importance of ISUP interworking, we describe the functioning of M3UA in more detail later in this chapter.

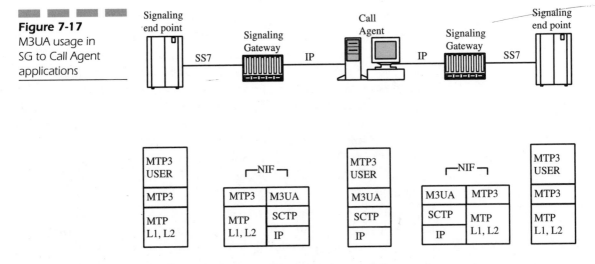

Figure 7-17
M3UA usage in
SG to Call Agent
applications

NIF = Nodal Interworking Function

- *SS7 SCCP-User Adaptation Layer* (SUA)   Provides an interface between SCCP user applications and SCTP. Applications such as TCAP use the services of SUA in the same way that they use the services of SCCP in the SS7 network. In fact, those applications do not know that the underlying transport is different in any way. Similar to other adaptation layers, SUA merely provides a transparent conduit whereby an application such as TCAP at an IP node can use the services of SCCP at an SG. Hence, there can be transparent peer-to-peer communication between applications in the SS7 network and applications in the IP network. At this writing, SUA has not been assigned a registered port number.

- *ISDN Q.921-User Adaptation Layer* (IUA)   In ISDN, user signaling such as Q.931 is carried by the Q.921 data-link layer. The equivalent Sigtran specification in an IP network is IUA. Thus, Q.931 messages can be passed from the ISDN to the IP network with identical Q.931 implementations in each network, neither of them recognizing any difference in the underlying transport. The registered port number for IUA is 9000.

SCTP

At this writing, SCTP is an Internet draft. As with all such documents, changes are possible at any time.

The primary motivation behind the development of SCTP is the fact that neither UDP nor TCP offer both the speed and reliability required of a transport protocol used to carry signaling. The design of SCTP is an attempt to make such reliability and speed available to SCTP users. In the SCTP specification, such a user is known as an *upper-layer protocol* (ULP). A ULP can be any of the protocols directly above the SCTP layer, such as M2UA or M3UA (as illustrated in Figure 7-15).

Some SCTP Concepts   Before attempting to explain how SCTP provides this reliable transport to various SCTP users, we must first become familiar with some SCTP terminology. Note that the following terms comprise just a selection of the most important terms—those that are necessary for an understanding of SCTP basics.

SCTP Endpoint   An SCTP endpoint is a logical sender or receiver of SCTP packets. In protocol terms, this endpoint is a combination of one or more

IP addresses and a port number. SCTP enables an endpoint to have multiple IP addresses, meaning that the endpoint can be multi-homed (i.e., spread over several physical platforms). This concept is important, because it enables fault tolerance in the network—something critical to carrier-grade performance.

Although a given SCTP endpoint can have multiple IP addresses, it can use only one port number. Thus, if an endpoint has several IP addresses, the same SCTP port number is applicable to each. The combination of an IP address and a port number is known as a transport address. Note that a given transport address can apply to only one SCTP endpoint, although that endpoint might have several transport addresses.

Association   SCTP works by establishing a relationship between SCTP endpoints. Such a relationship is known as an association and is defined by the SCTP endpoints involved and the current protocol state. Before applications at two endpoints can communicate, an association must be established. Once communication is complete, the association can be terminated. The association also can be terminated in error situations.

The upper-layer protocols such as ISUP, SCCP, or TCAP are not aware of such associations. After all, they are blind to the fact that the signaling is being carried by something other than standard MTP. Therefore, the task of instigating an SCTP association falls to the applicable adaptation layer.

Packets and Chunks   SCTP sits on top of IP. When SCTP wishes to send a piece of information to the remote end, it sends what is known as an SCTP packet to IP—and IP routes the packet to the destination.

The SCTP packet comprises a common header and a number of chunks, as depicted in Figure 7-18. The common header includes the source and destination port numbers which, when combined with the source and destination IP addresses, uniquely identify the endpoints. The header also includes a verification tag that is used to validate the sender of the packet. The verification tag is described in further detail later in this chapter.

The common header also includes an Adler-32 checksum, which is a particular calculation based on the values of the octets in the packet. This checksum is used to ensure that the packet has been received without corruption and provides another level of protection above the IP header checksum.

A number of chunks follow the common header, and each chunk is comprised of a chunk header plus some chunk-specific content. This content can be either SCTP control information or SCTP user information. In the case of SCTP user information from a ULP, the value of the chunk ID is 0,

Figure 7-18
SCTP packet format

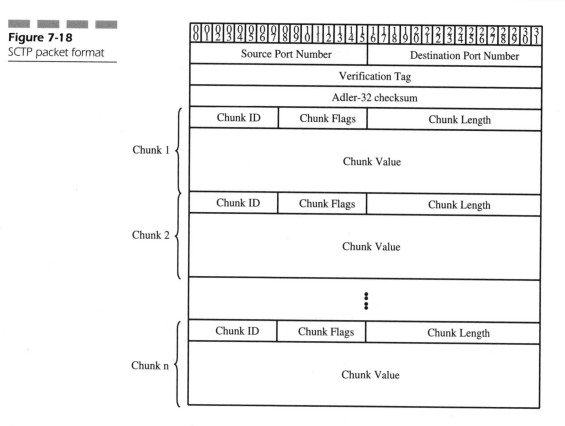

indicating user payload data. Otherwise, the chunk ID will have a value indicating a particular type of SCTP control information. The possible values for the chunk flags and chunk length depend upon the value of the chunk ID.

Streams A stream is a one-way logical channel between SCTP endpoints. You can also think of a stream as a sequence of user messages between two SCTP users. When an association is established between endpoints, part of the establishment of the association involves specifying how many streams the association is to support. If we think of a given association as a one-way highway between endpoints, then the individual streams are analogous to the individual traffic lanes on that highway.

To understand the stream concept in a network scenario, imagine a call agent that uses ISUP and that communicates with the outside SS7 network via an SG. The physical signaling links from the SS7 network terminate at the SG, and the actual speech circuits from the PSTN terminate at a media gateway that the call agent controls (as shown in Figure 7-19).

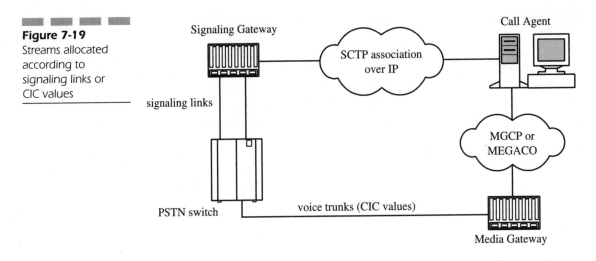

In the case where the call agent uses ISUP over MTP3 over M2UA, then the MTP3 function operating at the call agent will attempt to direct messages from ISUP to a particular signaling link, because that is one of the functions of MTP3. No physical signaling link exists at the call agent, however—there is only an SCTP association between the M2UA layer at the call agent and the M2UA layer at the SG. In order to handle this situation, the SCTP association should include N streams, where N is the number of signaling links terminated at the SG. In this manner, the M2UA function at the call agent can map a particular signaling link (as specified by MTP3) to a particular stream. Equally, the SG can map the stream to a particular signaling link to the destination.

In the case where the ISUP function at the call agent uses M3UA, then there is more flexibility in how streams are allocated in the SCTP association between M3UA at the SG and M3UA at the call agent. Because the choice of signaling link is made by the MTP3 function at the SG, it is not necessary for the streams to be allocated according to signaling link, although they could be. Alternatively, the streams could be allocated according to DPC/OPC combination or even according to OPC/DPC/CIC-range.

The use of these streams is an important tool to avoid head-of-line blocking and to ensure in-sequence delivery. In-sequence delivery is ensured, because each message sent for a given stream contains a stream sequence number. The receiving entity can make sure that messages are transmitted to the user application in order of stream sequence number. Secondly, each stream is processed separately so that message delivery for one stream is not held up while waiting for the next in-sequence message of a different stream.

Types of SCTP Chunks   Within an SCTP packet, a chunk ID identifies individual chunks. Though there are 65,536 possible values of chunk ID (and hence, 65,536 different chunk types), SCTP chunks can be grouped into four main categories. There are those that carry SCTP user data, those that carry SCTP control information, those reserved by IETF, and those that carry IETF-defined extensions. The correlation between chunk ID value and chunk type appears in Table 7-1.

SCTP Control Chunks   The INIT chunk is used to initiate an SCTP association between two endpoints. Unlike many other chunks, the INIT chunk cannot share an SCTP packet with any other chunk. In other words, a

Table 7-1

SCTP chunk types

ID Value	Chunk Type
00000000	Payload Data (DATA)
00000001	Initiation (INIT)
00000010	Initiation Acknowledgment (INIT ACK)
00000011	Selective Acknowledgment (SACK)
00000100	Heartbeat Request (HEARTBEAT)
00000101	Heartbeat Acknowledgment (HEARTBEAT ACK)
00000110	Abort (ABORT)
00000111	Shutdown (SHUTDOWN)
00001000	Shutdown Acknowledgment (SHUTDOWN ACK)
00001001	Operation Error (ERROR)
00001010	State Cookie (COOKIE ECHO)
00001011	Cookie Acknowledgment (COOKIE ACK)
00001100	Reserved for *Explicit Congestion Notification Echo* (ECNE)
00001101	Reserved for *Congestion Window Reduced* (CWR)
00001110	Shutdown Complete (SHUTDOWN COMPLETE)
00001111 to 11111111	Reserved by IETF, and IETF-defined chunk extensions

packet containing an INIT chunk must contain no other chunks besides the INIT chunk. The INIT chunk is further described later in this chapter.

The INIT ACK chunk is used to acknowledge the initiation of an SCTP association. As is the case for the INIT chunk, the INIT ACK chunk must not share a packet with any other chunk.

The SACK chunk is used to acknowledge the receipt of DATA (payload) chunks and to inform the sender of any gaps in the received chunks. Not every received chunk merits a SACK chunk in response. Instead, the SACK chunk works by indicating gaps. Let's suppose that a receiver has received chunks 1 through 5, plus chunks 8 and 9. The SACK message is used to indicate that chunks 6 and 7 are missing. These are the only chunks that need to be resent. This procedure is more efficient than TCP's retransmit mechanism.

The HEARTBEAT chunk is used to query the reachability of a particular endpoint. Let's assume that no chunks need to be sent from endpoint A to endpoint B during a particular period of time. In that case, endpoint A will send periodic HEARTBEAT messages to endpoint B just to make sure that endpoint B is still alive. The HEARTBEAT contains sender-specific information. The receiver of the HEARTBEAT chunk should respond with a HEARTBEAT ACK chunk containing heartbeat information copied from the received HEARTBEAT chunk.

The ABORT chunk is sent by an endpoint in order to end an association abruptly. The ABORT chunk can contain cause information regarding the reason for aborting the association. The ABORT chunk can be multiplexed with other SCTP control chunks into one packet. In such cases, however, the ABORT chunk should be the last chunk in the packet. If it is not the last chunk, then subsequent chunks in the packet are ignored.

The SHUTDOWN chunk is used for a graceful termination of an association. In the case that a higher-layer application or management application wishes to terminate an association, then the endpoint stops sending any new data to the far end. The endpoint will wait until all data sent to the far end has been acknowledged and will then send a SHUTDOWN to the far end in order to close the association. The SHUTDOWN will indicate the last data chunk received from the far end. The endpoint might, if necessary, retransmit user data to the far end before sending the SHUTDOWN chunk.

Upon receipt of a SHUTDOWN chunk, an endpoint will make sure that all of the user data that it has sent has been acknowledged, and if necessary, retransmit data to the far end. Once it is sure that everything sent in the past has been received, the endpoint will send a SHUTDOWN ACK chunk. The far end responds with SHUTDOWN COMPLETE. At this point, the association is ended.

The ERROR chunk is sent to indicate that the endpoint has detected some error condition. The chunk will include an error cause to provide further information about the type of error. For example, an endpoint might have received a chunk for a non-existent stream or might have received a chunk that is missing certain mandatory parameters.

The COOKIE ECHO chunk is only used during the initialization of an association. When an endpoint receives an INIT chunk and responds with an INIT ACK chunk, it includes a cookie parameter within the INIT ACK chunk. This parameter contains information that is specific to the endpoint and to the endpoint's view of the association, a timestamp, and a cookie lifetime value (60 seconds is recommended). When the far end receives the INIT ACK, it copies the cookie information and returns it in the form of a COOKIE ECHO chunk. The COOKIE ECHO chunk can be sent in a packet that also contains data chunks. The COOKIE ECHO chunk must be the first chunk in the packet, however.

The COOKIE ACK chunk is sent in response to a COOKIE ECHO chunk. Hence, the chunk is used only during the establishment of an association. Because the content of a COOKIE ECHO chunk is the same as what was sent in an INIT ACK, the sender of the INIT ACK can make sure that the initiator of the association has received the cookie information correctly. If the COOKIE ECHO chunk has been received without error within the lifetime specified for the cookie, the receiver of the COOKIE ECHO chunk sends a COOKIE ACK chunk. Otherwise, an ERROR chunk is sent. The COOKIE ACK chunk can be sent in the same datagram as DATA chunks but must be the first chunk in the datagram.

Payload Data (DATA) Chunk The DATA chunk is used to carry information to and from the ULP and has the format shown in Figure 7-20. The *Unordered* (U) bit means that the information in the chunk should be passed to the ULP without regard to sequencing. In other words, no stream sequence number is applied to the chunk.

SCTP might possibly segment a given user message. This situation could happen if the path MTU is smaller than the size of the message to be sent. The B and E bits are included because of such potential segmentation. The *Beginning* (B) bit indicates that this chunk contains the first segment of a user message. The *End* (E) bit indicates that the chunk contains the last segment of a user message. If a message is completely contained within one chunk, then both the B and E bits are set to zero. If the message contains more than two segments, then the first segment will have the B bit set to one and the E bit set to zero, while the last segment will have the B bit set to zero and the E bit set to one. All segments in between will have both the E and B bits set to zero.

The *Transmission Sequence Number* (TSN) is a 32-bit integer identifying this chunk in the context of the association. The TSN is independent of any

Figure 7-20

Payload data chunk format

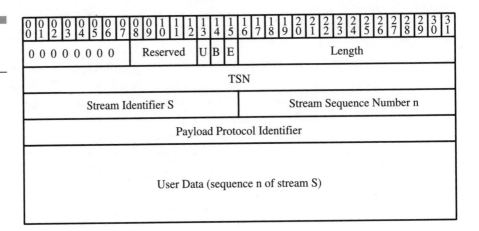

streams and is assigned by SCTP, rather than by any SCTP user. When an endpoint sends an INIT chunk, it includes a TSN value that corresponds to the first DATA chunk that it plans to send. Thus, the first DATA chunk sent will contain the same value of TSN as in the INIT chunk. Thereafter, the TSN is incremented for each new DATA chunk that the endpoint sends in this association.

The *Stream Identifier* (S) is a 16-bit integer that identifies the stream to which the data belongs. The *Stream Sequence Number* (n) is a 16-bit integer indicating the position of this message within the stream. For a given stream, the value of n begins at zero. This value increments for each message sent in a given stream. Note that a segmented message will have the same value of n in each segment.

The payload protocol identifier is passed from the user to the SCTP at the sending end and is passed from the SCTP to the destination user at the receiving end. This identifier is available to the users for passing further information about the chunk but is not examined or acted upon by SCTP.

Establishing an Association   An SCTP endpoint instigates the establishment of an association (i.e., an SCTP user instructs SCTP to establish the association). Note that an association is not established every time an application such as ISUP needs to send a message. Rather, an association is established in advance, such as when a new SS7 link is brought into service at an SG or when a new CIC range is brought into service.

The process appears in Figure 7-21 and begins by sending an SCTP packet containing an INIT chunk. The INIT chunk has the format shown in Figure 7-22.

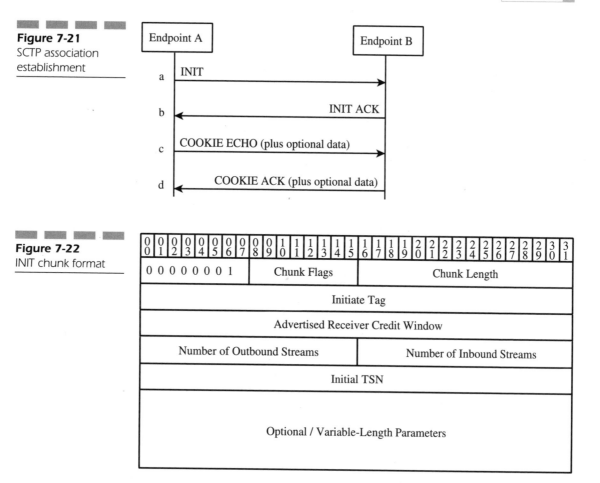

Figure 7-21
SCTP association establishment

Figure 7-22
INIT chunk format

The Chunk Flags field should be set to zero. The Initiate tag is a 32-bit integer and a random number that must not have the value zero. The end that receives this INIT chunk (the far end) will store the value of the Initiate tag. For every SCTP packet that the far end sends in this association, it will use the value of the Initiate tag as the Verification tag in the SCTP common header. The Advertised Receiver Credit Window (a_rwnd) represents the dedicated buffer space that the sender has allocated for this association. Although the buffer space available to the association might exceed this value during the lifetime of the association, it must never fall below this value. The number of *Outbound Streams* (OS) and the number of *Inbound Streams* (MIS) indicate the maximum number of streams that the initiator is willing to send and receive (respectively) for this association. The value zero is not allowed for either of these parameters. The Initial TSN is the TSN value that the sender expects to use when sending the first DATA chunk in the association.

The INIT chunk might also contain a number of optional parameters and/or variable-length parameters. This chunk might, for example, contain one or more IP version 4 or IP version 6 addresses.

When the other endpoint receives the INIT chunk, it responds with an INIT ACK chunk that has the identical format to the INIT chunk—with the exception that the optional/variable component must include a cookie. Importantly, the Verification tag in the SCTP common header must contain the same value as the received Initiate tag in the INIT chunk.

The INIT ACK chunk can also contain a number of IP version 4 or IP version 6 addresses. Given that the INIT ACK chunk also contains numbers of inbound and outbound streams, both ends of the association know the maximum number of streams that the other can send and receive. Each endpoint must respect the other's requirements and should not send more than what the other endpoint can handle.

Upon receipt of the INIT ACK chunk, the initiator of the association should send a COOKIE ECHO chunk—the content of which is copied from the cookie parameter of the received INIT ACK. In order to quickly start the message exchange for which the association is being created, one or more additional DATA chunks can be included in the packet. The COOKIE ECHO chunk must be the first chunk in the packet, however.

Finally, the receiver of the COOKIE ECHO chunk sends a COOKIE ACK chunk, which is a chunk containing the CHUNK ID value of 00001011, chunk flags all set to zero, and a length indicator of four. In other words, this chunk is a simple acknowledgement header with no additional information. The packet containing the COOKIE ACK chunk can also contain a number of DATA chunks, provided that the COOKIE ACK chunk is the first chunk in the packet.

Transferring Data   The reliable transfer of user data is achieved by the use of two SCTP chunks. The first is the DATA chunk described previously (depicted in Figure 7-20). The second is the SACK chunk, which is an SCTP control chunk (shown in Figure 7-23).

The easiest explanation of the SACK chunk is by example. Let's assume that an endpoint has transmitted data chunks 1 through 11. Let's also assume that chunks with TSNs 1 through 4 and chunks with TSNs 7, 8, 10, and 11 have been received. Hence, chunks 5, 6, and 9 are missing. Let's also assume that chunks with TSN 8 and TSN 11 have been received twice. Therefore, the received data appears as in Figure 7-24.

In this scenario, the Cumulative TSN ACK field would contain the value 4, because this field indicates the highest TSN value received without any

Figure 7-23
SACK chunk format

0 0	0	0	0	0	0	0	0	0	0	1	1	1	1	1	1	1	1	1	1	2	2	2	2	2	2	2	2	2	2	3	3

| 0 1 | 2 | 3 | 4 | 5 | 6 | 7 | 8 | 9 | 0 | 1 | 2 | 3 | 4 | 5 | 6 | 7 | 8 | 9 | 0 | 1 | 2 | 3 | 4 | 5 | 6 | 7 | 8 | 9 | 0 | 1 |

0 0 0 0 0 0 1 1	Chunk Flags	Chunk Length
Cumulative TSN ACK		
Advertised Receiver Credit Window (a_rwnd)		
Number of Gap Ack Blocks = N	Number of Duplicate TSNs = X	
Gap Ack Block #1 Start	Gap Ack Block #1 End	
Gap Ack Block #N Start	Gap Ack Block #N End	
Duplicate TSN 1		
Duplicate TSN X		

Figure 7-23
SACK chunk format

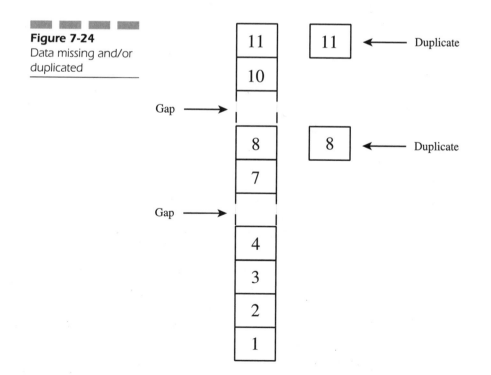

Figure 7-24
Data missing and/or
duplicated

gaps. The number of Gap Ack Blocks (N) indicates the number of fragments received after the unbroken sequence. In our example, N has the value 2 to indicate the fragment 7 to 8 and the fragment 10 to 11. The number of duplicate TSNs (X) indicates the number of TSNs that have been received more than once.

The Gap Ack Block number 1 start field indicates the offset of the first segment from the unbroken sequence. This offset means the difference between the TSN value in the Cumulative TSN ACK field and the lowest TSN value of the first segment. In our example, this offset is the difference between 4 and the value 7 (i.e., 3). The Gap Ack Block number 1 end field indicates an offset from the Cumulative TSN ACK and the highest TSN in the first segment. Thus, it has the value 4 in our example (the difference between 8 and 4). This process is repeated for each fragment. Thus, for the next fragment, the Gap Ack Block start and end values are 6 and 7, respectively.

Finally, we have a list of the duplicate TSNs. In this case, the values 8 and 11 would be listed.

The *Advertised Receiver Window Credit* (a_rwnd) indicates the updated buffer space of the sender of this Selective ACK. This a_rwnd indicates how much receive buffer space is free at the time the SACK is sent and enables the remote end to manage the amount of data sent, such that buffer overflow and consequent data loss can be avoided. SCTP specifies procedures based on the a_rwnd and the SACK chunk in general to ensure that congestion is avoided.

SCTP Robustness   Robustness is a key characteristic of any carrier-grade network. Robustness means that the network should implement procedures through which failures or undesired occurrences are minimized. Robustness means the capability to handle a certain amount of failure in the network without significant reduction in quality. Furthermore, the network should provide a graceful rather than a drastic degradation in the event of failures or overload. SCTP addresses these issues in a number of ways.

As mentioned previously, SCTP implements congestion-control mechanisms to ensure that one endpoint does not flood another with messages. Furthermore, SCTP incorporates Path MTU discovery so that messages are not sent if they are too long to be handled by the intervening transport network.

Recall that the SCTP common header contains source and destination port numbers. Also recall that the INIT and INIT ACK chunks might optionally include one or more IP addresses, which enable a given endpoint to be effectively multi-homed (i.e., to have multiple IP addresses). If the

INIT or INIT ACK chunk does not contain any IP addresses, then the IP address will default to the IP address from which the datagram is sent. Greater robustness is offered by having multiple IP addresses, however.

In the case that a given endpoint supplies several IP addresses, then the other end will choose one of those addresses as the primary destination address. In the case that the primary address fails, then one of the other addresses is chosen as the new primary address. Furthermore, if one or more DATA chunks is transmitted to a given address and is not acknowledged within the retransmission timer, then the sender should send the retransmission to one of the other addresses.

SCTP ensures that an endpoint is aware of the reachability of another endpoint through the use of SACK chunks in the case that DATA chunks have been sent and HEARTBEAT chunks in the case that an association is idle. Therefore, detection of path or endpoint failure is assured, and we can take appropriate action.

M3UA Operation

The following describes how M3UA over SCTP can be used in a VoIP network to provide seamless interworking with the external SS7 network for ISUP. First, let's become familiar with a number of terms that are applicable to M3UA. In fact, many of these terms apply to all of the adaptation layers:

- *Application Server* (AS) A logical entity handling signaling for a particular scope. For example, an AS could be a logical entity within a call agent that handles ISUP signaling for a particular SS7 DPC/OPC/CIC-range. Equally, an AS could be a logical entity within a database application that handles signaling for a particular *SCCP DPC/OPC/Subsystem Number* (SSN) combination. An AS contains a set of *Application Server Processes* (ASPs), which are those actual processes operating within the AS. In fact, we can consider the AS a list of ASPs, some of which are active and some of which are standby.

- *Application Server Process* (ASP) A process instance of an AS acting as an active or standby process. An ASP could be, for example, that process within an MGC that is currently handling ISUP signaling. The ASP has an SCTP endpoint; therefore, it can be spread across multiple IP addresses. In a robust network, there should be at least one active ASP and at least one standby ASP for a given application.

■ *Routing Key*   A set of SS7 parameters (such as SLS, DPC, OPC, and CIC) that identifies the signaling for a given AS. For example, if a given AS is to handle ISUP signaling for a particular combination of OPC/DPC/CIC- range, then that OPC/DPC/CIC-range is the routing key. Within a signaling gateway, a particular routing key will point to a particular AS. Within the AS, there will be at least one ASP actively handling the signaling.

■ *Network Appearance*   A mechanism for separating signaling traffic between an SG and an ASP, where all the traffic uses the same underlying SCTP association. Imagine, for example, an SG that is an international signaling gateway. This gateway has a national point code and an international point code and uses a national variant of MTP for communication within the national network. The gateway also uses the ITU-T variant of MTP for communication within the international network. For communication between an SG and call agent/MGC, the messaging exchanged must be placed in the context of the correct network appearance so that the appropriate link set can be used on the non-IP side of the SG.

Signaling Network Architecture   In order to ensure carrier-grade service, we must ensure that there is no single point of failure in the network between the SG and ASP. Consequently, SGs should be set up at least in pairs in a manner similar to STPs in the SS7 network. Furthermore, ASPs should be set up in a redundant or load-sharing configuration and spread over different physical platforms (hosts). Such a robust configuration is shown in Figure 7-25. Note that the requirements for such a robust configuration are not something unique to the usage of M3UA; rather, it applies to any system that needs to be fault tolerant (a requirement for carrier-grade operation).

Each ASP needs to have a point code. The allocation of point codes is completely flexible, however. For example, all ASPs connected to a given SG could share the same point code as the SG. In such a case, the combination of SG and ASPs appears to the SS7 network as a single signaling endpoint. Alternatively, all ASPs connected to a given SG could all share the same point code but have a different point code from the SG. In such a case, the SG would appear to the SS7 network as an STP, and the combined ASPs would appear as a single signaling endpoint located behind that STP. Another alternative would be for every ASP to have its own point code or for groups of ASPs to share a point code separate from the point code of the SG. In that case, the SG appears as an STP, and each ASP or group of ASPs appears as a unique signaling endpoint. The options available depend only

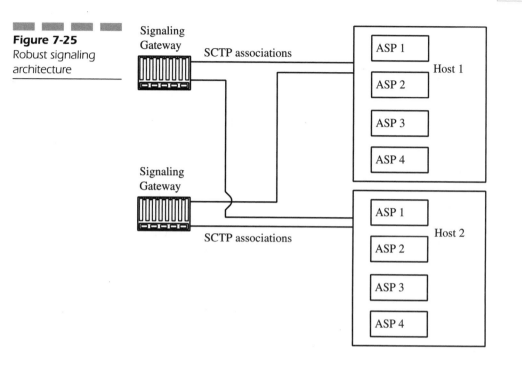

Figure 7-25
Robust signaling
architecture

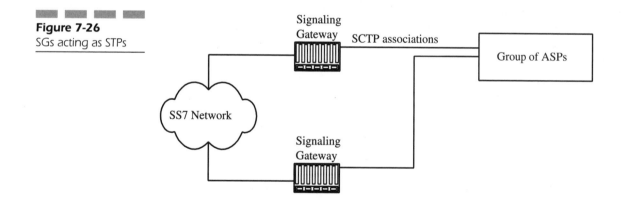

Figure 7-26
SGs acting as STPs

upon the translation and mapping capabilities of the SG. If a given ASP or group of ASPs can communicate with the SS7 network via more than one SG, then the ASP or group of ASPs must have a point code that is different from those of the two SGs. In such a scenario, the SGs operate as STPs, as shown in Figure 7-26.

NOTE: *M3UA uses a client-server model in which the SG is the server and the ASP is the client. The ASP takes responsibility for instigating an SCTP association with the SG.*

Services Provided by M3UA As mentioned, M3UA provides an interface between an application such as ISUP and the underlying transport. M3UA enables the application to utilize the services of the real MTP3, which is located at an SG. M3UA accomplishes this task in a transparent manner so that the application does not realize that the instance of MTP3 being used is at the SG, rather than local.

In order to provide services to the upper layer, M3UA must offer the same primitives to the upper layer as offered by MTP3:

MTP-Transfer request Sent from the upper layer to M3UA to request a message to be transferred to a particular destination

MTP-Transfer indication Used by M3UA to pass an incoming message to the upper layer

MTP-Pause indication Sent by M3UA to the upper layer to indicate that signaling to a particular destination should be suspended. This primitive is used, for example, when the destination is not reachable.

MTP-Resume indication Sent by M3UA to the upper layer to indicate that signaling to a particular destination can resume

MTP-Status indication Sent by M3UA to the upper layer to inform the upper layer of some change in the SS7 network, such as network congestion or a destination user part becoming unavailable

Transferring application messages from the application to the SS7 network is a relatively straightforward process. Let's take an example where an ISUP application at an MGC wishes to send a message to the SS7 network. The application issues an MTP-Transfer request to M3UA, and M3UA effectively sends this request to SCTP as a DATA chunk to be transmitted on a particular SCTP association and in a particular stream. SCTP ensures that it reaches the SG, where the content of the DATA chunk is passed up to M3UA, which passes it to the interworking function of the SG. The interworking function passes it to MTP3 on the SS7 side of the SG, which takes care of routing the message correctly to the SS7 network.

In the opposite direction, the process is similar. A message arrives at the SG and is passed from MTP3 to the interworking function and from the interworking function to M3UA. Based on the SS7 parameters of the message, such as DPC/OPC/CIC-range, the appropriate application server and ASP are chosen. In the case that more than one ASP is active, then one of the active ASPs is chosen (depending on the load-sharing algorithm). M3UA packages the message for transmission by SCTP as a DATA chunk on the appropriate SCTP association and within the appropriate stream. At the destination, the DATA chunk is passed to M3UA, which passes the information to the application by using the MTP-Transfer indication primitive.

M3UA Messages   M3UA includes a number of messages between peer M3UA entities. These have the generic format shown in Figure 7-27. This format contains a header followed by the M3UA message content. The header is common across all user adaptation layers. For M3UA, the protocol version is 0000 0001.

When M3UA passes user information between the SG and MGC (as described previously), it does so by sending M3UA data messages. These data messages are packaged as SCTP DATA chunks. When sending a data message, the message type field of Figure 7-27 has the value 0101.

In addition to packaging user messages in the form of M3UA data messages, M3UA includes other messages between M3UA peer entities so that the entities can communicate information regarding the SS7 network in general. For example, if a remote destination becomes unavailable, then the SG will become aware of this fact through SS7 signaling-network management messages. The ISUP application at a call agent must also be made aware of the event, however, so that it does not try to send messages that cannot reach their destination. M3UA at the call agent can indicate such an event through the use of the MTP-Pause indication primitive. But M3UA is not

Figure 7-27
Common message format

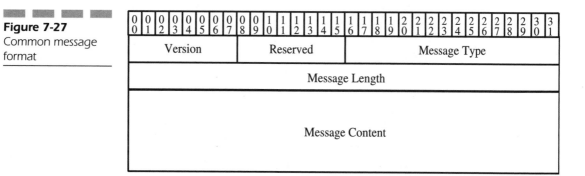

MTP3, and SS7 signaling-network management messages are not passed directly from the SG to the call agent. So how does the M3UA at the MGC know that the destination is unavailable? The answer is that the information is passed through other M3UA messages besides the DATA message. Table 7-2 lists the various types of M3UA messages.

Signaling Network Management Messages M3UA includes the following messages for signaling network management:

SS7 Network Isolation (S7ISO) This message is proposed for use when all links to the SS7 network have been lost. At this writing, its use is not yet fully defined.

Destination Unavailable (DUNA) This message is sent from the SG to all concerned ASPs, indicating that a destination within the SS7 network is not available. This message indicates the affected destination (DPC) and (optionally) the network appearance. Several destinations might be included in the message, so that one message can be used to indicate the

Table 7-2

M3UA message types

Message Name	Message-Type Value
Error (ERR)	0000
Notify (NTFY)	0001
Payload Data (DATA)	0101
SS7 Network Isolation (S7ISO)	0200
Destination Unavailable (DUNA)	0201
Destination Available (DAVA)	0202
Destination State Audit (DAUD)	0203
SS7 Network Congestion State (SCON)	0204
Destination User Part Unavailable (DUPU)	0205
ASP Up (ASPUP)	0301
ASP Down (ASPDN)	0302
Heartbeat (BEAT)	0303
ASP Up Ack (UP ACK)	0304
ASP Down Ack (DOWN ACK)	0305
ASP Active (ASPAC)	0401
ASP Inactive (ASPIA)	0402
ASP Active Ack (ACTIVE ACK)	0403
ASP Inactive Ack (INACTIVE ACK)	0404

All other values are reserved

unavailability of several signaling points. The message allocates 24 bits for each DPC. The network appearance is used when the SG is logically partitioned across multiple networks and serves to logically separate the signaling between SG and ASP, according to each of the logical portions of the SG. Appearance is also used to indicate the format of the DPC (e.g., 14-bit or 24-bit). Therefore, the same DUNA message can be used in ANSI, ITU-T, or other national-specific signaling environments.

DUNA is generated at the SG when it determines from MTP3 network-management messages that a destination is unavailable. The DUNA message is transmitted to the ASP, where M3UA uses it to create the primitive MTP-Pause indication (which it issues to the upper-layer protocol, such as ISUP).

Destination Available (DAVA)   This message is sent from the SG to all concerned ASPs when a destination becomes reachable. At the ASP, this message is mapped to the primitive MTP-Resume indication. The message contains the same information as the DUNA, with respect to the destinations in question and the network appearance.

Destination State Audit (DAUD)   This message is sent from an ASP to an SG when the ASP wishes to query the status of one or more destinations. The ASP could, for example, send the message upon expiry of a timer set upon receipt of DUNA or SCON. The SG responds with DAVA, DUNA, or SCON for the destination(s) in question.

SS7 Network Congestion (SCON)   This message is sent from the SG to all concerned ASPs to indicate that the route to a particular SS7 destination is congested. This message is generated at the SG and is sent to all concerned ASPs. At an ASP, the message is mapped to the primitive MTP-Status indication, which is issued to the upper-layer protocol.

Destination User Part Unavailable (DUPU)   This message is sent from the SG to all concerned ASPs to indicate that a given user part at a given destination is not available. The message indicates the DPC and the user part in question, such as ISUP, SCCP, or TUP. At the ASP, the message is mapped to the primitive MTP-Status indication. The MTP-Status indication can be used to indicate congestion on the network or the unavailability of a given destination user part. While in MTP3, these different events are indicated with the same primitive by using different cause codes, M3UA communicates these events between M3UA peers through the use of two different messages.

As we can see in Table 7-2, M3UA also includes a number of messages related to ASP management.

The *ASP Up* (ASPUP) message is used between M3UA peers to indicate that the adaptation layer is ready to receive traffic or maintenance messages. The *ASP Down* (ASPDN) message is used between M3UA peers to indicate that an ASP is not ready to receive traffic or maintenance messages. The *ASP Up Ack* (UP ACK) message is sent in response to ASPUP and the *ASP Down Ack* (DOWN ACK) message is sent in response to ASPDN. The *ASP Active* (ASPAC) message is sent by an ASP to indicate that it is ready to be used. This message indicates the mode of usage of the ASP, such as whether it is to receive all messages or whether it is to operate in a load-sharing mode. This message also can include a Routing Context. The Routing Context is related to the Routing Key described previously, and as such, it indicates to the SG what scope is applicable to the ASP (e.g., what DPC/OPC/CIC-range). The *ASP Inactive* (ASPIA) message indicates that the ASP is no longer active and should not be used. The *ASP Active Ack* (ACTIVE ACK) message is sent in response to ASPAC and the *ASP Inactive Ack* (INACTIVE ACK) message is sent in response to ASPIA. Figures 7-28 and 7-29 give examples of the usage of these messages.

The *Heartbeat* (BEAT) message is optional. This message is used between M3UA peers to ensure that they are still available to each other. When M3UA uses the services of SCTP, the BEAT message is not required at the M3UA level, because SCTP includes functions for ensuring that endpoints stay informed of each other's reachability.

Two management messages exist: the *Error* (ERR) message and the *Notify* (NTFY) message. ERR is used when a message is received with invalid contents. ERR includes a cause code to indicate the type of error discovered (and optionally, some diagnostic information). NTFY is used between M3UA peers to communicate the occurrence of certain events. A NTFY message can be used between the SG and ASP to indicate a change

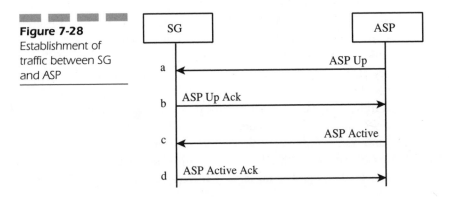

Figure 7-28

Establishment of traffic between SG and ASP

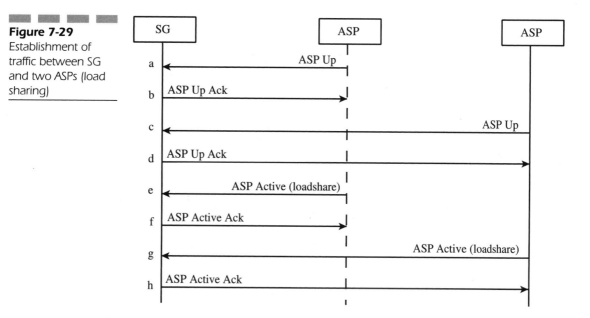

Figure 7-29
Establishment of traffic between SG and two ASPs (load sharing)

of status in another ASP. For example, if a given ASP uses the ASPIA message to indicate that it is inactive, then the SG could use a NTFY message to inform another ASP of the event. This notification would be used in the case where the two ASPs had been in load-sharing mode.

M2UA Operation

The previous discussion focuses on a situation where the call agent/MGC implements a function such as ISUP but does not implement any of the lower layers of the SS7 stack. In particular, MTP3 is not implemented at the call agent. The direct consequence of this approach is that the call agent does not have the same view of the SS7 network as an entity that implements MTP3. In particular, it does not send or receive signaling-network management messages.

If it were necessary for the call agent to be more involved in signaling network management, however, then it is possible to implement MTP3 at the call agent and use M2UA over SCTP to access the MTP2 functions at a signaling gateway. While this process gives the call agent more visibility of the SS7 network, another consequence is that the call agent can be considered more tightly coupled to the SG. The MTP3 function at the call agent is a normal MTP3 implementation that includes routing and distribution capabilities and that needs to know about the details of all signaling links that it might use. The main function of the SG is to take care of the MTP2 functions, which means that the SG acts as a signaling terminal for the MTP3 application at the call agent.

M2UA uses similar concepts to those used by M3UA, including the concept of an AS and an ASP. M2UA also includes similar messages, such as *ASP Up* (ASPUP), *ASP Down* (ASPDN), *ASP Active* (ASPAC), *ASP Inactive* (ASPIA), and *Error* (ERR). These messages are used in exactly the same way as M3UA, with the exception that the ASPs in M2UA perform different functions than those in M3UA. Whereas in M3UA, the ASP can be related to something like a DPC/OPC/CIC-range or some other set of characteristics, in M2UA, the ASP is an instance of MTP3 at the call agent.

The messages sent between M2UA peers use the same format as messages used by M3UA. The difference is in the Message Type field of the common header. The messages that are M2UA-specific have different message-type values than those messages that are M3UA-specific. Note that the ASP-management messages have the same message type in both M2UA and M3UA.

The following M2UA-specific messages are provided between M2UA peers:

DATA (Message type 0601) Used to carry an MTP2-user *Protocol Data Unit* (PDU)

ESTABLISH REQUEST (Message type 0602) Used to establish a link at the SG or to indicate that a link has been established. In setting up signaling links, the establishment of those links at the SG can take place in advance of management action on the call agent in order to establish links. In such a case, when the call agent sends an ESTABLISH REQUEST to the SG, the SG might simply return the ESTABLISH CONFIRMATION (Message type 0603).

RELEASE REQUEST (Message type 0604) Used by the call agent to request the SG to take a particular signaling link out of service. The request includes the reason why the link should be released. Once the link has been taken out of service, then the SG responds with a RELEASE CONFIRM (Message type 0605). The SG can also autonomously take a link out of service, which would occur, for example, in the case of a hardware failure at the SG. In such a case, the SG would send a RELEASE INDICATION (Message type 0606) to the call agent.

STATE REQUEST (Message type 0607) Sent from a call agent to the SG to cause the SG to perform some action on a particular signaling link. The actions include tasks such as requesting normal link alignment or flushing transmit and retransmit buffers. Upon completion of the action at the SG, the SG responds with a STATE CONFIRM (Message type 0608). The SG can also autonomously send a STATE INDICATION (Message type 0609) to the call agent to indicate the condition on a link. The message might be sent, for example, upon the occurrence of a remote processor outage or a change in the physical state of the link itself.

A number of messages are used during link changeover. In such an event, the call agent must retrieve certain information from the SG, particularly information that is stored at the SG for retransmission on a link. DATA RETRIEVAL REQUEST (Message type 060A) is used during link-changeover procedures. The call agent sends this message to request the *Backward Sequence Number* (BSN), to retrieve PDUs from the retransmit queue, or to flush PDUs from the retransmit queue. The SG responds with the DATA RETRIEVAL CONFIRM (Message type 060B). The SG can also just send a PDU to the call agent from the retransmit queue, in which case the SG will use a DATA RETRIEVAL INDICATION (Message type 060C). Multiple RETRIEVAL INDICATION messages can be sent from the SG to the call agent in the case of link changeover. When the final PDU from the retransmit queue is being sent, then the SG will use the DATA RETRIEVAL COMPLETE INDICATION (Message type 060D) instead. This message should be sent only once in a given changeover event.

Interworking SS7 and VoIP Architectures

The primary thrust of this chapter has been to describe how a VoIP network can interwork with an external SS7 network. Finally, we consider any differences that the architecture of the VoIP network might cause—specifically, whether the network follows the softswitch of the H.323 model.

Interworking Softswitch and SS7

The concepts behind Sigtran are aimed at interworking between SS7 and the softswitch architecture. The reason is because the softswitch architecture includes a number of media gateways placed close to the sources and sinks of media in the PSTN (i.e., at the edges of the IP network). The distributed softswitch architecture means that the call agents that control the media gateways might be far removed from the gateways themselves, however. Thus, having signaling gateways at the edge of the IP network makes sense.

We achieve interworking by using at least two signaling gateways that terminate SS7 links and pass SS7 messages to the actual applications at a call agent. The architecture would be as shown in Figure 7-30, and the protocols used include SCTP, M2UA, and M3UA (as previously discussed).

An Internet draft titled "Use SCTP as MEGACO Transport" discusses the use of SCTP as the underlying transport protocol for carrying MEGACO transactions between the MG and MGC. Merit exists to the argument that SCTP should be used for that purpose. This idea means that MEGACO messaging could make use of the reliable and quick transport that SCTP offers. Given that there is a semi-permanent relationship between an MG and an MGC, the SCTP relationships can be established at MG startup and can remain in place though the duration of the MGC-MG relationship. Whether such an approach becomes standardized remains to be seen. If it were to become standardized, then Figure 7-30 would be modified to show MEGACO/SCTP between the MGC and MC.

Ideally, SCTP could also be used as the transport protocol for carrying SIP messages. That procedure might not be so easy, however, because no semi-permanent relationship exists between a given SIP client and a remote server. In other words, the communication between a SIP client and a given server involves an ad-hoc relationship established only for the purpose of the call in question. Establishing an SCTP association every time a call is to be set up might not be efficient.

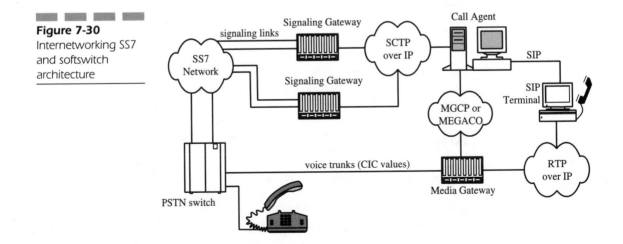

Figure 7-30
Internetworking SS7 and softswitch architecture

ISUP Encapsulation in SIP   SCTP and the associated adaptation layers provide the capability to convey SS7 application protocol information from an endpoint in the PSTN to a call agent in the center of the VoIP network. The call agent then has access to the signaling information needed to route calls on the VoIP network. SS7 application protocols themselves are not used internally in the VoIP network, however, or if they are, such use is proprietary. Rather, protocols such as ISUP are mapped to equivalent VoIP protocols, such as SIP. In fact, the SIP-ISUP mapping is the most common, and we already discussed this topic in Chapter 5, "The Session Initiation Protocol." Such mapping works well if one end of the call is a VoIP node and the other is a PSTN node. A potential problem exists, however, when the softswitch network provides a transit function between a circuit-switched network at the originating end and a circuit-switched network at the terminating end.

Because protocols are designed in different ways, you will rarely find a direct match between the messages and parameters of one protocol and those of another. In addition, interworking has to address the differences between state machines, timers, etc. The result, although workable, is often less than perfect. The reason is because protocol A might be slightly better in one aspect, while protocol B might be better in another. Consequently, interworking situations often lead to a lowest common denominator result, where the functionality offered is limited to that supported by both protocols. For example, SIP header fields enable the inclusion of information that simply does not exist in ISUP. A good example is the Retry-after header field that does not map to an equivalent in ISUP.

Equally, ISUP provides information that does not easily map to SIP headers (if at all). Often, such information might need to be mapped to the closest SIP equivalent (or perhaps discarded). This situation is not so bad if a call is placed from a PSTN user to an SIP user, because the SIP user agent server would not have the capability to interpret such information in any case. Networks exist today that provide long-distance service by using VoIP, however. They connect to the PSTN at numerous points. Long-distance calls from one PSTN user to another can transit via the VoIP network, entering the VoIP network at one point and leaving again at another. If ISUP is converted to SIP at the point of ingress and SIP is converted back to ISUP at the point of egress, the result might be that the ISUP messages leaving the network might be different to those that entered the network, as shown in Figure 7-31.

In order to counteract this problem, SIP enables the message body to encapsulate an ISUP message. Of course, SIP messages and responses are used between the gateways, and where appropriate, those messages must

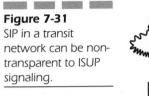

Figure 7-31
SIP in a transit
network can be non-
transparent to ISUP
signaling.

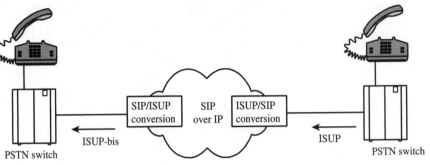

contain an SDP message body in order for the gateways to describe media to be sent between them. Some of the messages can also contain an ISUP message in binary form within the message body. Therefore, the message body can have multiple parts: an SDP session description used by the gateways and an encapsulated ISUP message carried transparently across the network.

Consider the scenario depicted in Figure 7-32. The incoming IAM is mapped to a SIP INVITE. The SIP INVITE contains an SDP session description and also contains the original ISUP message. At the egress point, the SDP information is used to set up the RTP session between the gateways, and the ISUP information is extracted to send to the PSTN. When the call is through-connected, the received ACM is mapped to the SIP 183 (session progress) response, which contains a temporary SDP media description and also contains an encapsulated ISUP ACM to pass to the originating PSTN node. A similar process occurs when the call is answered and an ANM message is returned from the called end. Note that the figure does not include a SIP message for the reliable delivery of provisional responses. We could assume, however, the use of such a message (PRACK). Although the 183 response is considered provisional, it should be delivered reliably.

In order to encapsulate an ISUP message with a SIP message body, the Content-Type: header field of a SIP message should not indicate `application/sdp`. Instead, the header field should have the following format:

```
Content-type: multipart/mixed; boundary = SDP-ISUP-boundary
```

The Content-type indicates that the message body contains different sets of information in different formats. The boundary field indicates the boundary between one description and the other within the message body. Its

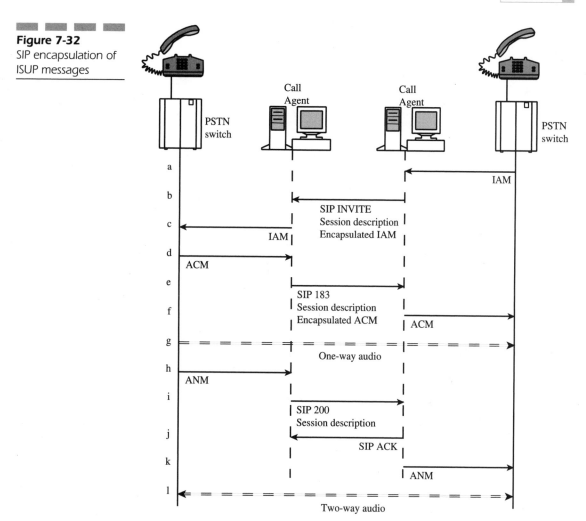

Figure 7-32
SIP encapsulation of ISUP messages

usage is described in RFC 2046, *"Multipart Internet Mail Extensions (MIME) Part two: Media Types."*

Within the message body, the beginning of each different payload is indicated by `--` `'boundary'`, where `boundary` is as specified in the Content-Type: header. Therefore, if a SIP message were to contain a Content-Type field as shown previously, then the each payload within the message body would be indicated by the string `--` `SDP-ISUP-boundary`. After the last payload, there is another line with the format `--` `'boundary'` `--`. In our example, the last line would be `--` `SDP-ISUP-boundary` `--`.

A SIP INVITE using ISUP encapsulation might look something like the following:

```
INVITE SIP: 9725551234@telco2.net SIP/2.0
From: SIP: 2145556789@telco1.net
Call-ID: 123456789@telco1.net
Content-Length: xxx
Content-Type: multipart/mixed; boundary = SDP-ISUP-boundary
MIME-Version: 1.0

-- SDP-ISUP-boundary
Content-Type: application/sdp

V=0
O=collins 22334455 123 IN IP4 444.333.222.111
S=network issues
C= IN IP4 444.333.222.110
T=0 0
M=audio 5555 RTP/AVP 0

-- SDP-ISUP-boundary
Content-Type: application/ISUP; version = ANSI
Content-Encoding: binary
1A 00 01 00 60 00 0A 03 06 0D 03 80 90 A2 07
03 10 18 27 85 31 48 0A 07 03 11 12 74 66 69
53 EA 01 00 00
-- SDP-ISUP-boundary--
```

Interworking H.323 and SS7

Interworking between an H.323 architecture and SS7 could work in a similar manner to interworking between softswitch and SS7. The approach would be to use signaling gateways that terminate SS7 signaling. The SS7 application information would be carried by using Sigtran from the SS7 gateways to one or more H.323 gateways.

Equally, however, the gateway itself could simply terminate SS7 links directly. In fact, such an approach has merit. In the H.323 architecture, the gateway itself contains a great deal of the application logic required for call setup, and the gateway is also the entity that performs the media conversion. The fact that it holds the application logic means that it must send and receive call control signaling information. The fact that the gateway performs the media conversion means that it should be placed at the edge of the IP network. Because the gateway is at the edge of the network, it is likely easier and more efficient for the gateway itself to terminate SS7 links from the PSTN directly. The architecture would be as shown in Figure 7-33.

In fact, interworking SS7 with H.323 is in some ways an easier proposition than interworking SS7 and softswitch (at least, for ISUP). The reason is because a close relationship exists between ISUP and Q.931. After all, ISUP is the ISDN User Part, the inter-switch network protocol correspond-

Figure 7-33

Interworking SS7 and
H.323 architectures

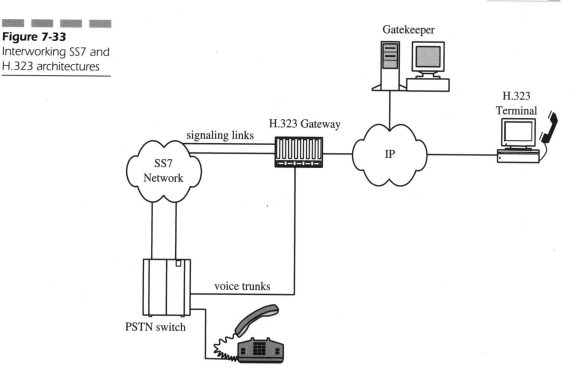

ing to the ISDN access protocol Q.931. Consequently, mapping is easy between Q.931 messages and ISUP messages. For example, Setup in Q.931 maps to IAM in ISUP, and ANM in ISUP maps to Connect in Q.931. Therefore, if an H.323 network were to receive a call from the PSTN using ISUP, then the signaling would appear as in Figure 7-34.

The example assumes the use of the H.323 Fast-Connect procedure. If the called H.323 terminal did not support that procedure, then the Connect message would not include a faststart element. In that case, the arrival of the Connect message at the gateway would not immediately lead to an ANM back to the PSTN. Rather, the logical channel procedures of H.245 would first be used to establish the media path between the gateway and the called terminal.

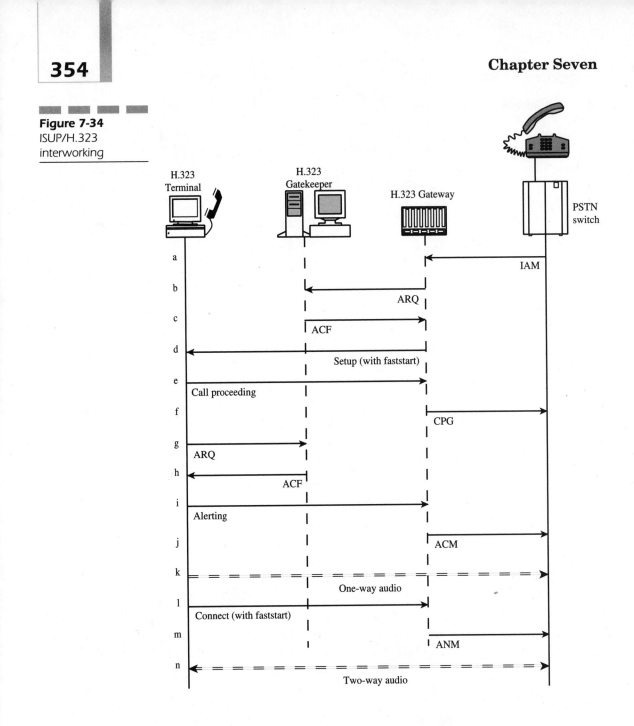

Figure 7-34
ISUP/H.323
interworking

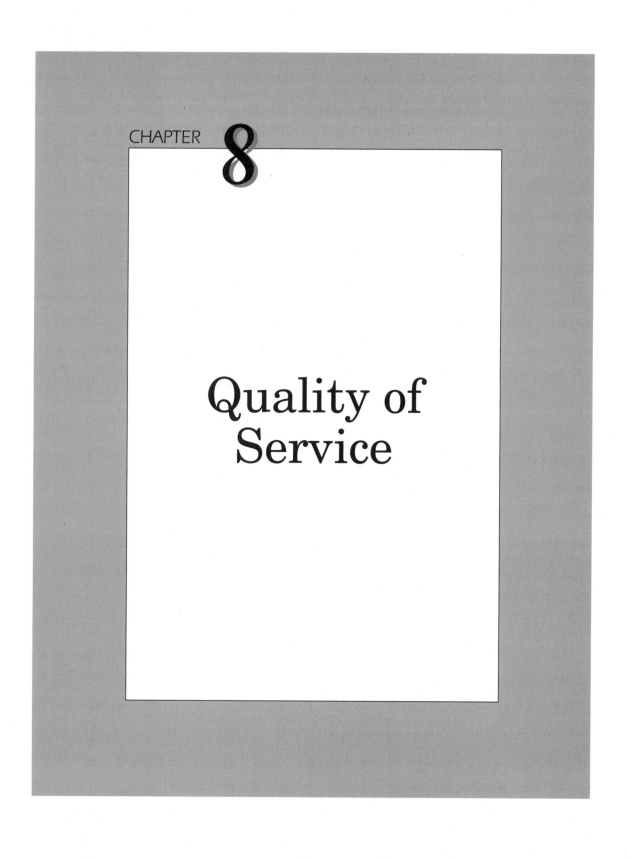

CHAPTER 8

Quality of
Service

Introduction

To many, *Quality of Service* (QOS) is one of the biggest, if not *the* biggest issue facing *Voice over IP* (VoIP). For this reason, many of the VoIP solutions on the market today that provide voice over the Internet are free services. The philosophy behind such offerings is that customers can hardly complain about poor quality when they are being provided a free service. Such a philosophy is fine if the business model is based upon advertising or some other revenue stream, rather than upon paying subscribers. Such VoIP providers might choose to continue providing free service by sending voice over the Internet and not worrying about the quality of the service. For many others, however, the massive revenue generated by *Public-Switched Telephone Network* (PSTN) operators every year is a large carrot. The plethora of competitive local and long-distance carriers attests to the fact that it is possible to successfully compete with the established PSTN operators and at least nibble at that carrot. The opportunity to take a big bite can be even greater if a network can be implemented at a lower cost than traditional circuit switching and if voice and data services can be integrated to offer a more complete service package. VoIP offers this opportunity. The big hurdle is QOS.

This chapter focuses on a number of technical solutions for providing QOS in a VoIP network. We describe the various solutions separately, followed by a description of how they can be combined in order to complement each other. First, however, we must consider what QOS means and why it is such an issue for VoIP networks.

The Need for QOS

QOS is a collective measure of the level of service delivered to a customer. QOS is considered the level of assurance for a particular application that the network can meet its service requirements. From a technical perspective, QOS can be characterized by several performance criteria, such as availability (low down time), throughput, connection setup time, percentage of successful transmissions, speed of fault detection and correction, etc. In an IP network, QOS can be measured in terms of bandwidth, packet loss, delay, and jitter. In order to provide a high QOS, the IP network needs to provide assurances that for a given session or set of sessions, the measurement of these characteristics will fall within certain bounds.

Unfortunately or fortunately (depending on your viewpoint), IP is by nature a best-effort service. IP itself offers no guarantees. For this reason, the *Transmission Control Protocol* (TCP) was developed as a layer above IP in order to ensure error-free, in-sequence delivery of information. Unfortunately, TCP provides this service at the expense of some delay, which is not a big problem for typical IP applications such as e-mail, Web browsing, file transfer, etc. The picture changes significantly, however, when real-time applications such as voice are added to the mix. These applications are time-sensitive to the extent that delays and packet loss can easily make a service unusable. When it comes to voice, delay and jitter are of particular significance. Consequently, TCP cannot be used, which is why UDP is used instead (as described in Chapter 2, "Transporting Voice by Using IP").

Using UDP for transporting voice is fine, provided that there is relatively low packet loss and relatively little congestion on the network. Traffic in an IP network can be bursty and unpredictable, however, which can lead to a situation where one application consumes network resources for a period of time (however brief), which forces packets from other applications to wait and/or to be discarded. Again, TCP can take care of this problem for non-real-time applications, but for other applications that use UDP, the application itself needs to take care of retransmission. For voice, however, neither TCP nor UDP can take care of these issues. TCP causes too much delay, and UDP is inherently unreliable. In the case of significant packet loss or delay, the damage can be such that it might be left to the application to take care of retransmission. In practical terms, a speaker may be forced to repeat what he or she just said—not exactly a desirable situation, and certainly not one that a paying customer will tolerate. Therefore, solutions must be found to provide better assurance of quality.

Why should we be concerned about all of these issues? The short answer is money. The long answer is that high-quality service is imperative if an operator is to attract and retain paying subscribers. If a provider is to charge for voice service, no matter how low the price, then the quality of that voice service must at least match the quality provided by today's circuit-switched network operators. When it comes to quality, however, circuit switching has a distinct advantage. Because a dedicated two-way transmission pipe is established between the parties in a call, and because there is no buffering of packets or packet loss, the quality is excellent. That quality is to be expected, because circuit switching was designed specifically for voice from the outset. The problem with circuit switching is that it is ill-suited to other forms of communication. IP, on the other hand, is well suited to non-real-time communication and can also be made suitable for real-time communication if solutions can be found for the QOS issue. Fortunately, solutions do exist, and others are being developed. In fact, QOS on IP networks is perhaps the area of greatest focus among those who are developing new standards in the IP world.

The technical solutions involve resource-reservation techniques, giving priority to interactive communications such as voice, QOS-based routing, and other mechanisms. The primary thrust of this chapter is to describe these technical solutions.

End-to-End QOS

Before delving into the technical solutions for QOS, we should point out that QOS is not just a technical matter on one operator's network. In general, a customer does not know or care about how a provider gets a call from one place to another, provided that the call gets there and that it meets the customer's quality expectations. In other words, QOS must be end-to-end and must have the support of all networks in the chain. A call might originate on one provider's network and terminate on another provider's network, perhaps traversing other networks in the meantime. Each of the networks must cooperate to ensure that the quality provided matches what is expected. This requirement raises issues of *Service-Level Agreements* (SLAs) between different operators. SLAs are agreements in which operators make commitments to each other regarding the type and quality of service to be offered and the penalties involved if such commitments are not met. We will discuss SLAs again later in this chapter.

Meanwhile, however, we must again remember that VoIP and Voice over the Internet are not necessarily the same. We must recognize that the Internet is a huge collection of different networks and is not managed by any one entity. For those parts that are actively managed, they are not necessarily managed in the same way or according to the same set of criteria. While SLAs might be possible between certain VoIP carriers, too many networks are involved with the Internet to place a call from one place to the other over the Internet and to always experience superior quality (at least, not in the near future). Many of the new operators readily agree that a particular QOS can be provided in managed networks, but they shake their heads at the suggestion that QOS is a realistic expectation for Voice over the Internet. For this reason, this book makes a significant distinction between VoIP and Voice over the Internet.

We do not mean that QOS can only be achieved when a call stays within a single network. Rather, SLAs at least offer the potential of ensuring quality across multiple networks. Until all *Internet Service Providers* (ISPs) own up to the same quality commitments and implement equivalent technical solutions, however, then anywhere-to-anywhere Voice over the Internet is likely to remain a best-effort, low-quality service.

Things will get better over time. To begin, a given VoIP provider might just provide long-distance service in one country, with connections to the circuit-switched PSTN at each end. A call might originate on the PSTN and terminate on the PSTN. Eventually, other operators come into play, and the different VoIP operators will connect to each other so that a given call can begin on IP and terminate on the PSTN. As time goes on, more and more VoIP providers will enter the scene so that calls can be carried over IP from source to destination. Many of these carriers will be the ISPs that currently provide us with Web hosting, e-mail, and other services. Provided that all of the providers in the chain embrace the same quality objectives and implement similar technical solutions, then they can provide the required quality. A significant amount of time is necessary, however, before that situation translates to excellent and reliable voice quality over the Internet—because the Internet will probably always involve networks that do not provide QOS guarantees.

It's Not Just the Network

Not only is QOS not confined to a single operator's network, but it is also not confined to network issues in general. In other words, a quality service means a lot more than just good voice quality. In the competitive environment of telecommunications, low prices and advanced features might attract customers, but the best way to retain customers is to provide a consistently high service across all facets of the business. Given that a potentially high cost may be associated with acquiring a customer in the first place, the last thing that a provider wants is to see a customer leave and go to a competitor. Therefore, QOS is provided not only by technical solutions in the network, but it is also provided through superior customer service, rapid service provisioning, 100 percent accurate billing, clear and concise product descriptions, etc.

Not long ago, for example, there was much complaining about several e-commerce companies that failed to adequately staff their customer-service operations. The ease and convenience of purchasing products online attracted the customers. When they needed help in executing those purchases, however, there was often no customer-service personnel available to provide that help. This situation resulted in many frustrated customers who were initially attracted by a new service but were left upset and disappointed. A quality service must extend to more than just the technology of the network, and serious consequences may result if it does not. While this book is a technical book that focuses on network solutions, we must remember that those network solutions can provide only one piece of the puzzle when it comes to overall service quality.

Overview of QOS Solutions

A number of solutions have been developed or are being developed to address the QOS issue. These solutions approach the problem from various angles. One approach is to ensure that resources for a given session are available and reserved for that session prior to establishing the session. Conceptually, this solution has certain similarities to circuit switching, where the bandwidth needed for a call is reserved for that call before the called party's phone rings.

Another approach is to categorize traffic into different classes or priorities, with higher-priority values assigned to real-time applications (such as voice) and lower-priority values assigned to non-real-time traffic (such as e-mail). In some ways, this approach is easier to implement, but it does require that no one application can totally shut out another. Just because voice exists on a network that supports voice and data applications does not mean that the data never obtains a reasonable share of resources. Therefore, a need exists to implement fair resource-allocation techniques.

Finally, perhaps the easiest technique to help solve the QOS problem is the provision of more bandwidth. After all, a lack of bandwidth is often the reason for packet delay or loss. Bandwidth costs money, however, and needs to be utilized efficiently.

More Bandwidth

Allocating more bandwidth might sound like a simplistic and expensive approach to QOS—simple because it does not require major system development, but expensive because it means significant overbuild. This overbuild would need to exist so that network resources would be available in times of traffic bursts. Unfortunately, this overbuild would remain unused for most of the time. That method would be an inefficient way of solving the QOS problem, but it does have some merit and should not be dismissed completely.

We live in an age where huge advances continue to be made in squeezing bandwidth from facilities that appeared to have reached their limit. Not long ago, a 9600-baud modem was considered almost a utopian situation for a modem connection. Now, 56kbps modems are standard for dial-up access, and DSL technology means that homes can have several megabits per second in each direction.[1] Such technologies deal with just the access part of the network. In the core of the network, we now have

Dense Wave-Division Multiplexing (DWDM),which offers the capability to support tens of gigabits per second today and hundreds of gigabits per second in the future on a single optical fiber. Consequently, as these new transmission techniques have become available, the cost of bandwidth has been decreasing and will continue to decrease.

On the other hand, however, Moore's Law says that computing power doubles roughly every 18 months, which means that applications are demanding bandwidth at an ever-increasing rate. In addition, the number of Internet users continues to increase. The result is that the demand for bandwidth is increasing rapidly. Historically, bandwidth availability and bandwidth demand have tended to move almost in lock-step. This relationship simply involves supply and demand. If bandwidth is scarce, then applications that require high bandwidth are either not developed because they would be impossible or too expensive to use or are deployed only on a limited scope. Once more bandwidth becomes available, however, then a major barrier to the deployment of such applications is removed, and application developers can proceed with implementing new ideas. They can do so without worrying about a bandwidth demand that might have been unconscionable a short time earlier.

Therefore, additional bandwidth is certainly needed, if only to support additional traffic as demand continues to grow. While additional bandwidth is a necessity for a network that is required to support voice in addition to the data traffic that it might have carried all along, additional bandwidth is not, by itself, a complete solution to the QOS issue. Mechanisms need to exist that manage the available bandwidth so that it is used most effectively to support the services provided.

QOS Protocols and Architectures

The following description is a high-level overview of the QOS solutions developed within the *Internet Engineering Task Force* (IETF). We provide details about these solutions in subsequent sections of this chapter.

---

[1]Many variants of DSL exist. The most common is Asymmetric DSL, which can theoretically provide up to 640kbps uplink and up to 9Mbps downlink. Whether a given customer experiences these rates is a different issue, with commercial as well as technical considerations.

Resource Reservation RFC 2205, the *Resource-Reservation Protocol* (RSVP), specifies resource-reservation techniques for IP networks. RSVP is part of the IETF integrated-services suite and is a protocol that enables resources to be reserved for a given session (or sessions) prior to any attempt to exchange media between the participants. Of the solutions available, RSVP is the most complex but is also the solution that comes closest to circuit emulation within the IP network. RSVP provides strong QOS guarantees, significant granularity of resource allocation, and significant feedback to applications and users.

RSVP currently offers two levels of service. The first level is Guaranteed, which comes as close as possible to circuit emulation. The second level is controlled load, which is equivalent to the service that would be provided in a best-effort network under no-load conditions.

Basically, RSVP works as depicted in Figure 8-1. A sender first issues a PATH message to the far end via a number of routers. The PATH message contains a *traffic specification* (TSpec) that provides details about the data that the sender expects to send, in terms of bandwidth requirement and packet size. Each RSVP-enabled router along the way establishes a path state that includes the previous source address of the PATH message (i.e., the next hop back towards the sender). The receiver of the PATH message responds with a *Reservation Request* (RESV) that includes a flowspec. The flowspec includes a Tspec and information about the type of reservation service requested, such as Controlled-load service or Guaranteed service.

The RESV message travels back to the sender along the same route that the PATH message took (but in reverse). At each router, the requested resources are allocated, assuming that they are available and

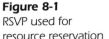

Figure 8-1
RSVP used for
resource reservation

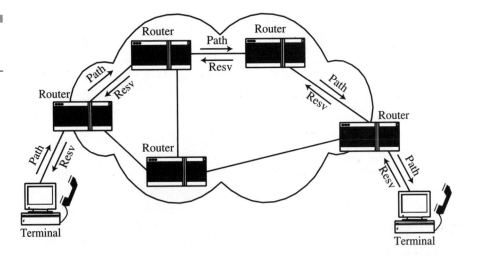

that the receiver has the authority to make the request. Finally, the RESV message reaches the sender with a confirmation that resources have been reserved.

One interesting point about RSVP is that the receiver, not the sender of the data, makes the reservations. This task is performed in order to accommodate multicast transport, where large numbers of receivers and only one sender might exist.

NOTE: *RSVP is a control protocol that does not carry user data. The user data (i.e., voice) is transported later by using RTP. This transport occurs only after the reservation procedures are performed. The reservations that RSVP makes are soft, which means that they need to be refreshed on a regular basis by the receiver(s).*

Differentiated Service (DiffServ) *Differentiated Service* (DiffServ) is a relatively simple means for prioritizing different types of traffic. RFC 2475, "An Architecture for Differentiated Services," describes the DiffServ protocol. Basically, DiffServ makes use of the IP version 4 *Type of Service* (TOS) field and the equivalent IP version 6 Traffic Class field. The portion of the TOS/Traffic Class field that DiffServ uses is known as the DS field. The field is used in specific ways to mark a given stream as requiring a particular type of forwarding. The type of forwarding to be applied is known as *Per-Hop Behavior* (PHB), of which DiffServ defines two types: *Expedited Forwarding* (EF) and *Assured Forwarding* (AF).

RFC 2598 specifies EF EF is a service in which a given traffic stream is assigned a minimum departure rate from a given node—one that is greater than the arrival rate at the same node—provided that the arrival rate does not exceed a pre-agreed maximum. This process ensures that queuing delays are removed. Because queuing delays are a major cause of end-to-end delay and are the main cause of jitter, this process ensures that delay and jitter are minimized. In fact, EF can provide a service that is equivalent to a virtual leased line.

RFC 2597 defines AF AF is a service in which packets from a given source are forwarded with a high probability, provided that the traffic from that source does not exceed some pre-agreed maximum. AF defines four classes, with each class allocated a certain amount of resources (buffer space and bandwidth) within a router. Within each class, a given packet can have one of three drop rates. At a given router, if there is congestion within the resources allocated to a given AF class, then the packets with the highest drop-rate values will be discarded first so that packets with a lower drop-rate value will receive some protection. In order to work well, the incoming traffic must not have packets with a high percentage of low drop rates. After all, the purpose is to ensure that the highest-priority packets get through in the case of congestion, and that cannot happen if all of the packets have the highest priority.

Label Switching   Label switching has gained significant interest in the Internet community, and significant effort has been placed behind the definition of a protocol called *Multi-Protocol Label Switching* (MPLS). In some ways, MPLS is similar to DiffServ in that it marks traffic at the entrance to the network. The primary function of the marking is not to allocate priority within a router, however, but to determine the next router in the path from source to destination.

MPLS involves the attachment of a short label to a packet in front of the IP header. This procedure is effectively similar to inserting a new layer between the IP layer and the underlying link layer of the *Open Systems Interconnection* (OSI) model. The label contains all of the information that a router needs to forward a packet. The value of a label can be used to look up the next hop in the path and forward to the next router. The difference between this routing and standard IP routing is that the match is exact and is not a case of looking for the longest match (the match with the longest subnet mask). This process enables faster routing decisions within routers.

The label identifies a *Forwarding Equivalence Class* (FEC), which means that all packets of a given FEC are treated equally for the purposes of forwarding. All packets in a given stream of data, such as a voice call, will have the same FEC and will receive the same forwarding treatment. Therefore, we can ensure that the forwarding treatment applied to a given stream can be set up such that all packets from A to B follow exactly the same path. If that stream has a particular bandwidth requirement, then that bandwidth can be allocated at the start of the session. This feature can ensure that a given stream has the bandwidth that it needs and that the packets that make up the stream will arrive in the same order as transmitted. Hence, this function provides a higher QOS.

In many ways, MPLS is as much of a traffic-engineering protocol as it is a QOS protocol. MPLS is somewhat analogous to the establishment of virtual circuits in *Asynchronous Transfer Mode* (ATM) and can lead to similar QOS benefits. MPLS helps provide QOS by helping to better manage traffic. Whether it should be called a traffic-engineering protocol or a QOS protocol hardly matters if the end result is better QOS.

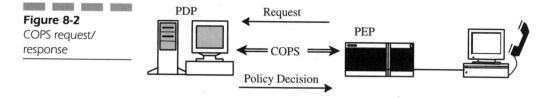

Figure 8-2
COPS request/
response

QOS Policies

Given that mechanisms exist to apply a higher QOS to certain streams compared to others, the issue arises as to which QOS levels should be applied to which types of traffic. This dilemma raises the issue of QOS policies. While protocols such as RSVP, DiffServ, and MPLS provide the mechanisms to differentiate traffic, QOS policies specify how those mechanisms are used.

One important part of QOS is the fact that certain users will have better service than others, and they might have to pay for that service. This situation creates an incentive for others to steal better service without paying for it. Therefore, a need exists for authentication functions to identify users and to ensure that a given user is who he or she claims to be. Furthermore, a given user might be entitled to a given level of QOS under certain circumstances but not under other circumstances. Therefore, certain rules must be established to specify which circumstances lead to which actions, and methods must exist for ensuring that those rules are applied. This function can be considered a type of policing function. Appropriately, the IETF developed a protocol known as the *Common Open-Policy Service Protocol* (COPS), which RFC 2748 defines. This protocol is a client-server protocol, as depicted in Figure 8-2. A *Policy Enforcement Point* (PEP), such as a router that needs to enforce certain rules, queries a *Policy Decision Point* (PDP) that makes the actual policy decisions. The PEP can be considered the policeman, and the PDP can be considered the judge. Given these analogies, the acronym COPS is particularly fitting.

The Resource-Reservation Protocol (RSVP)

The previous sections of this chapter gave a brief overview of QOS issues and solutions. Next, we delve deeper into the technical solutions and start with RSVP.

In order to use RSVP, we need to implement a number of functions in routers and hosts. These functions appear in Figure 8-3. Prior to data transfer, when a QOS reservation is being established, the admission control function determines whether sufficient resources exist to satisfy the requested QOS, and the policy control function determines whether the user has the authority to make the requested reservation. The policy control function could, for example, use COPS to communicate with a PDP to make the policy determination. During data transfer, the packet classifier determines the QOS to be applied to a given packet, and on the outgoing interface, the packet scheduler or other link-layer function determines when a given packet is to be forwarded. Together, the packet classifier, packet scheduler, and admission control are known as traffic control.

RSVP Syntax

Before describing the messages and functions of RSVP in greater detail, we must first mention a few points about the syntax of RSVP messages. The syntax of the information contained in RSVP messages is specified in RFC 2215, "General Characterization Parameters for Integrated Service Network Elements." Basically, RSVP is used to implement a number of services, and particular service numbers identify these services. For example, the Guaranteed Service has the service number 2, and the Controlled-load service has the service number 5. A number of parameters are associated with each service, and each parameter has a parameter number.

Both services and parameters are specified in messages with the format *Type-Length-Value* (TLV). Thus, a type field identifies each, followed by a length field, followed by the data itself.

Figure 8-3
RSVP within hosts
and routers

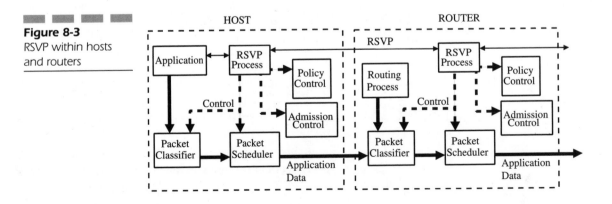

In some cases, data needs to be sent that is not associated with any particular service or that is common to a number of different services. For this purpose, service number 1 is defined. If a message is to be sent that contains both general information followed by service-specific information, then the message will specify service number 1 (containing the general information parameters), followed by service information (e.g., for service number 2) that contains service-specific information.

Establishing Reservations

As we previously described, RSVP reserves resources along a path from the receiver of user data back to the sender of that data. The reason is because RSVP is designed with multicast in mind. With multicast, there might be a limited number of senders and multiple receivers. Figure 8-4 depicts a scenario in which there is only one sender and many receivers. From sender to receiver, the data path branches. In the reverse direction, the path merges. Let's assume that Receiver 1 and Receiver 2 have different QOS objectives. In that case, the resources needed between Router 2 and Router 3 do not need to be the sum of the requirements of Receivers 1 and 2, just the maximum. Equally, the resources required between Routers 1 and 2 do not need to be the sum of the requirements of all four receivers, just the maximum.

RSVP deals with unidirectional data transfer only. In a typical two-party VoIP conversation, resource reservation must be carried out in both directions, assuming that both parties intend to speak.

The data sender sends a PATH message to the intended receiver of the data and includes in that message a traffic specification, TSpec, which describes the characteristics of the data to be sent. The receiver responds with an RESV message, which contains a flowspec indicating the type of resource reservation required. In the case of multiple receivers, the flowspec might be modified as it is passed back towards the sender(s) as a result of different receivers having different QOS requirements. This process is called merging flowspecs.

TSpec   A TSpec is sent from a sender in a PATH message and is also included in the information sent from a receiver to a sender in a RESV message. The Tspec takes the form of a token bucket specification plus a peak rate (p), maximum packet size (M), and minimum policed unit (m). The token bucket specification defines a bucket size (b) and a token rate (r), as depicted in Figure 8-5. This conceptual bucket has a capacity of b bytes and is filled with tokens at a rate of r bytes per second. Upon arrival of a packet,

Figure 8-4
RSVP used with
multicast

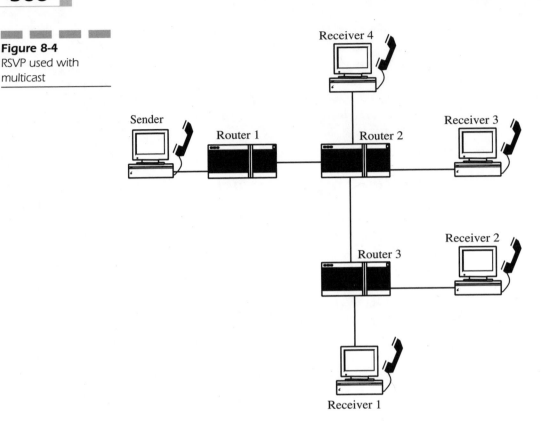

the packet will be transmitted onward only if the number of tokens in the bucket is at least as large as the packet. Otherwise, the packet might be dropped or queued until the bucket accumulates enough tokens. If the incoming flow is less than the token bucket rate r, then the bucket will eventually fill up—at which point the generation of tokens is suspended until a packet arrives to consume some tokens. Provided that the average rate flowing toward the bucket is less than or equal to r, then the token bucket should not constrain the flow. Furthermore, the rate of flow toward the bucket might exceed r for a period of time but cannot do so indefinitely without running out of tokens and being subject to delay or packet loss as a result.

NOTE: *A token bucket is not a physical container of packets but a numerical counter of tokens to schedule the releasing time of packets.*

Figure 8-5
Token bucket

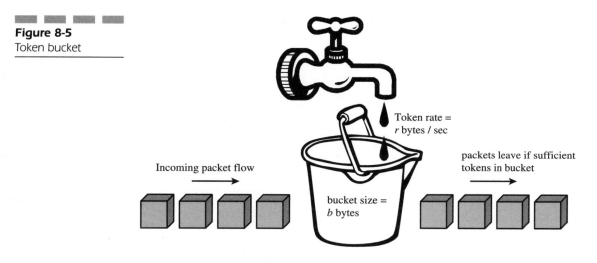

The peak rate (p), which should be greater than r, is the maximum data rate that the sender will use. The sender should not use that rate for any extended period of time, because to do so would be incompatible with the token bucket specification. The maximum packet size (M) is the maximum IP data packet that the sender will send. If any packets exceed that size, then they will be considered outside the limits of the traffic specification and will not be treated with the same QOS as packets within the specification. The minimum policed unit (m) is measured in bytes. All packets less than m bytes in length are considered to be m bytes long for two reasons. The first is to enable nodes in the network to estimate the amount of resources that will be needed to process each packet. The second is to bound the bandwidth overhead consumed by link-level headers. The maximum bandwidth overhead of link-layer headers will be the ratio of link-layer header size to m. Putting all of these components together, Figure 8-6 shows the format of a sender TSpec. The Token Bucket TSpec is defined in RFC 2215 as having parameter number 127.

Flowspec The flowspec, sent in an RESV message from a receiver, contains a TSpec of its own plus an indication of the QOS control service being requested. Currently, you can request two services: the Guaranteed service and the Controlled-load service. The flowspec for Controlled-load service appears in Figure 8-7, and Figure 8-8 shows the flowspec for guaranteed service.

The flowspec for Controlled-load service is simply a TSpec. The flowspec for Guaranteed service is a TSpec plus a Rate (R) term and a Slack (S) term. The R term indicates the bandwidth required and should be equal to or greater than the token rate (r). In the case that the rate is greater, then

Figure 8-6
Sender TSpec

3 1	3 0	2 9	2 8	2 7	2 6	2 5	2 4	2 3	2 2	2 1	2 0	1 9	1 8	1 7	1 6	1 5	1 4	1 3	1 2	1 1	1 0	0 9	0 8	0 7	0 6	0 5	0 4	0 3	0 2	0 1	0 0
V = 0				reserved												Overall length = 7 (excluding header)															
Service header = 1								0	reserved							Length of service 1 data = 6															
127												Flags (0)				Parameter length = 5 (excluding header)															
Token Bucket Rate (r) (32-bit IEEE floating point number)																															
Token Bucket Size (b) (32-bit IEEE floating point number)																															
Peak Data Rate (p) (32-bit IEEE floating point number)																															
Minimum Policed Unit (m) (32-bit integer)																															
Maximum Packet Size (M) (32-bit integer)																															

V = message format version

Figure 8-7
Flowspec for
Controlled-load
service

3 1	3 0	2 9	2 8	2 7	2 6	2 5	2 4	2 3	2 2	2 1	2 0	1 9	1 8	1 7	1 6	1 5	1 4	1 3	1 2	1 1	1 0	0 9	0 8	0 7	0 6	0 5	0 4	0 3	0 2	0 1	0 0
V = 0				reserved												Overall length = 7 (excluding header)															
Service header = 5								0	reserved							Length of service 5 data = 6															
127												Flags (0)				Parameter length = 5 (excluding header)															
Token Bucket Rate (r) (32-bit IEEE floating point number)																															
Token Bucket Size (b) (32-bit IEEE floating point number)																															
Peak Data Rate (p) (32-bit IEEE floating point number)																															
Minimum Policed Unit (m) (32-bit integer)																															
Maximum Packet Size (M) (32-bit integer)																															

V = message format version

extra bandwidth is being requested, which will reduce queuing delays. The Slack term is a number of microseconds and indicates the difference between the desired delay and the delay that would be achieved if rate R were used. A network element can use the slack to reduce the amount of resources reserved for this flow if necessary.

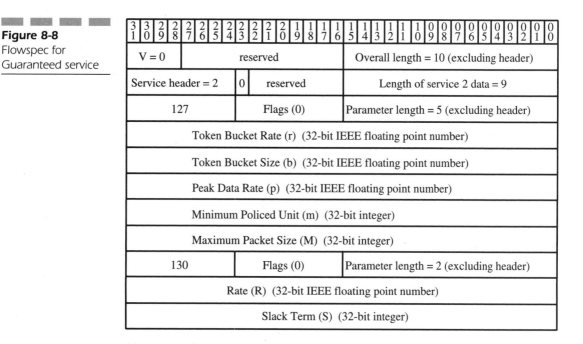

Figure 8-8
Flowspec for
Guaranteed service

| 3 3 2 2 2 2 2 2 2 2 2 2 1 1 1 1 1 1 1 1 1 1 0 0 0 0 0 0 0 0 0 0 |
| 1 0 9 8 7 6 5 4 3 2 1 0 9 8 7 6 5 4 3 2 1 0 9 8 7 6 5 4 3 2 1 0 |

V = 0	reserved	Overall length = 10 (excluding header)
Service header = 2	0 reserved	Length of service 2 data = 9
127	Flags (0)	Parameter length = 5 (excluding header)
Token Bucket Rate (r) (32-bit IEEE floating point number)		
Token Bucket Size (b) (32-bit IEEE floating point number)		
Peak Data Rate (p) (32-bit IEEE floating point number)		
Minimum Policed Unit (m) (32-bit integer)		
Maximum Packet Size (M) (32-bit integer)		
130	Flags (0)	Parameter length = 2 (excluding header)
Rate (R) (32-bit IEEE floating point number)		
Slack Term (S) (32-bit integer)		

V = message format version

Filter Spec Only a destination IP address and protocol ID identify an RSVP session. The session identity might optionally contain a destination port number. Thus, an RSVP session does not include any information about the sender of data. Given that the destination port is optional, however, and that the session is not identified by any sender-related information, there might be a problem with determining the specific data flow for which a given reservation is requested. For example, a given receiver might be receiving two streams of data from two different senders. We must indicate what QOS is being requested for the flow from each sender, but the session itself does not include sender information. At this point, the filter spec comes into play.

A filter spec defines the set of data packets (i.e., the flow) to which a particular QOS is to be applied, and as such, it is an input to the packet-classifier function. Its format depends on whether IP version 4 or IP version 6 is being used. At a minimum, the format contains a sender IP address and (optionally) a sender port number. The reason for the IP address is to identify the sender of the data to which the requested QOS is to be applied. The reason for the source port number is for situations such as a video conference, where there might be both a video and audio stream from a given sender to a given receiver with a different QOS requirement for each stream.

Given that the filter spec might contain a port number, each router in the path must have the capability to examine the details of the header to determine the port number, which raises a number of considerations. First, IP datagrams in the flow must not be fragmented, which in turn means that we must use path MTU discovery. Second, IP security might encrypt the header, which means that RSVP must include security functions in order to decrypt the header. Third, the IP version 6 header is of variable length, which might cause a significantly greater processing effort at the routers.

The filter spec is included in PATH messages from sender to receiver and in RESV messages from receiver to sender. When sent in a PATH message, this spec is known as a Sender template.

ADSpec   As previously described, a sender first sends a PATH message downstream towards one or more receivers. The TSpec in the PATH message indicates the type of data to be sent and is passed from sender to receiver without modification. The receivers respond with an RESV message to indicate the requested QOS, and the RESV is propagated upstream towards the sender. Having a receiver request a QOS that the intervening routers cannot meet is useless, however. For example, the RESV message might specify a bandwidth requirement that exceeds the bandwidth of one of the links in the path from sender to receiver. Therefore, a need exists for a receiver to be informed about the network between the sender and receiver, so that the receiver does not request something that the network cannot provide. To meet this need, RSVP includes the capability for the sender and for the routers to indicate their QOS capabilities within a PATH message from sender to receiver. This procedure is done through the use of an ADSpec, where the AD indicates advertising.

A sender constructs an initial ADSpec, indicating what it can support, and it includes this spec in the PATH message. At each router along the path from sender to receiver, the ADSpec is updated and passed on. By the time the PATH message gets to the receiver, the ADSpec provides a pretty good indication of what the receiver can reasonably request in terms of QOS from the routers and from the sender.

One interesting point about the ADSpec is that it can also indicate that one or more routers in the path from sender to receiver do not support RSVP. This scenario is perfectly possible. In the event that an RSVP-capable router sends a PATH message on a link to a router that does not support RSVP—or if an RSVP-capable router receives a PATH message on a link from a router that does not support RSVP—then the RSVP-capable router sets a bit in the ADSpec to indicate that RSVP is not supported end-to-end.

The general format of the ADSpec is depicted in Figure 8-9. This format includes general information (service number 1), which is optionally fol-

Figure 8-9

Format of ADSpec

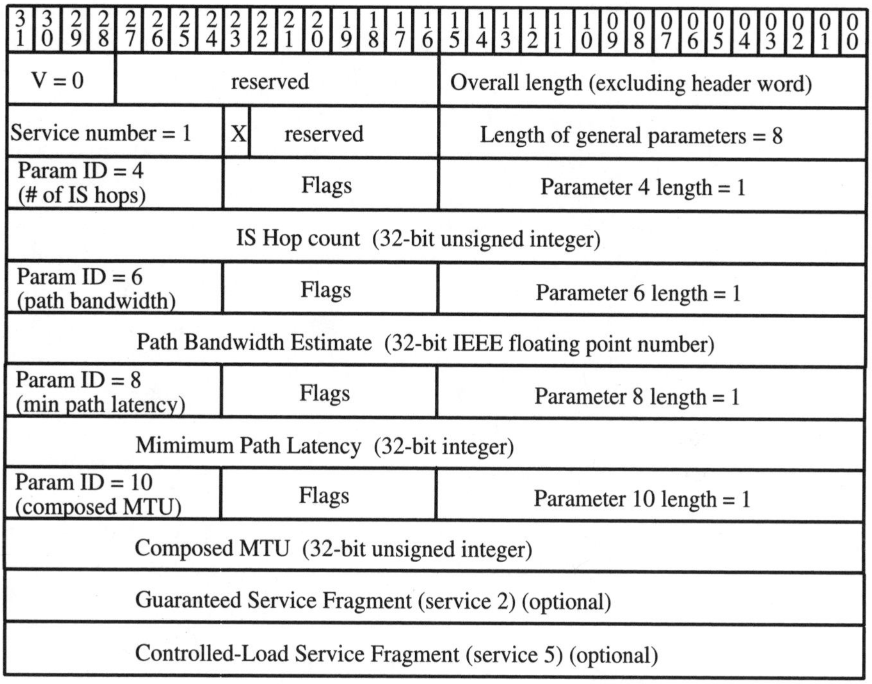

| 3 3 2 2 2 2 2 2 2 2 2 2 1 1 1 1 1 1 1 1 1 1 0 0 0 0 0 0 0 0 0 0 |
| 1 0 9 8 7 6 5 4 3 2 1 0 9 8 7 6 5 4 3 2 1 0 9 8 7 6 5 4 3 2 1 0 |

V = 0	reserved	Overall length (excluding header word)
Service number = 1	X reserved	Length of general parameters = 8
Param ID = 4 (# of IS hops)	Flags	Parameter 4 length = 1
IS Hop count (32-bit unsigned integer)		
Param ID = 6 (path bandwidth)	Flags	Parameter 6 length = 1
Path Bandwidth Estimate (32-bit IEEE floating point number)		
Param ID = 8 (min path latency)	Flags	Parameter 8 length = 1
Mimimum Path Latency (32-bit integer)		
Param ID = 10 (composed MTU)	Flags	Parameter 10 length = 1
Composed MTU (32-bit unsigned integer)		
Guaranteed Service Fragment (service 2) (optional)		
Controlled-Load Service Fragment (service 5) (optional)		

V = message format version

lowed by either or both of Guaranteed service information and Controlled-load service information.

The general information specifies whether a non-IS hop is involved in the path (the bit marked X in Figure 8-9). Non-IS means not Integrated Service capable—which, for the purposes of RSVP, means lacking RSVP support. If this bit is set, then the information in the rest of the ADSpec is no longer relevant, because QOS has to be end-to-end—and there is an obvious break in the QOS chain from sender to receiver.

The ADSpec also indicates the number of hops between IS-capable nodes, the path bandwidth, the path MTU, and the minimum path latency. The minimum path latency is the delay that would be experienced along the path if there were no additional queuing delays.

Following the general information in the ADSpec, there might be a Guaranteed service data fragment and/or a Controlled-load data fragment containing information related to those two services. These data fragments are inserted if the application might use one of these services for the session in question.

RSVP Messages We already discussed a number of RSVP messages, such as PATH and RESV, and we also looked at some of the information that these messages carry. We now look at the format of these messages and others in a little more detail.

Every RSVP message contains a common header, as shown in Figure 8-10. The various messages are Path (1), Resv (2), PathErr (3), ResvErr (4), PathTear (5), ResvTear (6), and ResvConf (7). The checksum is used to ensure that the message has been received without error. The `Send_TTL` is the IP TTL value of the message. With normal IP forwarding, RSVP can use this field to determine that a non-RSVP hop has been involved by comparing the value in the RSVP header with the IP TTL value, because a non-RSVP node will update the IP TTL value but not the RSVP `Send_TTL`.

Following the common header, the message contains a number of objects, such as a sender TSpec, an ADSpec, etc. Each object has a header and contents, as shown in Figure 8-11. The Class-num and C-Type identify the object. The Class-num identifies the object itself and the C-Type allows for different version of the object to exist. For example, the format of a filter spec varies depending on whether IPv4 or IPv6 is being used. In that case the Class-num indicates that object is a filter spec and the C-Type indicates whether it is the IPv4 or IPv6 variant.

The following list describes examples of some of the object types defined in RSVP. The list is not exhaustive, and the reader should consult RFC 2205 for a complete listing and associated definitions.

- *SESSION Class* Has a Class-num value of 1 and has a C-Type value of 1 for IP version 4 and a C-type value of 2 for IP version 6. This class includes the IP destination address, the protocol ID for the type of data being sent, and (optionally) the destination port.

- *FLOWSPEC Class* Has a Class-num of 9 and a C-type value of 2. We described the flowspec previously in this chapter.

- *SENDER_TEMPLATE Class* Has a Class-num value of 11. For IP version 4, the C-type value is 1, and for IP version 6, the C-type value is 2. The sender template is sent in a Path message and is, in fact, a filter spec that the sender sends.

Figure 8-10

RSVP common header

Byte 0		Byte 1	Byte 2	Byte 3
Vers = 1	Flags (0)	Message type	RSVP Checksum	
Send_TTL		Reserved	RSVP Length	

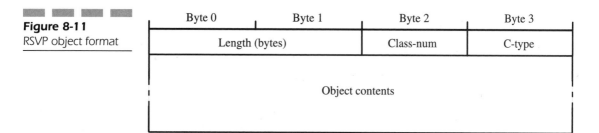

Figure 8-11
RSVP object format

Byte 0	Byte 1	Byte 2	Byte 3
Length (bytes)		Class-num	C-type
Object contents			

- *RSVP_HOP Class* Contains the IP address of the interface through which the last RSVP-capable node passed this message. This class is used in the Path message and is saved at each node along the route. The purpose is to ensure that an RESV message from receiver to sender follows the same path back through the network that the PATH message took. This class has the Class-num value of 3, a C-type value of 1 for IP version 4, and a C-type value of 2 for IP version 6.

- *TIME_VALUES Class* A timeout period in milliseconds that indicates how long the message is to be considered valid. This class has a Class-num value of 5.

- *ERROR_SPEC Class* An object that is included in RSVP error messages. This class includes the IP address of the node at which the error was detected, plus an error code identifying the type of error, plus some additional cause information. This class has the Class-num value of 6. The C-type value is 1 for IP version 4 and 2 for IP version 6.

- *STYLE Class* With this class, we can select different reservation styles in the case of multiple receivers and/or multiple senders. A Fixed-filter style means that a receiver uniquely identifies the individual sender and the corresponding data stream from that sender to which a given reservation is to apply. A Wildcard-Filter style indicates that the reservation request applies to all data streams from all senders in a given session. The Shared-Explicit style lists specific senders, indicating that a given request applies to data streams from those senders. The applicable Class-num value for the STYLE class is 8, and the C-type value is 1.

Example Reservations Figures 8-12 and 8-13 show two examples of successful reservations. Figure 8-12 shows a simple case where there is one sender and one receiver. In this case, a PATH message is propagated through the network from sender to receiver. The network does not modify the sender template and TSpec, but the ADSpec can be modified.

The receiver responds with a RESV message, which might optionally contain a request for the reservation to be confirmed. The message indicates the style of reservation to be applied and a list of flow descriptors. These are one or more flowspecs and or filter specs, depending on the reservation style in question.

In Figure 8-12, the receiver has requested confirmation of the success of the reservation. Consequently, a ResvConf message is returned. The

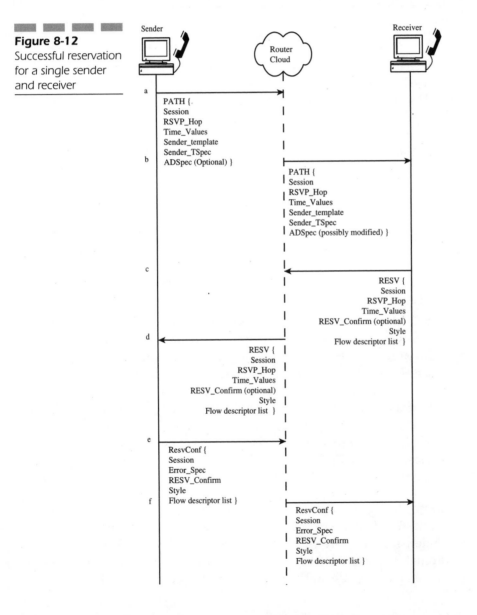

Figure 8-12

Successful reservation for a single sender and receiver

Error_Spec contains an error code of zero in the case of a successful reservation. The Resv_Confirm object is copied from the RESV message. The Style and flow descriptor indicate the particular reservations that are being confirmed. The reservations might not be identical to the information in the RESV message, because in the case of multiple senders, the reservation might not be successful for data streams from all senders.

In the multicast example in Figure 8-13, Receiver 1 sends an RESV message just before Receiver 2. In this example, the QOS requirement of Receiver 1 is stronger than Receiver 2. In that case, no need exists to modify the reservation that is already sent back to the sender, because

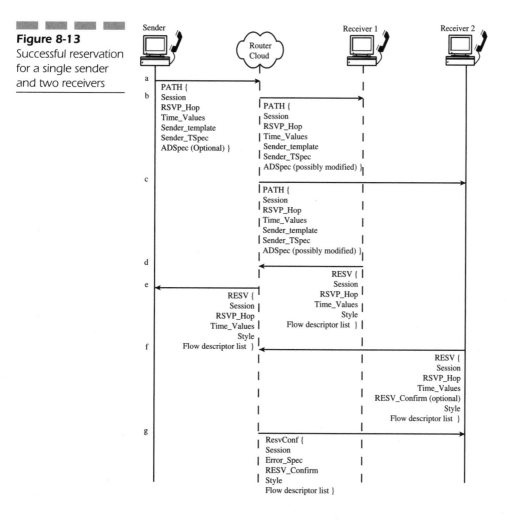

Figure 8-13

Successful reservation for a single sender and two receivers the reservation from Receiver 1 is good enough to support the requirements of both receivers. In the example, Receiver 2 requested a confirmation, but Receiver 1 did not. Note that the confirmation is sent from the router at which the reservations merge. This procedure is used in cases where a router receives a RESV message with a request for confirmation, and a reservation is already in place at the router that meets or exceeds the requirements of the reservation to be confirmed. In certain cases, this situation could lead to a false confirmation, particularly in the case where the existing reservation request fails at some point closer to the sender. Therefore, a reservation confirmation should only be considered as an indication of high probability that a reservation has succeeded.

Reservation Errors

Of course, a given resource reservation can fail—in which case an error message is returned. Two error messages exist: PathErr and ResvErr. PathErr messages are simply passed back to the sender whose PATH message led to the error.

ResvErr messages are sent to a receiver when a given reservation request fails. With merged flowspecs, a reservation request that fails for a given reason must not deny service for a different request that would have succeeded on its own. For example, imagine two receivers that both send reservation requests, and imagine that those requests meet at a single router that resides in the path from the sender to the receivers. If one receiver's request fails at the router due to an excessive bandwidth request, the other receiver's request should not fail if it had requested a smaller bandwidth that could be supported.

Guaranteed Service

RFC 2212 defines Guaranteed service. This service involves two elements: first, ensuring that there will not be any packet loss; and second, ensuring minimal delay. Ensuring that there will not be any packet loss is a function of the token bucket depth (b) and the token rate (r) specified in the TSpec. At a given router, provided that buffer space of value b is allocated to a given flow and that a bandwidth of r or greater is assigned, then we should experience no loss.

Delay is a function of two components. The first is a fixed delay due to processing within the individual nodes and is only a function of the path taken. The second component of delay is the queuing delay within the various nodes. Let's assume that a given node supports a token bucket model (b, r) and offers a given stream a bandwidth of R for a particular flow, where R is greater than or equal to r. Then, the queuing delay is less than $b / R + C / R + D$, where C and D are maximum deviations from an ideal fluid model. C is a rate-dependent error term and represents the delay that a flow might experience as a function of the flow rate. For example, the term could represent the delay involved in breaking up a datagram into the payload of ATM cells. D is a fixed-delay term within the node itself and is not dependent on the rate. C is measured in bytes, and D is measured in microseconds.

Recall that the ADSpec contains a Guaranteed Service object. This object has the form shown in Figure 8-14. One can see that the ADSpec includes cumulative values for C and D. In addition, there are values of C and D since the last traffic-reshaping point in the path. Traffic reshaping involves buffering at a node in the network to correct deviations from the original TSpec in the received data. This buffering is set up so that the non-conforming data can be brought back into conformance. Note that such non-conformance might not necessarily be the fault of the source application. This problem could also be the fault of queuing effects in routers.

When a receiver receives an ADSpec with the Guaranteed service data fragment, it can make a decision as to the value of R that should be requested in the flowspec within the RESV message. Because the path that the data will take fixes the Dtot, the receiver can do nothing to minimize the impact of the term. The delay impact of Ctot reduces as R increases, however. Therefore, the receiver can request a value of R such that the total delay will always be within acceptable limits.

Controlled-Load Service

Controlled-load service is a close approximation to the QOS that an application would receive if the data were being transmitted over a network that was lightly loaded. In other words, a high percentage of packets will be delivered successfully, and the delay experienced by a high percentage of packets will not exceed the minimum delay experienced by any successfully delivered packet. RFC 2211 describes Controlled-load service.

Figure 8-14

Guaranteed service
ADSpec data
fragment

| 3 3 2 2 2 2 2 2 2 2 2 2 1 1 1 1 1 1 1 1 1 1 0 0 0 0 0 0 0 0 0 0 |
| 1 0 9 8 7 6 5 4 3 2 1 0 9 8 7 6 5 4 3 2 1 0 9 8 7 6 5 4 3 2 1 0 |

Service number = 2	X	reserved	Length of service 2 data
Param ID = 133 (composed Ctot)		Flags	Parameter 133 length = 1
End-to-end composed value for C (Ctot) (32-bit integer)			
Param ID = 134 (composed Dtot)		Flags	Parameter 134 length = 1
End-to-end composed value for D (Dtot) (32-bit integer)			
Param ID = 135 (composed Csum)		Flags	Parameter 135 length = 1
Since last re-shaping point composed C (Csum) (32-bit integer)			
Param ID = 136 (composed Dsum)		Flags	Parameter 136 length = 1
Since last re-shaping point composed D (Dsum) (32-bit integer)			
Service - specific data (if present)			

The service specification does not identify specific characteristics of network elements that need to be minimized. The specification basically states that a node that offers Controlled-load service should ensure that it offers a given flow the necessary bandwidth and buffer space to support the TSpec indicated in a given request. For this reason, the flowspec for Controlled-load service is just a TSpec with no additional information. Equally, the Controlled-load data fragment of an ADSpec offers no additional information beyond that contained in the general part of the ADSpec.

One important point about the Controlled-load service is the QOS that would be experienced by packets that fall outside of the TSpec. No guarantee exists that such packets would obtain a QOS that is anything near the QOS of packets that fall within the TSpec. In fact, such packets are likely to experience significant delay or loss.

Removing Reservations and the Use of Soft State

You can remove reservations in two ways: explicitly by a sender or receiver, or as a result of a timeout.

Two messages exist for the explicit tear-down of a reservation state within network nodes: PathTear and ResvTear. A sender uses PathTear, which travels towards all receivers and deletes path states and reservation states in all nodes along the path. A receiver uses ResvTear, which effectively does the same thing but in reverse. A given receiver does not have the authority to delete a reservation state for other receivers, however. Therefore, in the case of merged reservations, a ResvTear message will propagate upstream in the network only to the extent that deletion of a reservation state applies only to the receiver in question.

RSVP uses a soft-state approach to managing reservations. In other words, reservations need to be refreshed on a regular basis or they will time out. Both PATH messages and RESV messages contain a refresh period (R) within the TIME_VALUES object. In other words, for example, to keep a reservation active, the receiver should send a new RESV message after R seconds. Within each node, there will be a timer L that is some function of R. If the node does not receive a refreshing message within L seconds, then the reservation state is deleted. L should be set to a value that is significantly larger than R, so that one or more RESV messages could be lost without deleting the reservation state. This setting is important, because RSVP does not include reliability functions for ensuring that RSVP messages have guaranteed delivery. Therefore, the possibility exists that refreshing messages could be lost.

DiffServ

RSVP is the most comprehensive QOS mechanism in existence for IP networks. Of the IP solutions available, RSVP comes closest to circuit emulation. RSVP does so at a significant cost, however. RSVP requires RSVP-enabled routers to examine IP headers in some detail, and it requires routers to maintain state for all existing reservations. Given that there might be millions of concurrent reservations, this situation translates to significant overhead and difficulty for RSVP to scale to large systems. In many cases, we can provide a good QOS to numerous applications without the complexity and overhead of RSVP. *Differentiated Services* (DiffServ) is one means for achieving this goal.

If we think of QOS purely in terms of bandwidth, then the only way to increase the QOS offered to a given data stream is to increase the bandwidth. We can perform this task by physically adding bandwidth or by offering one application extra bandwidth at the expense of another application. RSVP works by reserving a set bandwidth for a particular data stream. Once all of the bandwidth is used, further reservation attempts fail. The approach with DifServ is that it is acceptable to offer one application a greater QOS at the expense of another application if the first application really needs the QOS for acceptable performance and if the second application will not really notice a big difference. For example, we can offer a greater QOS to interactive traffic (such as voice) at the expense of non-real-time traffic (such as e-mail), because it matters greatly if voice is subject to delay. Conversely, it does not matter much if an e-mail message takes minutes instead of seconds to be delivered. DiffServ simply offers the capability to prioritize certain types of traffic and to offer certain types of handling in the network, depending on the priority of a given traffic stream.

DiffServ Architecture

Recall that the IP version 4 header has a *Type of Service* (TOS) field, and the IP version 6 header has a corresponding Traffic Class field. DiffServ renames these fields the DS field. As shown in Figure 8-15, DiffServ uses the least-significant six bits of the field to encode the *DS Codepoint* (DSCP) and does not currently use the remaining bits of the field. Within the network, the values of the DSCP in a given IP packet are used to handle the packet according to a certain PHB. Therefore, all that is required within the core of the network is the capability to examine the DSCP and act accordingly.

The DiffServ architecture is defined in RFC 2475, "An Architecture for Differentiated Services." In the DiffServ architecture, the packets of a given stream must be marked with the appropriate DSCP value so that routers in the network can provide the correct PHB. To avoid situations in which every packet from every stream is marked in order to obtain the highest priority, functions must exist at the edge of the network to ensure that only qualified packets are marked with a particular DSCP value. These functions include metering to measure packet rate, policing to ensure that traffic meets a particular agreed-upon profile, traffic shaping and dropping if necessary, and the actual marking itself. Collectively, these functions are known as traffic conditioning. Of course, the need also exists to classify packets in order to identify the stream to which they belong (or other characteristics that lead to the selection of a particular DSCP value). These functions work in a combined manner, as depicted in Figure 8-16.

Figure 8-15
DiffServ usage of the IP TOS/Traffic Class octet

Figure 8-16

Classifying and
conditioning traffic

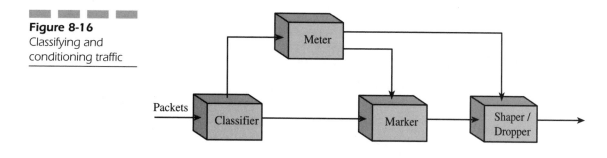

Figure 8-16

Classifying and conditioning traffic

One of the big advantages of the DiffServ architecture is that the classification, policing, and traffic-shaping functions are pushed to the edge of the network, while the requirements on routers within the network are kept to a minimum. In other words, the bulk of the effort and investment involved in using DiffServ is placed at or near the host that runs an application that wishes to use DiffServ. The amount of change required in the core of the network is minimal and does not change with the number of user applications that want to use differentiated services. Consequently, DiffServ scales extremely well.

The Need for Service-Level Agreements (SLAs)

The most critical component of successful DiffServ operation is at the edge of the network. Application packets *must* be correctly classified and marked, and applications *must* be prevented from abusing the system. In other words, a given network domain and packet source must agree on the rules related to packet classification and traffic conditioning, and these functions must be implemented at or near the source of the packets. The functions can be implemented at the source itself, such as within a customer's host(s), or they can be implemented in an edge router provided by the network operator (e.g., ISP or ITSP). In either case, a *Service-Level Agreement* (SLA) must be in place between the customer and network operator. Such an agreement should include a definition of the traffic profile to be sent from the source to the network, the classification and marking rules, and the behaviors to be applied within the network for specific DSCP values. The traffic profile might include a token bucket specification, as described previously. The classification rules could be based on combinations of source address, destination address, source port, destination port, protocol ID, time of day, or other criteria. The customer and provider must agree on the classification criteria, and just as important, the SLA should specify what is to happen if traffic is received that is outside of the traffic profile. This situation could include the dropping of packets, the marking of non-conformant packets with a different DSCP value to that used for conforming packets, traffic shaping, or additional charges to the customer.

Not only should there be SLAs between network operators and their customers, but there should also be SLAs between different network operators. In the context of this discussion, a network operator can be considered a particular DiffServ-capable domain with a common set of policies and PHB definitions. In the case where traffic needs to be passed from one operator to another, the two operators must agree on the rules related to the service that particular streams receive (and, in particular, the meaning to be applied to particular DSCP values). Only in this way can we achieve end-to-end QOS.

Per-Hop Behaviors (PHBs)

PHB is the treatment that a DiffServ router applies to a packet with a given DSCP value. Note that a router deals with multiple flows from many sources to many destinations. Many of the flows can have packets marked with a DSCP value that indicates a certain PHB. The set of flows from one node to the next that share the same DSCP codepoint is known as an aggregate. From a DiffServ perspective, a router operates on packets that belong to specific aggregates. When a router is configured to support a given PHB (such as dedicating bandwidth or policing incoming data flows), then the configuration is established with respect to aggregates, rather than to specific flows from a specific source to a specific destination.

Currently, two PHBs are defined: *Expedited Forwarding* (EF) and *Assured Forwarding* (AF). These are described in RFC 2598 and RFC 2597, respectively.

Expedited Forwarding The objective with the EF PHB is to provide a service that is low loss, low delay, and low latency, such that the service approximates a virtual leased line. The basic approach is to minimize the loss and delay experienced in the network by minimizing queuing delays. This job can be done by ensuring that at each node, the rate of departure of packets from the node is a well-defined minimum, and that the arrival rate at the node is always less than the defined departure rate. This statement implies that traffic-conditioning functions at the edge of the network are important in order to ensure that the incoming rate is always below the configured outgoing rate. The recommended DSCP value for the EF PHB is 101110.

We can implement the EF PHB in network nodes in a number of different ways (for example, by a priority queue). Such a mechanism could enable unlimited preemption of other traffic such that EF traffic always receives access first to outgoing bandwidth. This approach could, however, lead to unacceptably low performance for non-EF traffic. Therefore, the implementation of EF in a given node must be designed such that EF traffic does not inflict enormous damage to other traffic. One way would be to police the rate of incoming EF-traffic through a token bucket limiter. Another way to implement the EF PHB would be through the use of a weighted round-robin scheduler, where the share of the output bandwidth allocated to EF traffic is equal to a configured rate.

The specific implementation of the EF PHB within a network can have an impact on jitter. Appendix A of RFC 2598 includes a number of simulation results for different EF PHB implementations. This appendix specifically provides a comparison of performance for a *priority queue* (PQ) implementation versus a *weighted round robin* (WRR) implementation.

An Internet draft titled "The 'Virtual Wire' Behavior Aggregate" also exists, which discusses the traffic conditioning that should be applied at the edge of the network to ensure that the EF PHB can provide a service that mimics a dedicated wire from sender to receiver. The basic thrust of the draft is that the aggregate should have a well-defined minimum departure rate and that strict traffic shaping at the ingress to the DiffServ network can ensure that traffic is carried jitter free. (Jitter free actually means within a particular jitter window.) Basically, if a given packet is being received at a router, as soon as the last bit of the packet is received, the router starts to send the packet. Provided that that last bit of the next packet arrives at the router before the last bit of the first packet has departed, then there is no perceived jitter at the destination. In other words, a small jitter window is permitted for each packet. Provided that packets arrive within their jitter window, they can be carried to the far end with no perceivable jitter.

Assured Forwarding   The *Assured Forwarding* (AF) PHB is defined in RFC 2597. Its objective is to provide a service that ensures that high-priority packets are forwarded with a greater degree of reliability than packets of a lower priority.

Normally, the traffic into a DiffServ network from a particular source should conform to a particular traffic profile. In particular, the rate of traffic should not exceed some pre-agreed maximum. In the event that it does, the source of the traffic runs the risk that the data will be lumped in with normal best-effort IP traffic and will be subject to the same delay and loss possibilities. In fact, in a DiffServ network, certain resources will be allocated to certain behavior aggregates, which means that a smaller share is allocated to standard best-effort traffic. Thus, receiving best-effort service in a Diff-Serv network could be worse than receiving best-effort service in a non-Diff-Serv network. A given subscriber to a DiffServ network might want the latitude to occasionally exceed the requirements of a given traffic profile without being too harshly punished. The AF PHB offers this possibility.

The AF PHB is a means for a provider to offer different levels of forwarding assurances for packets received from a customer. Basically, the AF PHB enables packets to be marked with different AF classes, and within each class, to be marked with different drop-precedence values. Within a router, resources are allocated according to the different AF classes. If the resources allocated to a given class become congested, then packets must be dropped. The packets to be dropped are those that have higher drop-precedence values. From the point of view of a customer, this feature offers the possibility of saying something like, "Please don't drop my packets, but if you have to, then drop these particular ones before any others, because these are less important."

AF currently has four classes and three drop-precedence levels in each class. Therefore, a given packet has a class (i) and a drop precedence (j), and the associated codepoint is *AFij*. The recommended values of the codepoints are as shown in Table 8-1. We can see that the three most significant bits indicate the class, and the three least significant bits indicate the drop-precedence within the class.

Within a DiffServ network, the AF implementation must detect and respond to long-term congestion by dropping packets and respond to short-term congestion by queuing. This process implies a function that monitors short-term congestion, and which from that, derives a smoothed long-term congestion level. When the smoothed congestion level is below a particular threshold, then no packets should be dropped. If the smoothed congestion level is between a first and second threshold level, then packets with the highest drop-precedence level should be dropped. As the congestion level

Table 8-1	**Drop Precedence**	**Class 1**	**Class 2**	**Class 3**	**Class 4**
Assured forwarding codepoints	Low	001010	010010	011010	100010
	Medium	001100	010100	011100	100100
	High	001110	010110	011110	100110

rises, more and more of the high drop-precedence packets should be dropped until a second congestion threshold is reached, at which point 100 percent of the high drop-precedence packets are being dropped. If the congestion continues to rise, then packets of the medium drop-precedence level should also start being dropped.

The implementation must treat all packets within a given class and precedence level equally. Thus, if 50 percent of packets with a given class and precedence value are to be dropped, then that 50 percent should be spread evenly across all packets for that class and precedence. Put another way, a given flow should experience the same drop rate as every other flow with the same class and precedence level. Note that different AF classes are treated independently and are given independent resources. Thus, when packets are to be dropped, they are dropped for a given class and drop-precedence level. Packets of one class and precedence level might possibly experience a 50 percent drop rate while the packets of a different class with the same precedence level are not dropped at all. Regardless of the amount of packets that need to be dropped, a DiffServ node must not reorder AF packets within a given AF class, regardless of their precedence level.

Multi-Protocol Label Switching (MPLS)

We can see that RSVP and DiffServ take different approaches to solving the QOS problem. RSVP tackles the issue by reserving resources in advance of a given session, while DiffServ takes the approach of sharing the resources according to priority. Meanwhile, however, another approach exists that is extremely different: MPLS.

MPLS is not primarily a QOS solution, although it can be used to support QOS requirements. Rather, MPLS is best viewed as a new switching architecture. Standard IP switching requires every router to analyze the IP header and to make a determination of the next hop, based on the content of that header. The primary driver in determining the next hop is the destination address in the IP header. We compare the destination address with entries in a routing table and the longest match between the destination address and the addresses in the routing table determines the next hop. The approach with MPLS is to attach a label to the packet. The content of the label is specified according to a *Forwarding Equivalence Class* (FEC), which is determined at the point of ingress to the network. The packet and label are passed to the next node, where the label is examined and the FEC is determined. This label is then used as a simple lookup in a table that specifies the next hop and a new label to use. The new label is attached, and the packet is forwarded.

So what is the big difference between this label and standard routing based on the IP header? To begin with, the FEC is determined at the point of ingress to the network, where information might be available that cannot be indicated in the IP header. For example, the FEC could be chosen based on a combination of destination address, QOS requirements, the ingress router, or a variety of criteria. The FEC can implicitly indicate such information, and routing decisions in the network automatically take that information into account. In particular, a given FEC can force a packet to take a particular route through the network without having to cram a list of specific routers into the IP header. This feature can be important for ensuring QOS, where the bandwidth that is available on a given path has a direct impact on the perceived quality.

An important point about MPLS is that the label does not necessarily imply a new layer between layer 2 and layer 3. Obviously, the fact that a new label is attached to the packet can mean that a new layer exists, if only for the purpose of interpreting the contents of the label. The label itself can be carried at layer 2 if there is an appropriate field in which to carry it, however. For example, the label could be carried in ATM VPI or VCI fields. Similarly, the label could be carried in a Frame Relay DLCI field. Given that many IP implementations use ATM or Frame Relay at layer 2, this fit is quite neat.

MPLS Architecture

The requirements for MPLS are described in the Internet draft "A Framework for Multiprotocol Label Switching," and the architecture itself is described in the aptly named Internet draft "Multiprotocol Label Switching Architecture."

As already described, MPLS involves the determination (at the point of ingress to the network) of an FEC value to apply to a packet. That FEC value is then mapped to a particular label value, and the packet is forwarded with the label. At the next router, the label is evaluated, and a corresponding FEC is determined. A lookup is then performed to determine the next hop and a new label to apply. The new label is attached, and the packet is forwarded to the next node.

This process indicates that the value of the label can change as the packet moves through the network. What does not change is the FEC. Basically, the process works as shown in Figure 8-17. An incoming packet has a particular FEC based on some criteria. In the figure, the packet has an FEC of F. Note that this FEC is not carried in the packet. At the ingress router (R1), the FEC of F means that the packet must be sent to router R2 and that it should have a label value of L1. When the packet and label arrive at R2, R2 knows that the label value of L1 means an FEC of F for packets received from R1. R2 then proceeds to look up the next hop and label value. R2 determines that the packet should be forwarded to R3 with a label value of L2. The packet and label arrive at R3. R3 knows in advance that the label value of L2 means an FEC of F for packets from R2, and it uses this information for its routing decision (and so on).

We can see that the relationship between the FEC and the label value is a local affair between two adjacent *label-switching routers* (LSRs). If a given router is upstream from the point of view of data flow, then it must have an understanding with the next router downstream as to the binding between a particular label value and a particular FEC. For example, in Figure 8-17, the label value L1 is the outgoing label at R1 for packets with FEC F going to R2. At R2, the incoming label value of L1 is bound to FEC F for packets incoming from R1. Also at R2, the label value L2 is the outgoing label for

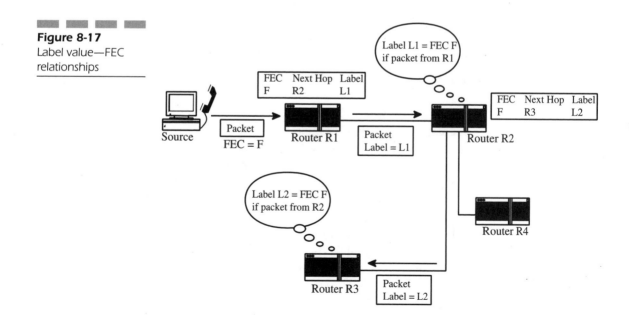

Figure 8-17
Label value—FEC relationships

packets with FEC F going to R3, etc. Thus, the binding between label value and FEC is limited to the two ends of an interface between adjacent LSRs for a particular direction of data flow. Also, if R2 were to receive packets with label value L1 on a link from a different LSR, such as R4, then a different FEC might apply.

Label Assignment and Distribution The foregoing discussion shows that adjacent LSRs need to agree on label-to-FEC bindings. In the MPLS architecture, the downstream LSR decides on the particular binding. The downstream LSR then communicates the binding to the upstream LSR, which means that there must be a label-distribution protocol between the two to support such communication. Several label-distribution protocols exist, in fact. In some cases, existing protocols such as RSVP have been extended to provide support for label distribution. In addition, a brand-new protocol called (not surprisingly) *Label-Distribution Protocol* (LDP) has been developed. If a given label-distribution protocol is used between two LSRs in order to exchange label/FEC binding information, then the two LSRs are known as label-distribution peers.

In general, we can perform label distribution in two ways. First, there is downstream-on-demand, whereby a given LSR can request a particular label/FEC binding from a downstream LSR. Second, there is unsolicited downstream, whereby a given LSR distributes label/FEC bindings to other (upstream) LSRs without having been explicitly requested to do so.

Although the MPLS architecture does not mandate that any particular label-distribution protocol should be used, we discuss LDP in this chapter.

FECs and Labels

Conceptually, we can use the value of an FEC to represent many things. In reality, however, the FEC takes the form of one or more IP addresses or IP address prefixes. For example, LDP specifies that an FEC is composed of a number of FEC elements, as shown in Figure 8-18. Each element is either a host address or an address prefix.

As mentioned, we can map a label to an ATM VPI/VCI or to a Frame Relay DLCI. In the case where the layer 2 protocol does not offer a field, then the label forms a shim layer between layer 2 and the network layer. This situation would be the case for Ethernet or the *Point-to-Point Protocol* (PPP). In such cases, the label is a 32-bit label, as shown in Figure 8-19. As we can see, the label value itself occupies 20 bits. The remainder is consumed by an eight-bit time-to-live field, a three-bit field for experimental

Figure 8-18
FEC format in LDP

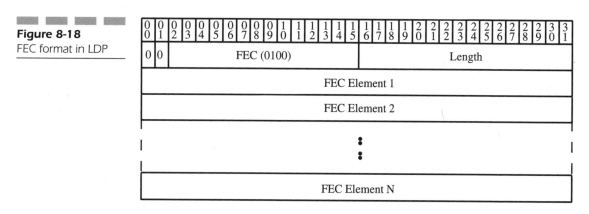

Figure 8-19
Generic label format

Figure 8-20
ATM label format

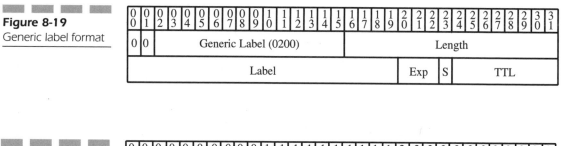

use (Exp), and a single bit (S) to indicate that this label is the last in the stack. We will discuss the concept of a label stack shortly.

In the case that ATM is used at layer 2, then the label can be mapped to the ATM VPI/VCI—in which case it takes the format shown in Figure 8-20. The V-bits indicate whether both the VPI and VCI are significant or just one of them. 00 indicates that both are significant. 01 indicates that only the VPI is significant and 10 indicates that only the VCI is significant. Where Frame Relay is used at layer 2, then the label is mapped to the DLCI as shown in Figure 8-21. The Len field indicates the number of bits in the DLCI. 0 indicates that the DLCI is 10 bits long. 2 indicates that the DLCI is 23 bits long. Values 1 and 3 are reserved.

Figure 8-21
Frame Relay label
format

0 0	0	0 0	0 0	0 0	0 0	0 1	1 1	1 1	1 1	1 1	1 1	2 2	2 2	2 2	2 2	2 2	3 3
0 1	2	3 4	5 6	7 8	9 0	1 2	3 4	5 6	7 8	9 0	1 2	3 4	5 6	7 8	9 0	1	

0 0	Frame Relay Label (0202)	Length
Reserved	Len	DLCI

The Label Stack   Much of our discussion so far has suggested that a given packet can have only one label. While this situation is valid, a given packet can have more than one label. In fact, MPLS describes a label stack in which there can be several labels, one after the other, in advance of the actual IP packet. In such a situation, an LSR that receives the labeled packet will base its actions on the first (top) label. The obvious question is, "Why might we need more than one label?" The answer is tunneling.

Let's suppose that a given packet with FEC F needs to follow the *label-switched path* (LSP) R1, R2, R3, and R4. Let's further assume that R2 and R3 are not directly connected but form the two ends of the tunnel R2, R2A, R2B, R2C, and R3. Therefore, the actual path taken is R1, R2, R2A, R2B, R2C, R3, and R4, as depicted in Figure 8-22. When the packet travels from R1 to R2, it has a single label. R2 determines that the packet P must enter the tunnel. Therefore, R2 attaches another label on top of the first, so that the label stack has a depth of 2. In fact, R2 might replace the first label with a label that is meaningful to R3, then place a new label on top. The packet P has two labels as it passes through the tunnel. When it reaches R2C, the LSR recognizes that it is the next-to-last LSR in the tunnel. (Both R2 and R3 are effectively part of the tunnel.) R2C pops the stack and forwards the packet to R3 with only one label. R3 uses the label to route to R4. In fact, R3, which is the next-to-last LSR in the overall LSP, might also pop the stack and pass the packet to R4 unlabeled.

The scenario depicted in Figure 8-22 is effectively an LSP nested within another LSP. The MPLS architecture enables multiple levels of nesting.

Actions at LSRs

A given LSR's actions obviously depend on the value of the label. In fact, the LSR's action is specified by the *Next Hop-Level Forwarding Entry* (NHLFE), which indicates the next hop, the operation to perform on the label stack, and the encoding to be used for the stack on the outgoing link. The operation to perform on the stack might mean that the LSR should replace the label at the top of the stack with a new label. The operation

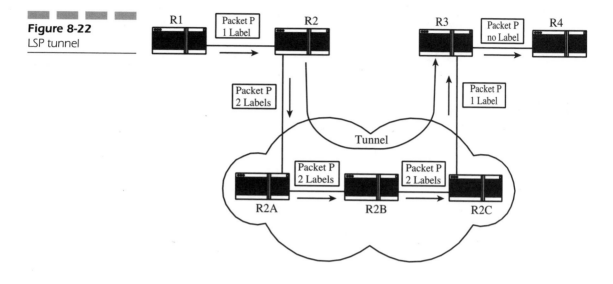

Figure 8-22
LSP tunnel

might require the LSR to pop the label stack or to replace the top label with a new label, then add one or more additional labels on top of the first label.

The next hop for a given labeled packet might be the same LSR. In such a case, the LSR pops the top-level label of the stack and forwards the packet to itself. At this point, the packet might still have a label to be examined, or it might be a native IP packet without a label (in which case, the packet is forwarded according to standard IP routing).

A given label might possibly map to more than one NHLFE. This situation might occur when load sharing is to take place across multiple paths. In such a case, the LSR chooses one of the NHLFEs to use, according to some internal procedures.

In general, if a router happens to know that it is the next-to-last LSR in a given path, then it should remove any labels and pass the packet to the final LSR without a label. The reason is because we want to minimize the amount of effort that the ultimate LSR needs to undertake. If the next-to-last LSR passes a labeled packet to the final LSR, then the final LSR must examine the label, determine that the next hop is itself, pop the stack, and forward the packet to itself. The LSR must then re-examine the packet to determine what to do with it. If, on the other hand, the packet arrives without a label, then the final LSR has one less step to execute. How a particular LSR determines that it is the next-to-last LSR for a given path is a function of label distribution and the distribution protocol used.

Label-Switched Paths

We have already introduced the concept of a *Label-Switched Path* (LSP)
and also the concept of the last and penultimate LSR in a given LSP. The
concept of an LSP needs to be viewed with the consideration that label
switching will be introduced in the form of islands within the IP network as
a whole, with each island comprising one or more IP domains. Consequently, there will be points of ingress and egress to the MPLS network. An
LSR that is a point of ingress to the MPLS network will be responsible for
choosing the FEC that should be applied to a given packet. Because label
distribution works in a downstream-to-upstream direction, an LSR that is
a point of egress is responsible for determining a label/FEC binding and
passing that information upstream. An LSR will act as an egress LSR with
respect to a particular FEC if the FEC refers to the LSR itself, if the next
hop for the FEC is outside the label-switching network, or if the next hop
means traversing a boundary (such as between OSPF autonomous systems).

We can consider an LSP as the path that applies to a given FEC from a
point of ingress to the egress LSR. The primary function of label-distribution
protocols is the establishment and maintenance of these LSPs. Note that
many points of ingress might exist, in which case the LSPs form a tree with
the egress LSR at the root. This structure can be termed a multipoint-to-
point tree.

MPLS Traffic Engineering

One of the most important applications of MPLS is in the area of traffic
engineering, which can be summarized as the modeling, characterization,
and control of traffic to meet specified performance objectives. As described
in RFC 2702, the performance objectives in question might be traffic oriented or resource oriented. The former deals with QOS and includes aspects
such as minimizing delay, jitter, and packet loss. The latter deals with the
optimum usage of network resources, particularly network bandwidth.

These two objectives are not necessarily mutually exclusive. For example, congestion avoidance is a major goal related to both network-resource
objectives and QOS objectives. From a resource point of view, we strongly
desire to avoid situations where one part of the network is congested while
another part of the network is underutilized, particularly if the underutilized resources could be used to carry some of the traffic that is experienc-

ing congestion. Equally, from a QOS point of view, we want to allocate traffic streams to available resources in order to ensure that those streams do not experience congestion and a consequent packet loss and delay.

Congestion is primarily caused in two ways. The first is simply because of a lack of sufficient resources on the network to accommodate the offered load. The second is the steering of traffic towards resources that are already loaded, while other resources remain underutilized. The expansion of capacity or the application of congestion-control techniques such as flow control can correct the first situation. Good traffic engineering can address the second situation.

The current situation with IP routing and resource allocation is that the routing protocols are not well equipped to deal with traffic-engineering issues. For example, a protocol such as *Open Shortest Path First* (OSPF) can actually promote congestion because it tends to force traffic down the shortest route, although other acceptable routes might be less loaded. In many network deployments, the solution to this problem has been to use traffic-engineering functions at layer 2 in the network. ATM, for example, enables virtual circuits to be easily rerouted in cases of congestion. Fortunately, MPLS lends itself to the application of traffic engineering above layer 2 so that the choice of link-layer protocol is a real choice.

Traffic Trunks RFC 2702 introduces the concept of a traffic trunk, which can be considered a set of flows that share specific attributes. In practice, these attributes include the ingress and egress LSRs, the FEC, and possibly other characteristics (such as average rate, peak rate, and priority and policing attributes). A traffic trunk is something that can be (and is) routed over a given LSP. In fact, we can explicitly specify the LSP that a traffic trunk should use. This capability has the immediate advantage of allowing us to steer certain traffic away from the shortest path, which is likely to become congested before other paths. Furthermore, we can change the LSP that a given traffic trunk will use, which enables the network to adapt to changing load conditions whether through administrative intervention or through automated processes within the network.

As described in RFC 2702, there are three main aspects of traffic engineering on an MPLS network:

- Mapping of packets to forwarding-equivalence classes
- Mapping forwarding-equivalence classes to traffic trunks
- Mapping traffic trunks onto the physical network topology through label-switched paths

How individual packets are assigned to a given FEC and how FECs are assigned to traffic trunks are functions to be decided at the ingress to the network. These decisions can be made based on any number of criteria, provided that the criteria are fully understood by both the MPLS network provider and the source of packets (either a customer or another network provider). This situation again emphasizes the need for SLAs between providers and between providers and customers.

Of the three mappings that need to take place, the third is at the heart of ensuring that the network provides the quality that is needed for a given type of traffic. This mapping can involve constraint-based routing, where traffic is matched with network resources according to the characteristics of the traffic and the characteristics of available resources. For example, one characteristic of traffic is a bandwidth requirement, and one characteristic of a path is the maximum bandwidth that it offers. Obviously, path selection should be limited to those paths that can offer at least the bandwidth that the traffic trunk needs.

Constraint-Based Routing and LDP   Constraint-based routing is described in the Internet draft "Constraint-Based LSP Setup using LDP" (CR-LDP). As the name suggests, CR-LDP offers a routing capability where the path that is to be used is chosen according to certain constraints that must be met. CR-LDP is a protocol that is based upon LDP. Therefore, before discussing the mechanisms of constraint-based routing (as described in the draft),we must describe some aspects of LDP.

The primary purpose of LDP is the establishment of LSPs with which particular FECs are associated. LDP offers four categories of messages:

Discovery messages   Used to announce and maintain the presence of an LSR on the network

Session messages   Used to establish, maintain, and terminate sessions between LDP peers

Advertisement messages   Used to create, change, and delete label mappings for FECs

Notification messages   Used to provide advisory information and to signal error information

Of these, the advertisement messages are used to set up the actual LSPs in the network. One of these, the Label Request message, is used by an upstream LSR to request a downstream LSR to assign and advertise a label for a given FEC. The message, as defined in LDP, has the format shown in Figure 8-23. The FEC TLV is as shown in Figure 8-18. Two optional parameters are available. The first is the Hop Count, which specifies the running total of the number of LSR hops along the LSP to be set up. The Hop Count can be used to determine whether a path contains too many hops, which might indicate the existence of a loop. The second is the Path Vector, which is a list of LSRs in the path. A major function of the Path Vector is to support loop detection.

The Label Mapping message is used to advertise a given label/FEC mapping. The format of the message is as shown in Figure 8-24. The Label TLV is a label as shown in Figure 8-19 in the case that layer 2 is Ethernet or PPP.

CR-LDP enhances LDP such that traffic parameters, resource requirements, and other characteristics can be incorporated in the establishment of LSPs. Furthermore, CR-LDP enables the establishment of explicit routes that can be considered a subset of constraint-based routes. The following sections describe some of the characteristics that CR-LDP addresses when establishing LSPs.

Explicit Routes A *Constraint-Based Routed LSP* (CR-LSP) is an LSP that is established subject to a number of criteria. The CR-LSP is established based on information that is available at the edge of the network. An

Figure 8-23
LDP Label Request message

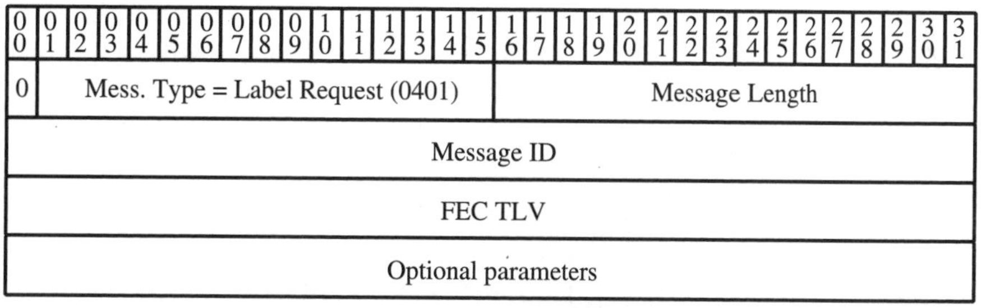

Figure 8-24
LDP Label Mapping message

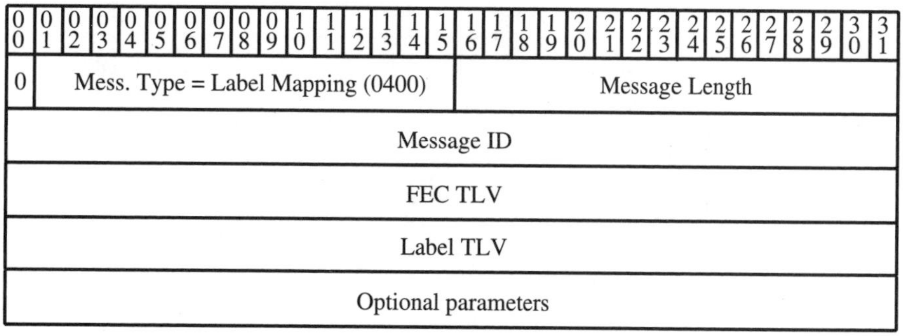

explicit route (ER) is one type of constraint-based LSP where some or all of the nodes to be used are specified. In a strict ER, all nodes in the path are specified, and the path is not allowed to include nodes that are not specified. In a loose explicit route, a number of nodes in the path are specified, but other nodes can also be used. CR-LDP enables explicit route information to be included in the LDP Label Request message. The overall idea is to define specific paths for traffic that has specific characteristics.

Traffic Characteristics   CR-LDP enables a number of traffic characteristics to be specified through the use of a traffic parameter, as shown in Figure 8-25. The peak rate indicates the maximum rate at which traffic should be sent to the CR-LSP and includes two components: the *Peak Data Rate* (PDR), measured in bytes/second, and the *Peak Burst Size* (PBS), measured in bytes. Each of these form the components of a token bucket specification, where the bucket is filled at a rate equal to PDR and the bucket size equals PBS.

The committed rate indicates the rate that the network commits to be available to the CR-LSP. The rate has the components *Committed Data Rate* (CDR) and *Committed Burst Size* (CBS). These also are used in a token bucket specification.

The *Excess Burst Size* (EBS) can be used at the edge of the MPLS domain for the purposes of traffic conditioning and can be used to measure the extent by which traffic exceeds the committed rate. This rate can also be used in a token bucket specification, where the token bucket size is EBS and the token rate is CDR.

Figure 8-25
CR-LDP Traffic
parameter

0 0	0 0 0 0 0 0 0 0 0 1 1 1 1 1 1	1 1 1 1 2 2 2 2 2 2 2 2 2 2 3 3
0 0	Traffic Parameter TLV (0810)	Length = 24 bytes
Flags	Frequency	Reserved	Weight
Peak Data Rate (PDR)		
Peak Burst Size (PBS)		
Committed Data Rate (CDR)		
Committed Burst Size (CBS)		
Excess Burst Size (EBS)		

The Frequency indicates how often the CDR should be made available. Frequent (1) means that the available bandwidth should average at least the CDR over any short interval. A short interval corresponds to a small number of the shortest packet times at the CDR. Very frequent (2) means that the available rate should average at least the CDR over any packet interval at the CDR.

The Weight determines the CR-LSP's share of any possible excess bandwidth above the committed rate. Its use is specific to a given MPLS domain.

Resource Classes CR-LDP enables the use of resource classes to specify what links can be used in a given CR-LSP. The purpose is to help limit the set of possible links to those that meet certain resource requirements. For example, a resource class could indicate OC-48, meaning that all links of OC-48 capacity belong to that resource class. Similarly, ADSL could be a resource class. Resource classes are sometimes known as colors, where resources of the same color belong to the same class. A given resource can belong to more than one resource class. Note that CR-LDP does not specify the specifics of resource classes; rather, it simply provides a means for indicating a resource class in LDP messages. CR-LDP does so by enabling a Resource-Class TLV. All LSRs within a given domain must have the same understanding of resource classes and the contents of the Resource Class TLV.

Preemption CR-LDP enables a CR-LSP to be allocated to a given traffic trunk and enables a CR-LSP to be established based on certain resource requirements. If a CR-LSP cannot be established due to a lack of available resources, then it is possible to reroute other traffic in order to make room. This procedure is called preemption.

Preemption involves the assignment of two priority levels to a given CR-LSP. The first, called setupPriority, indicates the authority assigned to a given CR-LSP to preempt (bump) another. The second, called holdingPriority, indicates how much authority is required by another CR-LSP to bump the CR-LSP that currently has the resources. B can only bump A if B's setupPriority is higher than A's holding priority. Actually, priority values are assigned in descending order. The value 0 is most important, and the value 7 is least important. For a given CR-LSP, its holdingPriority should be at least as good as its setupPriority. Otherwise, having bumped a less-important CR-LSP, it can be immediately bumped by a CR-LSP with the same priority as itself.

Modified LDP Messages for CR-LDP Using the new information defined in CR-LDP, the Label Request and Label Mapping messages are modified as shown in Figures 8-26 and 8-27. In addition to the parameters described

Figure 8-26
CR-LDP Label Request
message

| 0 0 0 0 0 0 0 0 0 0 1 1 1 1 1 1 1 1 1 1 2 2 2 2 2 2 2 2 2 2 3 3 |
| 0 1 2 3 4 5 6 7 8 9 0 1 2 3 4 5 6 7 8 9 0 1 2 3 4 5 6 7 8 9 0 1 |

0	Mess. Type = Label Request (0401)	Message Length
	Message ID	
	FEC TLV	
	LSPID TLV	(CR-LDP, mandatory)
	ER-TLV	(CR-LDP, optional)
	Traffic TLV	(CR-LDP, optional)
	Pinning TLV	(CR-LDP, optional)
	Resource Class TLV	(CR-LDP, optional)
	Pre-emption TLV	(CR-LDP, optional)

Figure 8-27
CR-LDP Label
Mapping message

| 0 0 0 0 0 0 0 0 0 0 1 1 1 1 1 1 1 1 1 1 2 2 2 2 2 2 2 2 2 2 3 3 |
| 0 1 2 3 4 5 6 7 8 9 0 1 2 3 4 5 6 7 8 9 0 1 2 3 4 5 6 7 8 9 0 1 |

0	Mess. Type = Label Mapping (0400)	Message Length
	Message ID	
	FEC TLV	
	Label TLV	
	Label Request Message ID	
	LSPID TLV	(CR-LDP, optional)
	Traffic TLV	(CR-LDP, optional)

previously, we notice the Pinning parameter and the LSPID. The Pinning parameter is used with loose explicit routes and includes a means of indicating whether or not a path can be changed at a given LSR if a better next hop becomes available later. Of course, such changes cannot apply to a strict explicit route. The LSPID parameter is a unique identifier for a CR-LSP within the MPLS network.

End-to-End QOS The previous discussion shows how MPLS and CR-LDP can be used to set up paths in the network based on the characteristics of the traffic to be carried. An important aspect of the solution, however, is how the traffic is classified and conditioned at the edge of the MPLS network. Given that QOS is an end-to-end effort, the actions taken at the edge are just as important as the actions taken in the core of the network (if not more so).

Combining QOS Solutions

The QOS solutions described in this chapter take different approaches, and each has its advantages and disadvantages. While RSVP is powerful, it has the drawback of requiring path states to be maintained in each router—something that can be difficult and costly on a large network. DiffServ is simpler but is more of a prioritization technique than a resource-guarantee mechanism. MPLS offers great promise as an overall solution but requires significant changes to all routers that want to use this method. Therefore, each of the solutions has its drawbacks, and no single solution is likely to be a silver bullet. Combining the different solutions in smart ways, however, can enable them to be used in different parts of the network where they can be put to best use. For example, if RSVP is used in one domain and DiffServ is used in another, then the possibility exists of mapping an RSVP service request (e.g., Guaranteed Service) to an appropriate DiffServ per-hop behavior (e.g., Expedited Forwarding). Similarly, we might be able to map a particular DiffServ behavior aggregate to an MPLS FEC. Figure 8-28 gives one example of how different solutions might be combined. Work is ongoing in the IETF and elsewhere on methods for combining the various QOS technologies.

Further Information

QOS is of major importance to the future of IP networking—a fact that most people in the industry recognize. We have seen that various solutions have been developed to address the QOS issue. Furthermore, a great deal of work is underway to improve these solutions and to enable the solutions to work together to provide end-to-end quality. You can find further information about this area on the IETF's Web site.

Another source of information is the QOS Forum, which is an international industry association dedicated to promoting the development and

Figure 8-28
Combining QOS
techniques for an
end-to-end solution

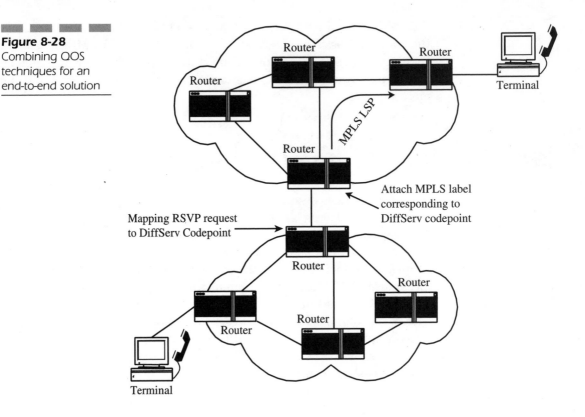

deployment of QOS solutions. Most of the industry's major players are involved in this organization. While it is not a standards-setting body, the QOS Forum does serve as a means for sharing ideas regarding QOS solutions. This forum also supports education, marketing, and testing related to QOS.

Accessing the Network

Introduction

Much of the discussion so far in this book has been about the functions and protocols within the network. While there has been some discussion about actions that need to take place at the network edge, we have not touched upon the technologies that enable an individual user to access the network in the first place. That is the subject of this chapter.

In any telecommunications network, the so-called "last mile" is the most expensive and most time consuming to implement. A user sitting at a device that is connected to a corporate *Local-Area Network* (LAN), which in turn is connected to the network via a DS3 or higher-speed facility, is in good shape. Many users do not have such a high-speed connection, however, and cannot justify the cost of leasing such facilities from the telephone company. Such users only have access to the copper wires that run from the local telephone exchange to their business or residence. Those facilities have been installed over a period of decades at an enormous cost. Furthermore, the majority of those lines were designed for carrying analog speech, and as described in Chapter 3 ("Speech-Coding Techniques"), speech is limited to less than 4kHz. In the new days of the Internet, with file transfer and multimedia applications, the demand for bandwidth is increasing drastically. While VoIP might not be a major bandwidth burden in its own right, one of its major attractions is its capability to integrate voice with other *Internet Protocol* (IP) applications that might well be more bandwidth greedy. Therefore, providing increased capacity over the last mile is a major concern.

In this chapter, we examine the issue of network access and describe some of the technologies used, particularly those that offer bandwidths that are suitable for supporting real-time multimedia communications. To start, we briefly discuss the latest modem technology that many people use for accessing the Internet. Subsequently, we address alternative technologies such as the *Integrated Services Digital Network* (ISDN), *Digital Subscriber Line* (DSL), cable modems, and wireless access.

Modems

Today, the majority of subscribers access the Internet through dialup modems. These modems operate by taking the stream of digital information from the user and converting it to an analog waveform that appears as analog speech to the telephone network. In other words, the bandwidth that this analog waveform uses is limited to that of human speech: less than 4,000Hz.

For some time, such modems offered low-speed connections (such as 9.6kbps or 14.4kbps). Today, however, the majority of modems in use are either V.34 or V.90 modems. These modems represent the latest in modem technology and offer speeds that far exceed the norm of a few years ago.

V.34 offers speeds of up to 33.6kbps in each direction. Analog-to-digital conversion imposes this limit. Recall from Chapter 3, "Speech-Coding Techniques," that quantization is the process of converting analog speech to a digital stream. This quantization introduces quantization noise, whereby the digital signal cannot represent the analog signal with 100 percent accuracy. The V.34 analog stream from the user undergoes this analog-to-digital conversion at the local exchange, because the local exchange treats a call from the modem like a speech call. Consequently, noise is introduced into the digital stream that is sent to the far end (e.g., an Internet Service Provider, or ISP). This noise is not a major problem for speech but is a significant problem when the analog stream actually represents digital bits. Consequently, we cannot achieve 64kbps over the analog line even though we are using 64kbps within the network. In fact, the maximum speed that we can achieve is 33.6kbps. Furthermore, this speed from a V.34 modem is an ideal situation. The actual speed that a user receives might be less, depending on both the distance of the line from the user to the local exchange and the condition of that line.

The latest and greatest is the V.90 modem, also referred to as a 56K modem (although it does not offer connections of 56kbps). In fact, the V.90 modem uses V.34 in the upstream direction (from user to network), with a maximum speed of 33.6kbps. In the downstream direction, it offers speeds of 53kbps or less. The obvious question is, "Why is it asymmetric?" Also, given that we can achieve a speed higher than 33.6kbps in the downstream direction, why can the higher speed not be supported in both directions? The answer is again related to analog-to-digital conversion.

Consider the scenario depicted in Figure 9-1. A customer dials in to an ISP via the *Public Switched Telephone Network* (PSTN). The connection from the customer to the PSTN is analog, but the connection from the PSTN to the ISP is digital (e.g., using an ISDN Primary Rate Interface, or PRI). Therefore, the analog-to-digital conversion happens only once from the customer to the network. This process occurs at the customer's local exchange. Of course, there is a digital-to-analog conversion at the same point, but in the opposite direction. While analog-to-digital conversion introduces noise, however, digital-to-analog conversion does not. This concept is the key to the V.90 modem.

The V.90 modem at the ISP is digital. When transmitting data, the modem effectively mimics the way that speech is transmitted in the PSTN,

Figure 9-1
V.90 modem
connection

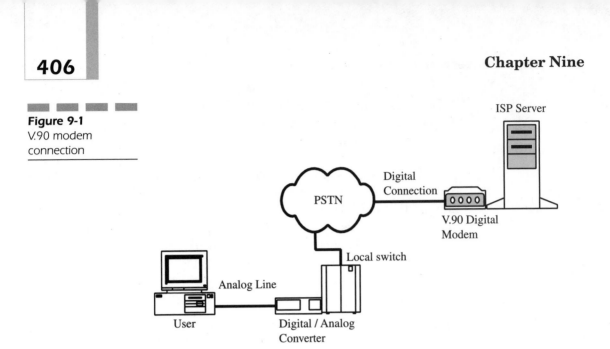

using eight-bit *Pulse-Code Modulation* (PCM) samples (values 0 to 255) at a rate of 8,000Hz. When the digital stream arrives at the user's local exchange, the eight-bit digital samples are converted to the exact corresponding analog voltage levels, which means that no quantization noise exists. In theory, we could achieve a downstream rate of 64kbps, but this situation is not the case in practice. To start, we cannot use all 256 levels —particularly if any noise exists on the analog line. In particular, the non-linear quantization in A-law or mu-law G.711 coding means that the voltage levels produced on the analog line for small PCM values are close. This situation provides little protection against noise on the line. Therefore, we use only a subset of the possible 256 values—those values that are less susceptible to noise. The use of only a subset of levels reduces the overall data rate that we can achieve.

ISDN

While 50kbps or a little higher from a modem might be considered a fast speed compared to the speeds that were available only a few years ago, this speed is still not sufficient to meet the demands of today's multimedia communications. Furthermore, even V.90 offers only 33.6kbps in the upstream direction, which is certainly not sufficient for an application such as decent-quality video in conjunction with voice. ISDN is one alternative for providing the higher bandwidth that such applications require.

Two main flavors of ISDN exist: *Basic Rate Interface* (BRI) and *Primary Rate Interface* (PRI). BRI is used over a local telephone line, and PRI is used over an E1 or T1 line.

BRI

ISDN is digital, as opposed to analog. Furthermore, the signaling uses a separate logical channel from the speech or media. With BRI, one signaling channel exists (known as the D-channel), and there are two bearer channels (known as B-channels). As a result, people often call BRI 2B+D, as depicted in Figure 9-2. The two B-channels operate at 64kbps and carry digital speech, video, or data. We can use the two B-channels in combination to give a combined rate of 128kbps in each direction, which is not bad. The D-channel operates at 16kbps and carries out-of-band signaling for call setup and tear-down. Combining the three channels gives a total bit rate of 144kbps, to which we add an overhead of 16kbps—resulting in a total line speed of 160kbps. The reason for the overhead is because information on the line is passed in frames. Each frame is 240 bits long and includes a synchronization field and a maintenance field, as shown in Figure 9-3. The maintenance field includes a CRC check and supports embedded commands for loopback testing.

The objective with ISDN BRI is that it should be useable up to a distance of 18,000 feet from the central office to the customer. Unfortunately, transmitting 160kbps would consume a bandwidth of 160,000Hz if we used one bit per baud, as is the case for standard *Alternate Mark Inversion* (AMI). This situation is unfortunate, because the higher the frequency used over

Figure 9-2
Basic Rate Interface
(BRI) — 2B+D

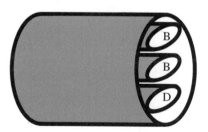

Figure 9-3
ISDN BRI frame

SYNC	B1 + B2 + D	Maintenance
18 bits	216 bits	6 bits

any transmission medium, the greater the attenuation of the signal. A signal with a frequency of 160kHz suffers too much attenuation over a typical copper wire to achieve the 18,000-foot range. Consequently, we need to send the 160kbps by using less frequency. In other words, we need to achieve more than one bit per second per Hz. The solution is line coding. In the United States, the coding scheme for ISDN BRI is *2 Binary, 1 Quaternary* (2B1Q), which is a type of *Pulse-Amplitude Modulation* (PAM) and is shown in Table 9-1. A pulse has more than one amplitude available. In fact, each pulse can represent two bits instead of one, effectively cutting the frequency requirement in half. By using this coding, we reduce the attenuation, and we can achieve the requirement of 18,000 feet from the local exchange to the customer without having to resort to repeaters.

Multiple devices (up to eight) can share a given ISDN BRI line. In order to use BRI, a particular device must either be ISDN-capable or must be connected via a *Terminal Adapter* (TA). ISDN-capable devices are known as *Terminal Endpoint 1* (TE1), and those that are not ISDN-capable (and therefore need to use a terminal adapter) are of the type *Terminal Endpoint 2* (TE2). The various devices are connected to a *Network Termination 1* (NT1), which interfaces to the actual line from the local switch, as shown in Figure 9-4. The fact that multiple devices can share a line is useful. Unfortunately, these advantages come at a price in the real commercial world. Because an ISDN line uses different equipment at the central office, the charge that most local phone companies levy for ISDN BRI service is significantly greater than the charge for a regular analog line.

	Bits	Quaternary Symbol	Voltage Level
Table 9-1			
2B1Q coding	00	−3	−2.5
	01	−1	−0.833
	10	+3	+2.5
	11	+1	+0.833

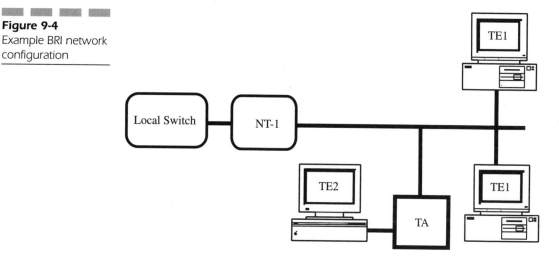

Figure 9-4
Example BRI network
configuration

Primary Rate Interface (PRI)

With PRI, in North America there are 23 B-channels and 1 D-channel (23B+D), all operating at 64kbps to occupy a full T1. In most of the rest of the world, however, there are 30 B-channels and one D-channel to occupy a full E1. Because PRI offers much more bandwidth, it is far more expensive and is usually used by businesses. In fact, the most common usage of PRI is to connect *Private-Branch Exchanges* (PBXs) to a local exchange, which is a standard way of doing things. Unfortunately, the word *standard* might be a poor choice. Although PRI is standardized in the *International Telecommunications Union* (ITU) Q.93X series of recommendations, different vendors have implemented the recommendations in different ways. Consequently, different flavors of PRI exist. For example, the Nortel PRI and the Lucent PRI are different. When implementing PRI, we must ensure that the flavor of PRI that the PBX supports can be supported by the local exchange (and vice-versa).

The Future of ISDN

ISDN has been around since the early 1980s and was initially hailed as the future of telecommunications for the foreseeable future. While PRI is widely deployed, particularly for connecting PBXs to the PSTN, BRI has not had as much success—at least, not in the United States. We could say that ISDN was a technology that arrived before its time. When ISDN was first introduced, some users thought it was great to be able to use 128kbps, but not many people actually needed that type of bandwidth. No widespread applications existed, and the Internet was still something that the U.S. Department of Defense mainly used. Therefore, this new technology had been invented and much noise was made about it in the telecommunications industry, but there was no significant market for this technology. For a long time, relatively few network implementations of BRI existed to the extent that joke names were given to the acronym ISDN, such as "the Invention Subscribers Don't Need" and "It Still Does Nothing." The Internet changed all of that, however. Whereas it used to be difficult to justify the extra cost of BRI because few people needed 128kbps, that bandwidth is now certainly needed for anyone who uses wishes to use the multimedia applications available today. Consequently, customers have been subscribing to BRI service at a rapid rate since the mid-1990s.

ISDN (and BRI in particular) took a long time to fulfill their promises, largely because a significant demand did not exist. Now that people need ISDN, many feel that 128kbps is not enough bandwidth given the content-rich multimedia applications that are now available. Furthermore, there are newer alternative technologies to BRI that offer higher speeds, often at a lower cost to the consumer. Among these are DSL technologies and cable modems. Apparently, BRI (after many years in waiting) has just started to take its place in the market, only to be surpassed by newer, better, and cheaper alternatives.

Digital Subscriber Line (DSL)

DSL technology has really come to the forefront recently, particularly since the Telecommunications Act of 1996. Regulations now enable new carriers to have access to the local loop that the *Incumbent Local-Exchange Carrier* (LEC) owns. This situation has introduced competition in local access in the same way that we have seen major competition in long-distance services since the mid-1980s. The new *Competitive Local-Exchange Carriers* (CLECs) can access the local loop and use DSL to provide a range of new services over those local facilities. Those services include high-speed Internet access and data and voice services, all over a standard copper telephone line. Of course, the incumbent LECs can offer the same as what the CLECs offer, and most of them are doing so. Only since new competitors have arrived on the scene have incumbents started to offer these services, however. The effect of competition has been that customers now have a range of new services to choose from and a range of providers from which to purchase these services. The incumbent LECs have a new revenue opportunity, and CLECs have a new business. Competition benefits everyone. What makes this competition possible, besides the regulatory environment, is DSL.

The basic concept with DSL is to use higher frequencies on the copper line than the standard voice frequencies. These frequencies are used to carry additional information, which could be voice or data. Higher frequencies enable faster data rates. As previously described, however, higher frequencies attenuate faster with distance. Therefore, the highest data rates are available only to those users who are located nearest the central office.[1]

DSL Variants

DSL actually comes in a range of different flavors, with differing speeds and transmission characteristics. The variant of DSL that is most appropriate for a given customer depends on the data rates required in each direction and the physical location of the customer premises. Furthermore, not every service provider offers all variants of DSL. In fact, most offer a limited selection.

High Bit-Rate DSL (HDSL)   HDSL was the first version of DSL to be deployed. HDSL was initially a means for providing DS1 service over standard copper facilities without the need for frequent repeaters. Standard T1 service using AMI over a copper pair requires repeaters to be placed every 3,000 to 5,000 feet to overcome attenuation of the signal. Although these repeaters are relatively simple devices, labor and other factors associated with implementing standard T1 AMI has meant that the provision of T1 service to a customer used to take weeks and the cost of the T1 service used to be high, at least compared to today's prices. The primary objective with HDSL was to make T1 service easier and faster to provision, and cheaper for the customer.

HDSL uses two copper pairs (four wires), and on each pair, 2B1Q coding is used to provide two bits per baud. In order to achieve the 1.544kbps speed of a T1 line, half of the data (772kbps) is sent on one pair, and half is sent on the other pair. The fact that fewer bits are sent over each copper pair and

---

[1]Strictly speaking, this location is where the line terminates—either at the central office or at a remote office.

that more efficient line coding is used means that the frequency band occupied by the HDSL signal is much less than that occupied by standard T1 AMI. As a result, the signal suffers much less attenuation and can propagate over much longer distances. In fact, HDSL enables T1 speeds to be offered over distances of 12,000 feet without repeaters.

HDSL is extremely popular and is widely used. In many cases where a customer orders T1 service from the telephone company, HDSL is actually provided. A new version of HDSL (HDSL2) has now been developed to enable T1 speeds in both directions over distances of 12,000 feet with just a single copper pair.

You should note that unlike some of the other DSL variants, HDSL does not enable a given copper pair to be used simultaneously for digital data services (such as IP traffic) and analog *Plain Old Telephone Service* (POTS). The reason is because POTS uses frequencies of up to 4kHz, while HDSL uses the frequencies from zero (DC) up to about 400kHz (which means that an overlap exists). This overlap prevents the two services from being used simultaneously. For that matter, POTS cannot be used with ISDN either. Figure 9-5 shows the frequencies that POTS and T1 AMI use, and Figure 9-6 shows the frequencies that ISDN and HDSL use. Note that Figure 9-6 shows the main lobes for ISDN and HDSL. Both technologies also have smaller sidebands beyond what is shown.

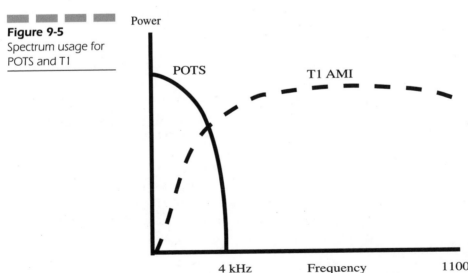

Figure 9-5
Spectrum usage for POTS and T1

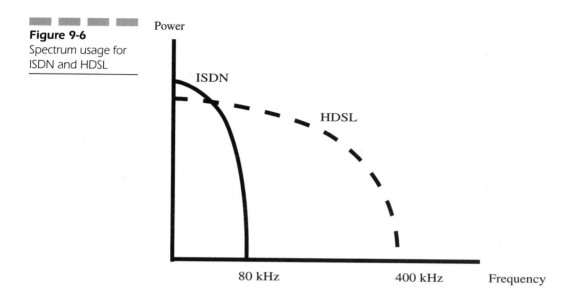

Figure 9-6
Spectrum usage for
ISDN and HDSL

Asymmetric Digital Subscriber Line (ADSL)   ADSL is perhaps the most widespread DSL variant deployed for high-speed Internet access. ASDL is a service in which the downstream rate (from network to user) is far greater than the upstream rate (from user to network). Unlike HDSL, ADSL uses a wider frequency range in the downstream than in the upstream, and this wider frequency range enables greater bandwidth in the downstream direction. ANSI T1.413 specifies the upstream band as 25kHz to 138kHz, with the downstream band occupying the band above (up to a maximum of about 1,100kHz).

Figure 9-7 shows the spectrum usage for ADSL. To begin, we can see that ADSL operates outside the POTS frequencies, which enables POTS to be carried simultaneously with ADSL. We also see two ways in which we can achieve ADSL: *Frequency Division Multiplexing* (FDM) and *Echo Cancellation* (EC). With FDM, the upstream frequencies are separate from the downstream frequencies. With EC, the downstream frequencies include frequencies that are also used in the upstream, as well as higher frequencies up to 1,104kHz. In both cases, we can see that the downstream has a far greater frequency allocation than the upstream, which is one of the main reasons why a far greater speed can be achieved in the downstream compared to the upstream. Another major factor in the downstream speed of ADSL is that it uses frequencies in the downstream that are not much used by other technologies, except for T1 AMI and VDSL. This factor is a major consideration when we consider cross-talk, as discussed later in this chapter.

Figure 9-7
Spectrum usage for
ADSL

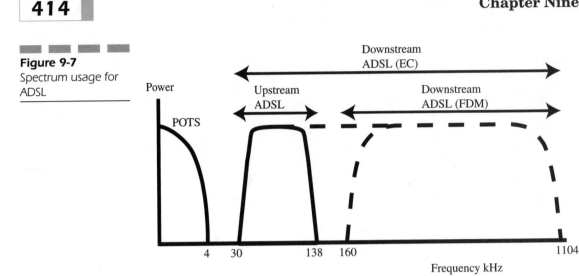

Most implementations of ADSL use FDM, which is easier (and therefore cheaper) to implement within modems. EC enables more bandwidth to be used in the downstream, because it has a broader frequency range. Furthermore, the usage of lower frequencies enables a longer reach, because lower frequencies do not attenuate as quickly. On the negative side, EC is more difficult to implement because it requires extremely accurate echo cancellation. EC also suffers from self-NEXT, as described later.

ADSL provides for downstream speeds of up to 8Mbps with upstream speeds of up to 640kbps, depending on the distance from the customer to the local switch and the characteristics of the physical copper in use (i.e., the gauge of the copper). Whether a given carrier actually offers such speeds to a given customer, however, is another matter. Unlike HDSL, ADSL enables a copper line to be used simultaneously for high-speed data traffic and analog POTS traffic. One can have a PC connected to the Internet over the phone line and still use the phone as normal.

A version of ADSL exists that is known as *Rate-Adaptive ADSL* (RADSL). This version has the advantage of adapting the speed of the service according to the length and condition of the copper wire and taking these uncertainties out of the equation when establishing the service.

Very High Bit-Rate DSL (VDSL)   VDSL offers the highest rates of connection, with a blistering speed of up to 52Mbps in the downstream direction and up to 2.3Mbps in the upstream direction. VDSL achieves this speed by using far higher frequencies than other DSL variants. The VDSL band begins at 300kHz and extends up to 11MHz. Because signals at higher frequencies attenuate more rapidly than lower-frequency signals, however, the reach of VDSL is not far. In fact, these speeds can be achieved

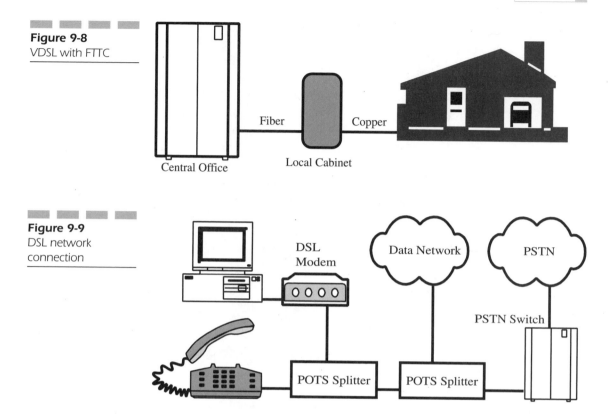

Figure 9-8
VDSL with FTTC

Central Office

Fiber

Local Cabinet

Copper

Figure 9-9
DSL network
connection

DSL Modem

Data Network

PSTN

PSTN Switch

POTS Splitter

POTS Splitter

over only about 300 meters. At 1,000 meters, the downstream speed that can be supported drops to 15Mbps.

Given that most people do not have the luxury of living within 300 meters of a central office, offering the highest speeds that VDSL can provide is not a practical proposition in most cases. The suggestion exists that *Fiber to the Curb* (FTTC) can be used to help overcome this situation, however. In such a scenario, fiber facilities are provided to a cabinet in a residential or business area, and copper line extends from this cabinet to the individual users, as shown in Figure 9-8. This setup, of course, means a large investment. To date, this type of investment has not been widely done, which means that VDSL is not yet widely deployed.

The ADSL Architecture

Figure 9-9 shows a generic network connection for ADSL. At the customer premises, there is an ADSL modem to which a PC or other digital device connects. In fact, depending on the modem in question, the modem might form an integrated-access device with several devices connected to it. In addition, there will be a splitter. The purpose of the splitter is to separate the POTS frequencies from the higher frequencies that ADSL uses.

At the central office location, we find another splitter. At this splitter, the POTS signal is extracted and sent to the PSTN switch, while the DSL signal is sent to a data network (typically, to an ISP).

In real-world ADSL configurations, each individual ADSL connection is not sent individually to an ISP. Instead, the incoming ADSL lines from the customer are collected at a *DSL Access Multiplexer* (DSLAM) from which the connection to the data network originates (as shown in Figure 9-10). In the formal language of the ADSL reference architecture, the DSLAM contains the ATU-C (ADSL Transmission Unit, Central Office), which is the ADSL termination at the CO. The DSL modem at the customer premises contains the ATU-R (ADSL Transmission Unit, Remote). In Figure 9-10, the POTS splitter is shown separately from the DSLAM, although it can be contained within the DSLAM. A variant of ADSL exists, known as ADSL Lite (also known as G.Lite). ADSL Lite removes the need for a POTS splitter at the customer premises, which has the advantage that installation of DSL service at a customer premises is extremely easy.

G.Lite is specified in ITU-T G.992.2. The basic requirement for splitterless operation is to minimize the power injected by ADSL within the POTS band. The major side effect, however, is that the data rates that can be achieved are significantly reduced. For a typical copper pair over a distance of 15,000 feet, ADSL Lite can offer an upstream rate of 256kbps and a downstream rate of 1,024kbps.

Switching in an ADSL Network   Figure 9-10 shows the DSLAM as an ATM device. This arrangement is the most common, although other interfaces are possible (such as Frame Relay). While the choice of ATM is a straightforward one in most cases, the choice of switching technique is less simple and varies from one provider to another. Given that IP will be used over the DSL link and that IP over ATM is most likely to be used on the network side of the DSLAM, the issue is whether switching is to be done at layer 2 or layer 3 (in other words, whether the DSLAM switches traffic purely at the ATM layer or performs IP forwarding).

The choice is one to be made between the service provider (e.g., an ISP) and the access provider (i.e., the owner of the DSLAM). In the case where the two are the same company, then it is quite common for the DSLAM to perform IP forwarding and for there to be routers on the network between

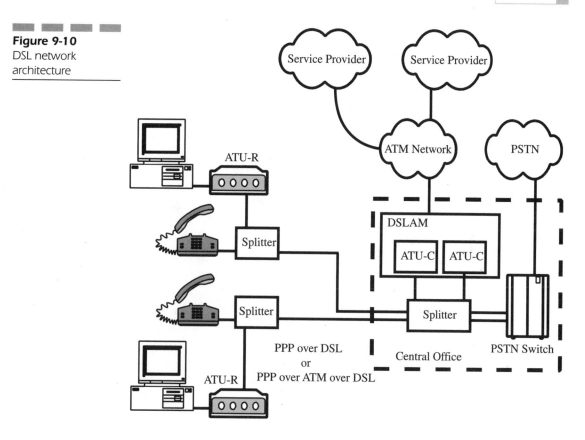

Figure 9-10
DSL network architecture

the DSLAM and the service provider. In fact, the DSLAM can be considered a part of the service-provider network. In the case where the two are separate companies, it is more common for switching to be performed at the ATM layer only. Thus, the network between DSLAM and service provider might contain ATM switches but not routers. In this case, ATM *Permanent Virtual Circuits* (PVCs) are established between the DSLAMs and the service providers. The interworking between the IP access (IP over PPP over DSL) and the ATM network is handled at the DSLAM.

A third choice also exists, which involves extending the ATM network to the user premises. In this case, the ATM PVC from the service provider does not terminate at the DSLAM. Instead, the PVC runs between the service provider and the ATU-R.

DSL Modulation

As previously described, both ISDN and HDSL use 2B1Q coding. This coding effectively provides two bits/s per Hz. If this coding were to be used for ADSL, then a maximum speed of about 2Mbps would be achieved in the downstream by using the frequencies specified for ADSL. We know, however, that ADSL is capable of much faster speeds. The coding scheme used enables this faster speed.

Two main coding schemes have been implemented for ADSL. The first is *Carrierless Amplitude Phase* (CAP) modulation, and the second is *Discrete Multitone* (DMT). Of these, DMT has been standardized by both ANSI and ETSI and is described here.

DMT is similar to *Orthogonal Frequency Division Multiplexing* (OFDM), which is used in some digital RF broadcasting systems. DMT works by breaking the available frequency spectrum into a large number of small chunks, effectively creating many narrow-band carriers. These carriers, also called tones, are all transmitted in parallel—with each carrying a small fraction of the total information to be sent. ANSI T1.413 specifies 256 of these subcarriers, each with a bandwidth of 4kHz (as shown in Figure 9-11). Each of the subcarriers can be modulated to carry from 0 to 15 bits/sec/Hz, enabling up to 60kbps per tone.

DMT is robust and adapts well to line conditions. The modulation can be varied such that there can be more kbps per tone in good line conditions and fewer kbps per tone in poor line conditions. Furthermore, we can use more dense modulation at different frequencies. Given that lower frequencies suffer less attenuation than higher frequencies, the lower-frequency tones can be made to carry more kbps than higher-frequency tones. Also,

Figure 9-11
Discrete Multitone

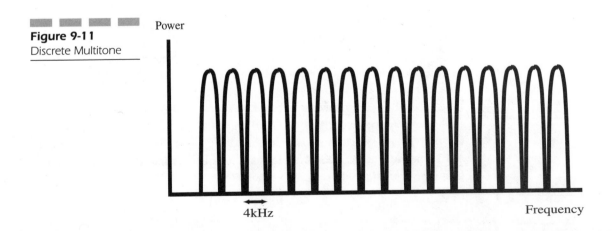

there might be interference in a particular band (e.g., from an AM radio station), and DMT can ensure that this band is used to carry less information.

DSL Challenges

The figures quoted previously for the speed of various DSL solutions should be considered the optimum, particularly for ADSL and VDSL. In most cases, we cannot achieve the maximum throughput that these technologies can support. One of the main reasons is the fact that simple copper lines are used—lines that have been installed for many years and that might not be in the best condition for carrying high-frequency signals. Furthermore, many of the lines have bridged taps.

A bridged tap is an unterminated stub off a line, as depicted in Figure 9-12. The simplest example of a bridge tap is where there are several extension sockets in a residence, but not all of them are used. That is not the only kind of bridge tap, however. In fact, a given line might have many such taps along its length. Historically, when a telephone company provided service to a subscriber, the line from the subscriber was simply tapped into a line in a main distribution system. While bridged taps are not a problem at voice frequencies, they can be a significant problem at higher frequencies. The fact that a stub is not properly terminated means that signals can be reflected. This situation can cause significant loss at certain frequencies. Fortunately, DMT enables these frequencies to be identified and used less than those that are unaffected.

Figure 9-12
Bridged taps

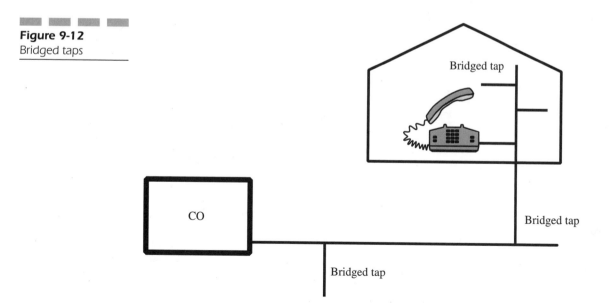

Not only are bridged taps an issue, but the copper lines also might have splices that are less than perfect (they might have been chewed by rats or squirrels, or water might have leaked in). All of these factors impede the performance of DSL.

Perhaps the biggest issue that DSL encounters is cross-talk. Cross-talk occurs when the signal on one line interferes with the signal on another line, and this problem is quite common on copper lines where there might be many lines together in a single conduit. Cross-talk comes in two forms: *Near-End Cross-Talk* (NEXT) and *Far-End Cross-Talk* (FEXT), both of which appear in Figure 9-13. NEXT occurs where a strong signal from a transmit source is coupled to a nearby receiver. FEXT occurs where the signal from one source couples into another line and is received at the far end, along with the desired signal. Of the two types of cross-talk, NEXT is the more damaging. For example, near a central office, the downstream transmit signals are strong—whereas the upstream receive signals are weak. Cross-talk from the transmit side to the receive side can distort the received signal. In the case where the technology uses the same frequencies in the downstream and the upstream, such as with EC ADSL, then there is self-NEXT. An EC ADSL transmitter causes interference to EC ADSL receivers nearby. Self-NEXT is a potential problem for any system where the same spectrum is used in both the transmit and receive directions. Of course, an FDM system does not suffer from self-NEXT because the downstream spectrum does not overlap the upstream spectrum (but it is still subject to cross-talk from other wires in a cable bundle, such as those carrying HDSL, ISDN, or T1AMI). When it comes to the achievable speed that ADSL can offer a customer, cross-talk is one of the limiting factors (but not a show-stopper).

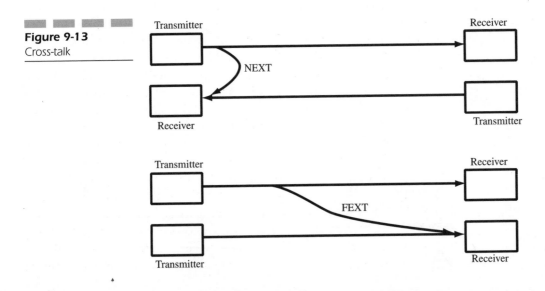

Figure 9-13

Cross-talk

What *is* a show-stopper, however, is the presence of loading coils. These are coils that are placed on copper lines, particularly on long lines of more than 18,000 feet. They are effectively inductors that work with the natural capacitance of the copper pair and help to reduce attenuation at low frequencies, which is fine for voice. The combination of inductor and capacitor effectively creates a low-pass filter and blocks the higher frequencies of DSL, however. In order for DSL to operate over such lines, the loading coils must be removed. Loading coils might exist on some lines that are shorter than 18,000 feet, but fortunately, they are the exception rather than the norm.

Using the Cable TV Infrastructure

The copper phone lines are not the only source of bandwidth for many residences and businesses. In many parts of the United States and other countries, *Cable TV* (CATV) service is available. As standard POTS copper wires provide the opportunity to enable other services besides POTS, so does the CATV infrastructure provide the opportunity to offer other services besides television.

Hybrid-Fiber Coax

CATV systems have been around for many years. For most of that time, they have been based on coaxial cable and have been used to offer TV service only in a broadcast mode. In other words, they have supported one-way transmission only. Obviously, such an infrastructure is not suitable for carrying VoIP-based multimedia services. Recent developments have significantly changed the infrastructure in two ways, however. First, much of the transmission is now conducted over fiber facilities. The architecture is generally *Hybrid Fiber-Coax* (HFC), whereby the first part of the transmission network uses fiber, with coax used for local distribution. Second, most CATV providers have spent a great deal of effort and money to upgrade the infrastructure in order to support transmission in both directions—from the service provider to the customer and from the customer to the service provider. These two enhancements enable the support of many services besides TV. These other services include voice, video, and data.

Figure 9-14 depicts an HFC network. The root of the network is known as the head-end. Here, the TV signals received by antennas enter the network. If the facilities are to be used for communication with other networks

Figure 9-14
Hybrid-fiber coax

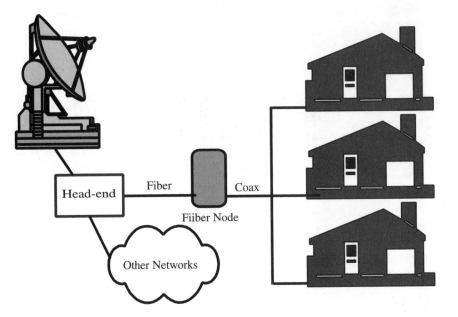

as well as for TV service, then the connection to those other networks is made at the head-end. The facilities from the head-end pass over fiber to a fiber node within a neighborhood. Coax is used from the fiber node to the local residences and business. Note that the HFC architecture is not unique to CATV companies. In fact, in Figure 9-14, a central office could be depicted instead of a CATV head-end, because an LEC or CLEC can also use the HFC architecture as an alternative to DSL. The CATV companies, however, are most actively pursuing the HFC approach.

Data Transmission over the HFC Network

The manner in which the CATV network is to be used for two-way transmission is specified in the *Data-Over-Cable Service Interface Specifications* (DOCSIS). These specifications have been developed by the CATV industry in the United States and are considered the standard.

DOCSIS specifies that data transmission in the upstream direction occupies the band from 5MHz to 42MHz, while downstream transmission takes place at higher frequencies, somewhere between 50MHz and 860MHz. Figure 9-15 gives a pictorial representation of the spectrum usage. Above the spectrum allocated for upstream transmission are a number of TV channels, each occupying a 6MHz channel within the CATV system (8MHz per

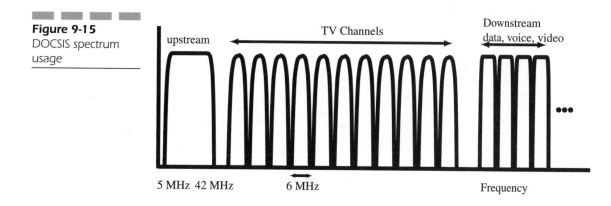

Figure 9-15
DOCSIS spectrum
usage

channel in Europe). Above these channels, the spectrum is available for downstream transmission of video, voice, and data. This downstream information is also transmitted by using a number of 6MHz bands. According to DOCSIS, the point in the spectrum at which the transmission of downstream information may begin is variable, depending on how occupied the spectrum is by TV channels.

The modulation used for transmission of information is either QPSK or 16QAM. QPSK is usually used in the upstream direction, while QPSK or 16QAM can be used in the downstream direction. The reason for using QPSK in the upstream direction is because this piece of the spectrum is more susceptible to interference from outside sources, such as radio stations, electronics, and appliances. While QPSK is not as efficient as 16QAM, it is more robust in the presence of interference.

Data over cable can offer upstream and downstream speeds of tens of megabits per second. Unlike DSL, however, the HFC is a branched structure in which many users might be tapped into the same piece of coax. Consequently, the bandwidth on the coax needs to be shared among the various users, and all users cannot take advantage of top speeds at all times. In fact, DOCSIS specifies mechanisms whereby the available bandwidth can be shared among the users. Because of this requirement to share the available spectrum, practical commercial implementations of the HFC architecture provide subscribers with upstream speeds in the order of 256kbps to 512kbps and downstream speeds of 2Mbps to 3Mbps.

The Packet Cable Architecture

While the DOCSIS specifications describe the usage of the CATV infrastructure (i.e., the functions required of cable modems, etc.), another effort

Figure 9-16
Packet Cable
architecture

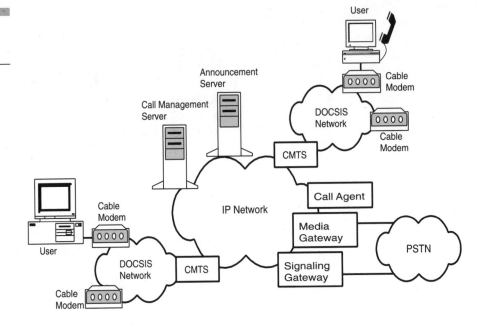

has been focused on the protocols required for the support of voice and multimedia applications over a DOCSIS-compliant infrastructure. This effort is known as the Packet Cable initiative. The corresponding network architecture is shown in Figure 9-16.

The cable modem provides the interface between the user device and the packet cable network. The modem contains a *Media Terminal Adapter* (MTA), which supports media conversion, audio codecs, packet network signaling, interfaces to user devices (including an RJ11 interface for a standard telephone), and analog line signaling. The MTA contains many of the functions associated with a media gateway (described in Chapter 6, "Media Gateway Control and the Softswitch Architecture"). In fact, the signaling between the MTA and the packet network uses MGCP.

At the other end of the DOCSIS access network (e.g., the head-end) is the *Cable Modem Termination System* (CMTS). The CMTS provides access from a number of cable modems to the IP network and can be considered the edge of the IP network. As such, the CMTS handles a number of QOS-related functions, such as allocating bandwidth, classifying packets, and traffic shaping and policing.

The *Call-Management Server* (CMS) contains the call-handling intelligence. From the point of view of an MTA, the CMS appears as a call agent. For example, when the MTA wishes to establish a call, it uses MGCP to notify the CMS. The CMS performs digit analysis and call-routing functions and then uses MGCP to tell the MTA how to establish the media path.

For interfacing with the PSTN, the packet cable architecture uses a call agent, media gateway, and signaling gateway, borrowing heavily from the softswitch architecture. Interestingly, at this writing, the most recent Packet Cable specifications do not elaborate on the signaling used between CMS entities or between a CMS and a call agent. Given the approach used in the rest of the network, however, SIP is the obvious choice for this type of signaling.

Fixed Wireless Access

In many parts of the world, the telephone infrastructure is much less developed than in the United States, and CATV is extremely limited. Nevertheless, a growing demand exists for high-speed communications services. In such environments, however, technologies such as ISDN, DSL, or cable modems are non-starters—simply because the facilities do not exist and the cost to build them would be excessive. To fill the gap, fixed wireless access is an attractive alternative. Furthermore, even in those countries where the communications are better developed, fixed wireless access has certain advantages. This method offers an entryway for those companies that wish to compete in the communications business but that do not have access to existing facilities. Fixed wireless access also offers a means to serve those customers who cannot be served by other technologies for various reasons. In many cases, this method can be deployed faster than other alternatives.

Numerous alternatives exist for fixed wireless access, using various frequency bands. In the United States, there is *Local Multi-Point Distribution Service* (LMDS), which is a point-to-multi-point microwave system. A number of proprietary systems also operate in the unlicensed band, and there is *Multi-Channel Multi-Point Distribution Service* (MMDS). Of these, the one that has seen the most activity and success so far in the United States is LMDS.

LMDS

In 1998, the FCC auctioned a total of 1,300MHz of spectrum in the 28GHz and 31GHz bands for use in providing LMDS service. This service is a fixed wireless service that can be used to offer voice, video, and data services (i.e., multimedia). In addition, one company in particular has been granted permission to use frequencies in the 24GHz band for the provision of LMDS services.

LMDS is usually implemented as a point-to-multi-point microwave system. Customers in a given geographical area are provided with a radio antenna, which provides connectivity with a central hub, as depicted in Figure 9-17. A single hub can serve multiple customers, with the exact ratio depending on the frequency spectrum available to the operator and the bandwidth contracted to each customer. Within a city, dozens of hubs might exist or as little as one hub, depending on the demand for service, the geographical spread of customers, and the topology of the city.

No formalized standards for LMDS exist, so the modulation schemes, access techniques, and other characteristics can vary from vendor to vendor. A combination of *Frequency-Division Multiplexing* (FDM) and *Time-Division Multiplexing* (TDM) is commonly used. A common approach is to use TDM in the downstream direction and FDM in the upstream direction. *Code-Division Multiple Access* (CDMA) is also a possibility. When it comes to modulation schemes, QPSK, 16QAM, and 64QAM are used in various deployments.

Within the customer premises, a network interface unit provides connectivity between the user devices and the external radio network. This system can serve as an integrated access device, supporting a range of devices and applications (as depicted in Figure 9-18).

Figure 9-17
LMDS

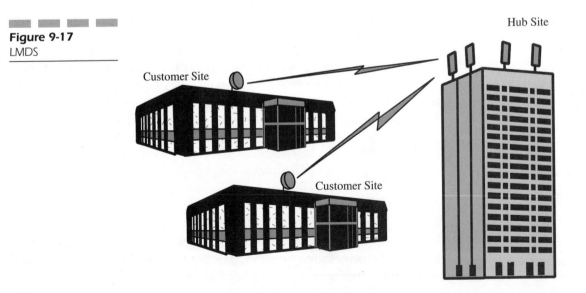

Hub Site

Customer Site

Customer Site

Figure 9-18
LMDS customer
premises

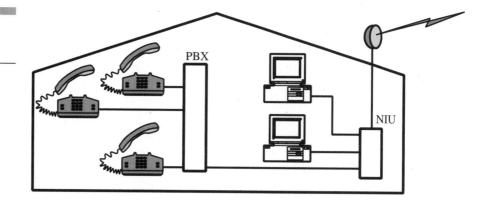

The individual hubs are usually connected to an ATM network. At each hub location, there is an interface to a fiber network. This fiber network normally provides backhaul to an ATM network. Therefore, any IP applications, including VoIP, are carried over IP over ATM.

LMDS can offer speeds of tens of megabits per second in each direction. Such speed is rarely (if ever) encountered in commercial applications, however, because it would consume too much bandwidth and limit the bandwidth that is available to other customers. Commercial offerings are usually in the order or a T1 or E1 worth of bandwidth, which is still pretty good.

Mobile Services

Mobile telephony services have been in widespread use since the mid-1980s. Initially, the systems were analog, and a significant number of these first-generation systems are still in service. Since the early to mid-1990s, however, second-generation digital systems have been implemented by using technology such as IS-95 CDMA, IS-136 TDMA, and GSM. These digital systems have helped fuel an amazing increase in the penetration of mobile service. They have enabled clearer speech, smaller and sleeker handsets, longer battery life, and new services. Among these new services are *Short Message Service* (SMS), which is a means for transmitting and receiving short text messages and data services. Unfortunately, however, the data services use circuit-switched data at low speeds (9.6kbps to 14.4kbps). Consequently, the main application for mobile service is still circuit-switched voice. But things are about to change.

Having seen the amazing growth of the Internet and the demand for multimedia applications, including voice, many in the telecommunications industry have recognized the need for high-speed access from mobile devices and the need to support packet-based communication. They see a world in which mobile devices are no longer just the simple mobile telephones that we have become used to; rather, they see advanced communications devices supporting voice, Web browsing, video, and high-speed data with advanced, intuitive user interfaces. Such groups include network operators, equipment vendors, and standards bodies such as the ITU, ETSI, TIA, and others. Consequently, a great deal of effort has been placed on developing technology for third-generation (3G) mobile services. As part of this effort, three main approaches have been taken: *Wideband CDMA* (W-CDMA), CDMA2000, and UWC-136. The first approach represents the upgrade path for GSM systems; the second represents the upgrade path for IS-95 CDMA systems; and the third represents the upgrade path for IS-136 TDMA systems. In each case, the overall objective is the same—to provide high-speed packet-based communications over the air in support of voice and data applications.

When it comes to network operators, we see a willingness to spend enormous amounts of money to gain access to the necessary spectrum in order to offer 3G services. At this writing, an auction of 3G spectrum in the United Kingdom has recently finished. The net result is that five licenses have been awarded, and the British treasury is richer to the tune of more than $35 billion. A similar auction is planned for the United States in March 2001, although at a different frequency band. Nevertheless, the result in the United States is likely to be similar: huge amounts of money committed for the opportunity to offer advanced multimedia services to mobile subscribers.

The following section provides a brief overview of the changes being made to bring high-speed access to mobile networks in order to enable fast multimedia communications. Note that these are broad issues, and we would need a separate book to address them in significant detail.

The Evolution to 3G for GSM

The evolution from GSM as we know it towards 3G is happening in steps. The first step along this road is the introduction of packet data support over the air interface, using a technology known as *General Packet Radio Service* (GPRS). This service provides the capability to support packet-switched data instead of circuit-switched data applications over the air.

Standard GSM uses 200KHz channels divided into eight time slots. These time slots are normally used to carry a single voice conversation in a circuit-switched mode. They can also be used to carry circuit-switched data, typically at 9.6kbps, although a channel coding supporting 14.4kbps has been developed. All that GPRS does is enable some of these time slots to be used in a packet-switched mode. This function offers efficiency in carrying data applications such as Web browsing, because a given subscriber only uses a time slot while something is being sent. Therefore, a user is not occupying bandwidth while just contemplating a Web page, as would be the case with circuit switching. Thus, GPRS enables more efficient bandwidth usage but does not magically increase the available bandwidth. Therefore, if a given 200KHz channel with eight time slots is completely dedicated to GPRS, then the maximum rate that can be achieved is either 76kbps or 115kbps, depending on channel coding. In most cases, the maximum is 76kbps. In reality, this rate is unlikely to be provided to any one user, because it consumes significant air interface capacity. The next step makes such speeds and higher much more feasible in real networks, however.

The next step is *Enhanced Data Rates for GSM Evolution* (EDGE), which changes the modulation scheme that is used over the air. In standard GSM, the modulation scheme is *Gaussian Minimum Shift Key* (GMSK). With EDGE, the modulation scheme is changed to 8PSK, which is more bandwidth efficient. GMSK can provide for more than a three-fold increase in bits/s/Hz. With both GPRS and EDGE, customers can achieve rates of perhaps 384kbps at the top end. Again, however, such rates are unlikely to be offered to any single customer in a practical network implementation. Both GPRS and EDGE could be considered part of generation 2.5. They represent significant improvements over existing deployed systems but do not meet the goals of 3G. A great advantage of both is that they still use the existing 200kHz channel spacing in GSM, which means compatibility with deployed systems. In fact, deployed systems can be (and are being) upgraded with GPRS and EDGE capabilities.

While EDGE and GPRS developments are certainly wonderful from a data communication point of view, they do not mean a great deal for voice communication, because voice is still expected to use circuit switching even when EDGE and GPRS are in place. Although GPRS is a packet technology, it was not originally designed to accommodate the delay sensitivities of voice and is not planned to carry voice initially.

The real deal when it comes to 3G for evolved GSM systems occurs when a number of things change. The first and most important of these is the air interface. Instead of using a TDMA access technique, the 3G solution will use W-CDMA with a 5MHz channel width, as opposed to today's TDMA

200KHz channel. The second major change is that circuit switching will ultimately be abandoned completely in favor of packet switching and transport. In other words, for IP and for voice, we mean VoIP. At that time, speeds of up to 2Mbps will be available over the air interface to support voice and a range of multimedia applications. Also, of course, the real killer application will still be there: mobility.

The 3GPP All-IP Architecture The network architecture for development of the W-CDMA architecture is being handled within an organization called the *Third-Generation Partnership Project* (3GPP). The group has already developed a set of specifications known as Release 99, and work is ongoing for an updated set of specifications known as Release 2000. Included in this work is an all-IP architecture that takes the general form shown in Figure 9-19. This system can be considered an evolution of the GPRS architecture, and a number of elements will be familiar to those who are familiar with GPRS. Note that it is possible for some of the network elements shown in Figure 9-19 to be combined on the same physical platform.

The central call-handling function is the *Call State Control Function* (CSCF). CSCF manages calls to and from mobile subscribers and has some similarities to a proxy server, as used in SIP architecture. In order to manage subscribers' calls, CSCF interfaces with other networks (such as a subscriber's home network) and with external IP networks. The CSCF also interfaces with a number of other functions on the network.

The *Gateway GPRS Support Node* (GGSN) is the interface to the outside world for media streams to and from the mobile subscriber. The SGSN is the serving GPRS support node, and it provides the interface between the core network and the *Radio Access Network* (RAN). The SGSN keeps track of the location of mobiles as they move about the network and controls access to and from those mobile subscribers.

The *Home Subscriber Server* (HSS) is the master database for subscribers and contains all subscriber service and feature data. The HSS also keeps track of subscribers' locations, although not to the same granularity as the SGSN.

The MGC and MGW are media gateway functions, as described in Chapter 6. They provide a means to interface with other networks, such as the PSTN.

The *Roaming Signaling Gateway* (R-SGW) supports signaling to legacy mobile networks for roaming purposes. R-SGW supports the Sigtran protocols, such as M2UA, M3UA, and SUA, to enable conversion from IP-based signaling transport within the all-IP 3G network to external SS7 MTP-based transport. The *Transport Signaling Gateway* (T-SGW) deals solely with signaling conversion for call-related signaling.

Figure 9-19
3GPP All-IP network
architecture

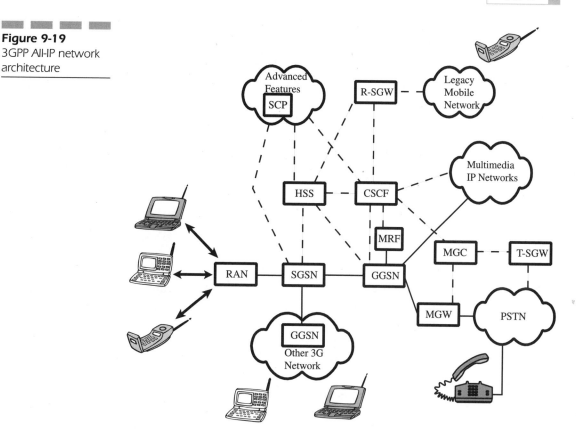

The *Multimedia Resource Function* (MRF) offers multi-party and multi-media services, such as conferencing, and performs a similar function to an MCU in the H.323 architecture.

The Evolution to 3G for IS-136 TDMA

For existing IS-136 operators, the evolution to 3G is somewhat similar to that for GSM. The approach involves the use of GPRS/EDGE technology as a stepping stone for higher-speed services. This approach will enable speeds of up to 64kbps as a first step. The subsequent step is to change the air interface. Currently, the IS-136 channel bandwidth is 30kHz. With the changes planned as part of 136 HS (high speed), two choices will exist for channel bandwidth. The first, 200kHz, will enable speeds of up to 384kbps and is aimed for use in outdoor mobile environments. The second, 1.6MHz, will facilitate speeds of up to 2Mbps and is aimed at indoor, low-mobility environments (e.g., non-vehicular).

Unlike the other technologies described here, the evolution from existing IS-136 TDMA involves a continued use of TDMA as the access technique on the air interface. The core of the network, however, will be much like that of the GSM/WCDMA camp (as shown in Figure 9-19) because of the GPRS and EDGE evolution paths. By using the same overall approach in the evolution of the non-radio part of the network, the network architecture of an evolved TDMA network becomes almost identical to that of an evolved GSM network.

The Evolution to 3G for IS-95 CDMA

For existing IS-95 CDMA network operators, the path to a third generation is also a step-wise effort and is aimed at maintaining compatibility with existing systems. The first step is called 1XRTT, which continues to use the existing channel bandwidth of 1.25MHz but can offer speeds of up to 144kbps over the air. The second stage, known as 3XRTT, involves an increased channel bandwidth of 3.75MHz, providing a speed of up to 2Mbps. Further steps using greater multiples of 1.25MHz are also envisioned for the future, with bandwidths of 6×, 9×, and 12× 1.25Mhz.

From the point of view of the core network architecture, some work has been done within the *Third-Generation Partnership Project 2* (3GPP2) towards defining a packet network architecture. At this writing, this architecture addresses a packet transport, but only for packet data services and not for voice. 3GPP2 intends for voice to still be circuit switched. Given that the network architecture is still in draft form, this situation might change.

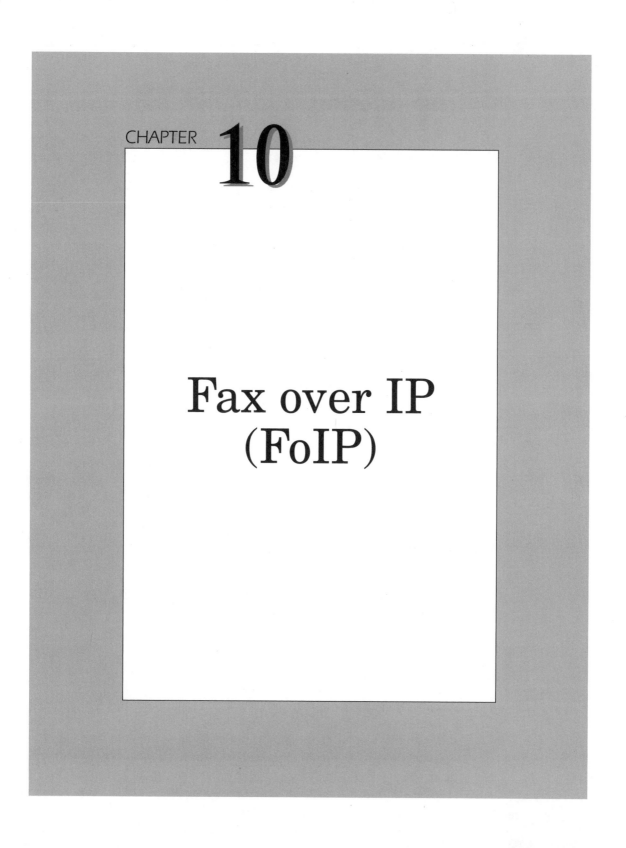

CHAPTER **10**

Fax over IP (FoIP)

Introduction

Facsimile (fax) service has been around for many years. Most offices and many homes have fax machines. While fax service might be considered an essential business tool, this service is not something that usually generates much excitement. Many people do not think much about a fax machine until it needs paper or toner. This attitude is understandable. After all, intelligent services, conference calling, and other advanced features that are common in the telephony world usually do not apply to fax service, so there is not much to get excited about (unless, of course, you start to look at the costs and revenues associated with fax service). Recall Figure 1-6 from Chapter 1, "Introduction," and recall the fact that *Fax over IP* (FoIP) is projected to be a major source of revenue for IP networks in the future. This statement suggests that FoIP deserves some consideration. That is the purpose of this chapter.

Today's Fax Service

Before discussing FoIP in any detail, we must recognize that FoIP is used to send faxes from IP-enabled devices to traditional fax machines and vice-versa. Furthermore, situations will arise in which traditional fax machines are connected to an IP network via gateways and communicate with each other over the intervening IP network without realizing that the network is packet-switched, rather than circuit-switched. Therefore, interoperating between traditional circuit-switched fax service and IP-based fax service will be of major importance. Hence, before proceeding with any description of FoIP, we should review the workings of circuit-switched fax service so that we can view the interworking issues from the appropriate perspective.

T.4 and T.30

Fax service works according to two main specifications: ITU-T Recommendation T.4 and ITU-T Recommendation T.30. T.4 addresses the formatting of fax images and their transmission from sender to receiver. To this end, the protocol addresses how the fax machine scans the documents, the coding of scanned lines, and the modulation and transmission scheme used. Depending on the speed of transmission, we will use one or another of the ITU-T V.series recommendations. For transmission at 14,400bps, V.17 applies.

T.30 is a control protocol between the machines and deals with communication that is not related to the actual transfer of scanned information. This communication includes call establishment, capabilities exchange, end-of page and multi-page signaling, and call release. For example, T.30 defines the tones that fax machines generate at the start of a call. Additionally, T.30 defines a number of messages sent between fax machines during a call. These messages are used for one fax machine to inform the other of its capabilities, to signal the end of a page or the end of a document, etc. The signals are sent at a speed of 300bps and are modulated according to ITU-T Recommendation V.21. In other words, although the signals begin as digital messages, they are modulated to analog form by using V.21.

Figure 10-1 shows a typical fax call. Once the user presses the start key and the number is dialed, the sending machine starts sending a *calling tone* (CNG). This tone is a short, periodic tone of 1,100Hz. The tone transmits for a duration of 0.5 second every 3.5 seconds. The main purpose of the tone is to let the called terminal know that a fax machine is calling. If the called number is a combined telephone/fax machine, then the person who answers can put the called device into fax mode. Note that the CNG tone is only sent from an automatic fax machine—one that does not require human intervention. In the case that the calling machine is a combined telephone/fax machine where the user must use a handset to dial, then CNG is not transmitted. The simple reason is because the person might wish to place a voice call instead of a fax call and would not want the called party to think otherwise.

Once the call is answered, the called device returns a *Called Terminal Identification* (CED) tone, which lasts somewhere between 2.6 seconds and 4.0 seconds. This 2,100Hz tone is continuous. The called device then sends a T.30 *Digital Identification Signal* (DIS), which indicates its capabilities (e.g., the capacity to receive data at 14,400bps). The device might also send a *Called Subscriber Identification* (CSI) message, which contains the phone number of the called terminal (not shown). Equally, the called terminal might send a *Non-Standard Facilities* (NSF) message to indicate any proprietary capabilities that the device might support.

Upon receipt of the DIS, the calling terminal sends a *Digital Command Signal* (DCS), which indicates the mode of transmission to be used (e.g., the speed), which will align with the capabilities of the called terminal as indicated in the received DIS message. The terminal might optionally send a *Transmitting Subscriber Identification* (TSI) message that contains the telephone number of the calling terminal (not shown). The terminal then sends a *Training Check* (TCF), which involves a training sequence sent on the high-speed (T.4) channel. The sequence is a series of zeros transmitted for 1.5 seconds.

Figure 10-1
Basic Group 3 fax call

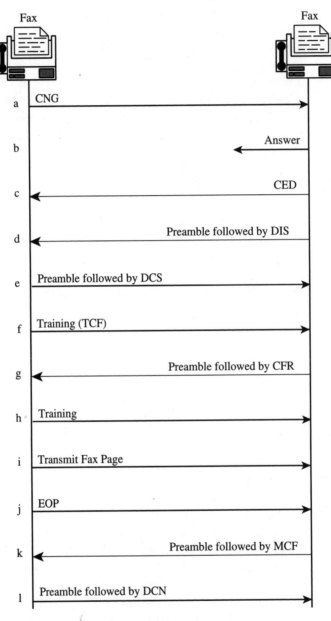

Once the training sequence has been successful, the called terminal sends a *Confirm to Receive* (CFR) message, indicating that the training has occurred correctly and that the terminal is ready to receive the faxed document. At this point, a training sequence is sent, followed by a page of the document (according to whatever speed/modulation scheme that the two terminals have agreed upon).

Once a page has been sent, and if there are no more pages to be sent, then the sending terminal sends an *End of Procedure* (EOP) message. Not only does this message indicate the end of the page, but it also indicates that no more pages are to be sent—whether from this document or from any other document. Upon receipt of the EOP, the called terminal responds with a *Message Confirmation* (MCF) message, confirming that the page was received. The sending terminal then sends a *Disconnect* (DCN) message, and the two terminals hang up.

If we had needed to send more than one page, then the calling terminal would have sent a *MultiPage Signal* (MPS). The called terminal would still respond with MCF to indicate that the signal was received and that it was ready to receive more pages. The calling terminal would have then sent a training sequence followed by the next page.

All messages sent on the T.30 (low-speed) channel use an HDLC frame structure. During the exchange of capabilities, end of procedure/multi-page signaling, and disconnection, a preamble precedes all messages sent on the T.30 channel. This preamble is a series of HDLC flags that are sent for a period of 1.5 seconds. The purpose of the preamble is to ensure that all elements of the channel (e.g., echo cancellers) between the two terminals are appropriately conditioned to ensure unhindered passage of the subsequent message.

The actual page transmission is performed according to Recommendation T.4. For a simple black-and-white transmission, the image is scanned one line at a time, with a density of at least 3.85 lines/mm and a minimum of 1,728 black or white picture elements (pels) per line for a typical letter size page. This value corresponds to 200 pels/25.4mm. These are minimum requirements, however, and many machines exceed these minimums.

T.4 specifies a number of code words that can be transmitted. Each code word indicates a number of contiguous black or white pels, and a special code word is used to indicate the end of a line. Thus, if a given line contained two black sections separated by a white section, then all that would be transmitted is a code word for a black run of a certain length, followed by a code word for a white run of a certain length, followed by a code word for a black run of a certain length, followed by an *end-of-line* (EOL) marker.

The maximum transmission time for a single line must be fewer than 13 seconds, according to the most recent version of T.4. Earlier versions of the recommendation have this limit set at five seconds, however, which means that current devices need to comply with the five-second limit in order to ensure compatibility with older devices.

At the end of a page, the device sends the *Return to Control* (RTC) signal, which is a sequence of six EOL code words. The RTC is used to indicate to the far end that it should expect a message on the T.30 channel (for example, EOP or MPS).

NOTE: *T.4 enables a great deal more to be transmitted than just indications of black and white. Over the years, T.4 has been enhanced through a number of annexes that describe several optional functions. For example, a T.4 fax can operate in character mode, whereby representations of Roman characters can be sent. Also, it is possible to send a binary file by using T4, you can send images using gray-scale or even color instead of just pure black and white. In the case that documents are sent using grayscale or color, images are encoded according to ITU-T Recommendation T.81 (JPEG) and are sent as binary data to the far end.*

You should note that with the exception of the tones generated at the beginning of a call, all of the information sent by a fax machine starts off in digital format. This situation applies whether the information is a T.30 signal or the actual information on the page. For example, scanned white and black segments are stored as digital code words for transmission. With a regular Group 3 fax, the information is modulated to an analog format for transmission over the regular telephone network. The information is then converted back to digital at the far end. The fact that it starts as digital and ends up as digital, however, suggests that with a digital network all the way through, the information could be sent in digital format from sender to receiver without ever being converted. FoIP can provide that function.

Real-Time Fax

ITU-T Recommendation T.38 was developed to address real-time communication between fax devices over an IP network. This recommendation is basically aimed at supporting real-time fax without regard to the IP network architecture in place. The approach assumes that native Group 3 fax devices are connected to gateways, where those gateways provide the interface between the IP network and a standard fax line. IP-enabled fax devices might also exist, which interface directly with the IP network. Such a device could be, for example, a PC with fax send-and-receive capability or a pure-IP fax device. For such IP-enabled fax devices, the gateway function is considered embedded within the device itself. Figure 10-2 shows the basic network model.

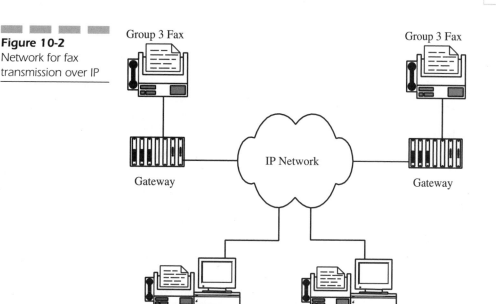

Figure 10-2
Network for fax
transmission over IP

The primary function of the gateways is to ensure that the real (packet-based) characteristics of the transport network are hidden from the native G3 fax devices. In other words, the gateways must demodulate the voice-band analog signals, such as CNG and CED, and transmit an equivalent IP-based signal to the far end. At the far end, the process works in reverse so that the original voice band signal is presented to the native G3 fax machine. In addition, the gateways must convert the T.30 control messages to and from native HDLC format to a packet format and must, of course, similarly convert the actual T.4 message data itself. Furthermore, a gateway must deal with the fact that packets might be lost or delayed on the IP network. This process involves generating HDLC flags locally, which prevent a fax machine from timing out while waiting for some response from the far end. Finally, the gateway must buffer packets to avoid the effects of jitter on the IP network. As mentioned in Chapter 1, jitter is not a factor in circuit-switched networks, which means that a native G3 fax device is not prepared to handle jitter.

Note that a gateway can be directly connected to a fax machine or can be located within the network and connected to the fax machine only on a per-call basis. For example, the gateway function might reside within a more general-purpose PSTN-IP gateway, with the fax functionality of the gateway invoked only when a fax call is to be carried.

Sending Fax Messages over IP

The information to be sent from one gateway to another is sent in a number of IP packets, each of which can have two parts: a message type field and (optionally) a message data field, as shown in Figure 10-3. The message type field values fall into two main categories: T30_Indicator or T30_Data. T30_Indicator messages do not carry a data field; rather, the message is just an indication and does not require additional data. T30_Indicator messages indicate CNG or CED tones, a T.30 preamble, or training according to the modulation scheme and speed chosen (e.g., V.17 14,400bps or V.29 9600bps). T30_Data type messages do contain a data field. Depending on the message in question, the data field carries either a T.30 HDLC command or a T.4 fax image segment. The actual value of the T30_Data message type indicates the type of information being carried within the data field.

We can further divide the data field into field-type and field-data parts. The field type indicates whether the information contains a T.30 HDLC signal or a T.4 image segment. In the case of T.30 messages, such as CFR and MCF, the field type indicates a T.30 HDLC message and the field data carries the corresponding T.30 HDLC frame.

Not all parts of the HDLC frame are sent in a given field-data element, however. To start, HDLC flags are not included, and the HDLC frame-check sequence is sent separately. In fact, the HDLC frame-check sequence, used to indicate the end of an HDLC frame in standard T.30 signaling, is sent as a field-type element with no field-data associated. To illustrate how this process works, consider Figure 10-4, which shows an IP packet that contains the information related to two T.30 HDLC frames. The message type (V.21 channel 2) indicates that the packet contains T.30 control information. The message data includes several concatenated pieces of HDLC information. The first portion of the data indicates the data field type as being an HDLC frame. This information is followed by the actual HDLC frame itself, which is structured according to T.30 and would indicate a signal such as CFR, EOP, etc. The *Frame-Check Sequence* (FCS), however, which is normally sent with the HDLC frame in standard T.30, is not sent as part of the HDLC data but as a subsequent field-type element. This field type does not

Figure 10-3
Packet structure for IP Fax

Message Type	Data (optional)

Figure 10-4
Concatenation of fields

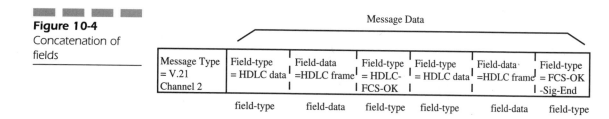

require any field-data information. In Figure 10-4, the field type HDLC-FCS-OK indicates that the FCS of the previous HDLC frame was correctly received at the gateway from the transmitting fax machine. This field type also indicates that more HDLC information is to follow. Furthermore, another field type appears that indicates another HDLC frame. The second HDLC frame follows, again without the FCS. This information in turn is followed by the HDLC-FCS-OK-Sig-End field type, which indicates that no more HDLC information is being sent for now.

In the case where an actual T.4 image segment is to be sent, then the message type indicates the speed and modulation type being used (for example, V.17 14,400bps), and the message data contains the actual image information.

Figure 10-5 provides an example of the message exchange between two standard fax machines that are communicating with each other via two gateways and via an intervening IP network. The message exchange, as shown in the figure, begins upon answering of the incoming call at the called fax machine. By comparing the signal flow in Figure 10-5 with that in Figure 10-1, you can see how we achieve the mapping from native fax-related signaling to and from IP signaling.

a–b. The called fax machine generates a CED tone. The gateway detects the tone and maps it to the T30_Indication equivalent to the CED.

c. The CED indication is received at the gateway of the calling fax machine, and the gateway generates a CED tone towards the fax terminal. Meanwhile, at the called end, the called fax device is generating a V.21 preamble in preparation for sending a T.30 DIS signal.

d. The gateway detects the preamble and generates the T30_Indication that corresponds to a V.21 preamble.

e. The gateway at the calling end receives the preamble indication and forwards the T.30 preamble to the calling fax. Meanwhile, at the called end, the fax device sends the T.30 DIS signal.

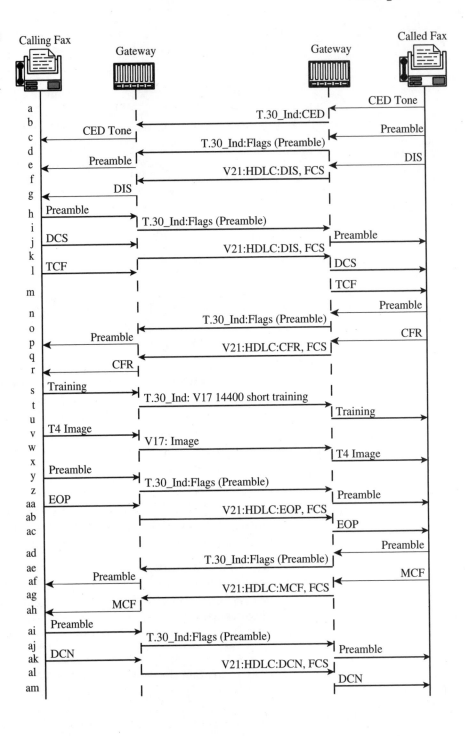

Figure 10-5
FoIP call using
gateways

f–g. The gateway at the called end places the DIS and a frame-check sequence into an IP packet and sends the packet to the calling gateway. The calling gateway receives the IP packet and sends a complete HDLC frame for the DIS signal, including frame flags and the frame-check sequence.

h–i. The calling fax prepares to send the T.30 DCS message, and like all such messages, it first issues a preamble. The gateway at the calling end receives the preamble and sends a T30_Indication to the called gateway, indicating the preamble.

j. The gateway at the called end receives the preamble indication and sends a native T.30 preamble towards the called fax device. Meanwhile, having sent a preamble, the calling device proceeds to send the DCS signal.

k. The gateway at the calling end receives the DCS signal from the calling fax device and sends an IP packet to the gateway at the called end, which contains the DCS HDLC command and a frame-check sequence.

l. The gateway at the called end sends a native T.30 DCS signal to the called fax device. Meanwhile, having sent the DCS command, the calling fax machine begins a training process that involves sending a series of zeros, modulated in the same manner as will be used for transmission of the T4 fax data. The gateway receives the training sequence but does not transmit anything in turn.

m. Having sent a DCS signal to the called fax device, the gateway at the called end sends a training sequence to the called fax device. The gateway performs this procedure without any particular indication from the calling side. Basically, the gateway knows that a training sequence will follow the DCS and does not need any external stimulus as a reminder to generate the sequence.

n–o. Having received the training sequence, the called fax device prepares to send the CFR signal. As with all T.30 control signals, however, the device must first send a preamble. The gateway receives the preamble and sends a corresponding T30_Indication for the preamble to the calling side.

p. The gateway at the calling side receives the IP preamble indication and sends the corresponding native T.30 preamble to the calling fax device. Meanwhile, having sent a preamble, the called fax sends the T.30 CFR signal to the gateway.

q–r. Upon receiving the T.30 CFR signal, the called gateway sends an IP packet containing the CFR and an indication of the HDLC frame-check sequence. At the originating end, this IP packet causes the gateway to send a CFR signal to the calling fax device.

s–u. Having received the CFR signal, the calling fax machine first begins the training process to prepare the receiver for reception of the T4 fax data on the fast channel. The training sequence is sent to the local gateway, which simply sends an indication to the far gateway that the training has occurred. The gateway at the called end uses this indication to generate a training sequence towards the called fax.

v–x. Having sent the training sequence, the calling fax machine proceeds to send the actual fax data on the fast channel that was agreed upon by the two fax devices. This data is passed in native form to the local gateway, which packages the information into one or more IP packets for transmission to the remote gateway. At the called end, the gateway extracts the data from the IP packets and sends the information to the called fax. The gateway must make sure that the information is played out to the called fax with correct timing, compensating for any jitter on the IP network.

y–z. Once the page has been sent, the calling fax prepares to send the T.30 EOP message but must first send the preamble. The local gateway receives the preamble and sends an IP T30_Indication for the preamble to the far end.

aa. The gateway at the called end uses the T30_Indication for the preamble, in order to trigger a preamble towards the called terminal. Meanwhile, after having sent a preamble, the calling device sends a T.30 EOP signal to indicate the end of the fax.

ab–ac. The gateway at the calling end sends an IP packet containing the EOP signal and an HDLC frame-check sequence. The gateway at the called end uses this information to generate the EOP signal locally in native T.30 format.

ad–ae. Having received the EOP, the called fax generates a preamble in preparation for sending the T.30 MCF message. The gateway generates an IP indication for the preamble and sends it to the calling gateway.

af. The gateway at the calling end uses the received indication to generate a preamble locally. Meanwhile, after having sent a preamble, the called fax proceeds to send the T.30 MCF signal.

ag–ah. The called gateway generates an IP packet, including the MCF signal and an HDLC frame-check sequence. This packet is sent to the calling gateway, which generates the T.30 MCF signal locally.

ai–aj. The calling terminal, upon receipt of the MCF signal, sends a preamble in preparation for sending a DCN. The local gateway receives the preamble and sends an IP T30_Indication representing the preamble.

ak. The called gateway receives the indication and triggers a T.30 preamble to the called fax device. Meanwhile, the calling fax device, having sent the preamble, sends a DCN.

al–am. The local gateway receives the DCN and generates an IP packet containing a DCN signal and a frame-check sequence, which it sends to the remote gateway. Upon receiving the packet, the remote gateway sends a native T.30 DCN signal to the called terminal. The call is now complete, and the two fax terminals hang up.

Fax and H.323

The previous descriptions explain how various T.30 messages or T.4 data can be structured into IP packets for transport across an IP network. We should recall, however, that a fax is basically a type of phone call that requires call establishment and release signaling in the same way that a voice call requires such signaling. In any case, each end of the FoIP call must know the IP address and port number to which the fax messages should be sent. To this end, in 1999, an amendment was added to recommendation T.38. This amendment specifies the call-establishment procedures for a fax call that uses T.38. Basically, the amendment documents how real-time FoIP can be accommodated within the H.323 architecture and describes the mechanisms to be used in H.323 fax call setup and some extensions to H.245 for support of fax. We should note that H.245 already has built-in support for T.38 fax service. Therefore, the H.245 extensions specified in Amendment 1 to T.38 are quite minor.

Basically, call establishment is performed in the same manner as a voice call in H.323, with the requirement that the Fast-Connect procedure of H.323 must be used. The procedure is described in Chapter 4 ("H.323") for voice calls and is quite similar for fax calls. The originating fax-enabled device or gateway sends a H.225.0 Setup message containing a faststart element. This element contains a number of OpenLogicalChannel structures specifying the options that the sending end wishes to use. The called end responds with a faststart element in one of the H.225.0 response messages, up to including the Connect message. The faststart element of the response indicates the choice of OpenLogicalChannel structures chosen by the responding gateway from those offered by the calling gateway.

In the faststart element of the Setup message, the calling terminal specifies a TCP/UDP port that it wishes to use for the fax transfer and indicates a preference for either TCP or UDP (but should have the capability to handle either). The receiving terminal makes the actual choice of transport, and the receiver indicates its choice of either TCP or UDP in the faststart element of the selected H.225.0 response message. In general, the receiver should comply with the preference of the sender. Of course, the response from the receiver also includes a port number to which the fax message and T.30 signals should be sent.

Note that the logical channel signaling that the Fast-Connect procedure supports does not involve any negotiation of fax capabilities (for example, V.17 at 14,400bps) between the two fax machines. This function is performed subsequently by using the messages of T.30, as described previously. In fact, once the call is established, it proceeds in the manner shown in Figure 10-5. Once the fax transmission is complete and one terminal hangs up, then the H.225.0 RLC message is used to release any signaling channels between the two gateways.

In addition to the amendment to T.38, a new Annex, Annex D, has been prepared for H.323 to address the same issue. The approach is quite similar, although it does not have the mandatory requirement for the Fast-Connect procedure to be used. Thus, this approach also enables the use of a separate H.245 control channel. In addition, the new Annex enables fax and voice to be carried in the same call. As of this writing, Annex D to H.323 has not yet been published.

TCP or UDP

Unlike VoIP, FoIP (as described in the preceding sections) can use either TCP or UDP. The reason why TCP becomes an option is that fax is not as delay sensitive as real-time voice. Granted, we do not want a page to disappear into one fax machine and come out of the destination fax machine many minutes later. Small delays measured in seconds are not a problem, however. On the other hand, fax is far more loss sensitive than voice. In terms of sensitivity to delay and loss, fax is much closer to data than to voice. For that reason, TCP is the preferred choice of transport protocol.

Fax and SIP, SDP, and RTP

The IETF's approach toward real-time fax service is different from the ITU's approach. The approach suggested by the *PSTN and Internet Interworking* (PINT) group of the IETF involves the addition of fields to SDP to enable SDP to support an indication that fax transmission is to occur. The extensions to SDP would be included with the SDP description in an SIP INVITE.

Recall that the various pieces of information that we send over a fax call start their life in digital form but are transmitted in analog form (really, just a series of tones). Also recall that in Chapter 3, "Speech-Coding Techniques," we discussed a new RTP profile for the support of the various tones generated in the PSTN, including fax tones. Using that profile, gateway devices can send an RTP representation of a tone received from a fax machine, and the remote gateway can regenerate the tone at the far end. This process would accommodate the various messages that go back and forth during a fax call, without the need for additional SIP signaling. Thus, once a call is established by using SIP and is answered, then the tones used between the fax machines can be communicated by using the applicable RTP representation of those tones.

The problem, however, is that it might not be obvious that a call is a fax call until it is answered and a CED tone is generated. A CNG tone at the beginning of the call might let the gateway on the transmitting side know that a fax call is being attempted, but the CNG tone is not guaranteed. Therefore, CED might be the first indication of a fax call, and by the time the CED is generated, session descriptions have already been exchanged between the two ends of the call (e.g., gateways controlled by separate MGCs). Those session descriptions might not have included any information regarding a fax call. This area is one in which the SIP INFO method could be used in conjunction with a MEGACO notification. A gateway connected to a called fax machine could detect CED and issue a notification to its MGC. That notification could be used to trigger an SIP INFO message to the calling MGC to indicate that the CED has been detected. The MGC at the calling end could then instruct the MG to switch over to a fax modem, as opposed to a voice-specific modem.

The IETF has not yet fully fleshed out the full details of real-time fax machine-to-fax machine communication, and we can expect to see further work in this area. In fact, RFC 2542 ("Terminology and Goals for Internet Fax") addresses the issue and specifies the requirements to be met. RFC 2542 also addresses another issue, however: an e-mail based fax service. This issue has garnered most of the focus within IETF to date and is discussed next.

E-Mail-Based Fax

While an overall ambition exists within the IETF to support fax in many different guises, as documented in RFC 2542, the approach has been to use existing IP-based protocols rather than to invent something new to support FoIP. Consequently, a great deal of the effort has been focused on a type of fax service that uses existing e-mail technology. In other words, the solutions to date have involved sending and receiving fax documents in the form of e-mail.

RFC 2305 describes a means of sending fax by using Internet e-mail. The basic goal is that an e-mail user should be able to send a message to a fax recipient, and a fax user should be able to send a fax message to an e-mail recipient. The specification describes a G3Fax device, which is a standard fax device as specified in ITU-T T.4, and an IFax device, which is an IP-enabled device that can send or receive Internet fax messages. In the context of RFC 2305, an Internet fax message is basically an e-mail. Thus, an IFax device is accessed via an e-mail address. An IFax device could be a regular PC or other workstation, but it could, however, be a device much like a standard fax machine in appearance (with the exception that it has an IP interface, rather than a standard analog-line interface). Therefore, when the device receives a message, it could proceed to print the message immediately rather than store it (as is normally done by an e-mail application).

Figure 10-6 illustrates the overall network architecture. Two types of gateways exist: an on-ramp gateway and an off-ramp gateway. An on-ramp gateway provides a connection from a G3Fax device towards the IP network for calls from the G3Fax device to an IFax device. This gateway converts standard fax information to an e-mail format for sending to one or more IFax devices. An off-ramp gateway handles calls in the other direction: from an IFax device to a G3Fax device. This gateway converts an e-mail to fax format and sends it to a G3Fax device via standard T.30 and T.4 communication.

In terms of image transmission, the approach is to use the *Tag Image File Format* (TIFF) for the transmitted fax. Basically, a fax is sent to an e-mail recipient as a TIFF image, and an e-mail is sent to a fax off-ramp gateway as a TIFF image.

Perhaps the biggest issue to be overcome is addressing, and significant effort has been applied in this area. Two specifications have been developed that have particular significance in this area. The first, RFC 2303, describes the requirements for PSTN addressing formats to be used with e-mail. The second, RFC 2304, focuses on the inclusion of fax addresses within e-mail.

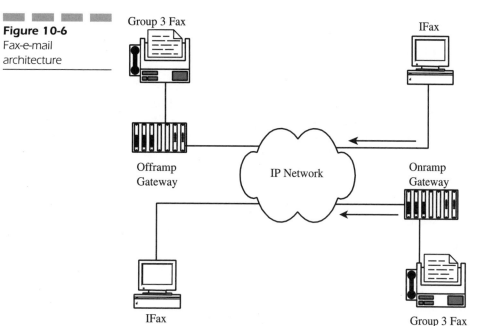

Figure 10-6
Fax-e-mail
architecture

An e-mail system that implements the specification would enable the inclusion of a fax address in the TO: field of an e-mail. This address would have the following format:

```
FAX=+12145559999@faxesrus.org
```

An e-mail sent to such an address would be routed to an appropriate off-ramp gateway. Upon receipt of the e-mail, the gateway would outdial to the fax number that is included in the e-mail address. Upon answer of the terminating fax device, the off-ramp gateway would act much like a standard T.4 fax machine and transmit the TIFF image. The gateway could easily add information such as a cover page, page count, and other items that are typical of a standard fax transmission.

Going back the other way (i.e., from a standard fax machine to an e-mail recipient) is not so easy. The problem is addressing. A standard Group 3 fax device has the capability to address any telephone number but not any other types of addresses. Therefore, it cannot indicate a recipient address of the form `mailuser@domain`. If the fax call arrives at an on-ramp gateway, the on-ramp gateway might only have a phone number as the destination address. Therefore, it cannot route the e-mail version of the fax. At this stage, the problem has not been resolved, and e-mail to fax is standardized but fax to e-mail might not be. ITU-T Recommendation T.37, which describes store-and-forward fax on the Internet, borrows a great deal from RFC 2305 and is effectively an ITU recommendation that refers the reader to RFC 2305. Interestingly, T.37 makes reference to an off-ramp gateway but does not mention an on-ramp gateway. Thus, T.37 addresses only the e-mail-to-fax direction, leaving the other direction for further study.

All is not lost, however. The fact that solutions are not standardized has not prevented some companies from providing a fax to e-mail service. The service involves a mapping between a subscriber's fax number and the subscriber's existing e-mail address. When a subscriber signs up for the service, the subscriber specifies an e-mail address and the service provider provides a fax number for the subscriber to use. This fax number actually terminates at a gateway device, which is programmed with the subscriber's e-mail address. When a fax is received at the gateway, it is stored and then forwarded to the subscriber's e-mail address in image form as an attachment to an e-mail.

REFERENCES

Chapter 2

Bradner, S. October 1996. "The Internet Standards Process—Revision 3," RFC 2026.

Postel, J. September 1981. "Internet Protocol," RFC 791.

Mogul, J. and Postel, J. August 1985. "Internet Standard Subnetting Procedure," RFC 950.

Mogul, J. October 1984. "Broadcasting Internet Datagrams," RFC 919.

Postel, J. and Reynolds, J. October 1984. "Domain Requirements," RFC 920.

Braden, R. October 1989. "Requirements for Internet Hosts—Communication Layers," RFC 1122.

Braden, R. October 1989. "Requirements for Internet Hosts—Application and Support," RFC 1123.

Baker, F. June 1995. "Requirements for IP Version 4 Routers," RFC 1812

Malkin, G. November 1994. "RIP Version 2 Carrying Additional Information," RFC 1723.

Rekhter, Y. and Li, T. March 1995. "A Border Gateway Control Protocol 4 (BGP-4)," RFC 1771.

Postel, J. September 1981. "Transmission Control Protocol," RFC 793.

Postel, J. August 1980. "User Datagram Protocol," RFC 768

Schulzrinne, H. et al. January 1996. "RTP: A Transport Protocol for Real-Time Applications," RFC 1889.

Schulzrinne, H. January 1996. "RTP Profile for Audio and Video Conferences with Minimal Control," RFC 1890.

Schulzrinne, H. and Casner, S. "RTP Profile for Audio and Video Conferences with Minimal Control," work in progress.

Perkins, C. et al. September 1997. "RTP Payload for Redundant Audio Data," RFC 2198.

Mills, D. March 1992. "Network Time Protocol (Version 3) Specification, Implementation and Analysis," RFC 1305.

Deering, S. August 1989. "Host Extensions for IP Multicasting," RFC 1112.

Fenner, W. November 1997. "Internet Group Management Protocol, Version 2," RFC 2236.

Moy, J. March 1994. "Multicast Extensions to OSPF," RFC 1584.

Hinden, R. and Deering, S. July 1998. "IP Version 6 Addressing Architecture," RFC 2373.

Kent, S. and Atkinson, R. November 1998. "Security Architecture for the Internet Protocol," RFC 2401.

Kent, S. and Atkinson, R. November 1998. "IP Authentication Header," RFC 2402.

Kent, S. and Atkinson, R. November 1998. "IP Encapsulating Security Payload (ESP)," RFC 2406.

Mogul, J. and Deering, S. November 1990. "Path MTU Discovery," RFC 1191.

Chapter 3

ITU-T. August 1996. "Methods for subjective determination of transmission quality," Recommendation P.800.

ITU-T. 1988. "Pulse code modulation (PCM) of voice frequencies," Recommendation G.711.

ITU-T. October 1988. "32kb/s adaptive differential pulse code modulation (ADPCM)," Recommendation G.721.

ITU-T. December 1990. "40, 32, 24, 16 kbit/s Adaptive Differential Pulse Code Modulation (ADPCM)," Recommendation G.726.

ITU-T. December 1990. "Coding of speech at 16 kbit/s using low-delay code excited linear prediction," Recommendation G.728.

ITU-T. March 1996. "Dual rate speech coder for multimedia communications transmitting at 5.3 and 6.3 kbit/s," Recommendation G.723.1.

ITU-T. March 1996. "Coding of speech at 8 kbit/s using Conjugate-Structure Algebraic-Code-Excited Linear Prediction (CS-ACELP)," Recommendation G.729.

TIA/EIA. March 1998. "High Rate Speech Service Option 17 for Wideband Spread Spectrum Communications Systems," IS-733.

ETSI. December 1999. "Enhanced Full Rate (EFR) speech transcoding," EN 300 726 (GSM 06.60).

ETSI. April 2000. "Adaptive Multi-Rate (AMR) speech transcoding," EN 301 704 (GSM 06.90).

Schulzrinne, H. and Petrack, S. "RTP Payload for DTMF Digits, Telephony Tones and Telephony Signals," work in progress.

The complete volume of the ITU material, from which the texts reproduced are extracted, can be obtained from:

International Telecommunications Union
Sales and Marketing Service
Place des Nations
CH-1211 GENEVA 20 (Switzerland)
Telephone: +41 22 730 61 41 (English)/+41 22 730 61 42 (French)/+41 22 730 61 43 (Spanish)
Telex: 421 000 uit ch/Fax: +41 22 730 51 94
X.400: S = Sales; P = itu; A = 400net; C = ch
E-mail: sales@itu.int
Web address: http://www.itu.int/publications

Chapter 4

ITU-T. February 1998. "Packet-based multimedia communications systems," Recommendation H.323.

ITU-T. February 1988. "Call Signaling protocols and media stream packetization for packet-based multimedia communication systems," Recommendation H.225.0.

ITU-T. September 1988. "Control protocol for multimedia communication," Recommendation H.245.

ITU-T. March 1993. "ISDN user-network interface layer 3 specification for basic call control," Recommendation Q.931.

ITU-T. March 1993. "Video coded for audiovisual services at p x 64 kbit/s," Recommendation H.261.

ITU-T. March 1993. "Generic procedures for the control of ISDN supplementary services," Recommendation Q.932.

Chapter 5

Handley, M. et al. March 1999. "SIP: Session Initiation Protocol," RFC 2543.

Handley, M. and Jacobson, V. April 1998. "SDP: Session Description Protocol," RFC 2327.

Schulzrinne, H. et al. April 1998. "Real Time Streaming Protocol (RTSP)," RFC 2326.

Handley, M. et al. "Session Announcement Protocol," work in progress.

Handley, M. et al. "SIP: Session Initiation Protocol," revised version, work in progress.

Donovan S. et al. "SIP 183 Session Progress Message," work in progress.

Donovan S. "The SIP INFO Method," work in progress.

Rosenberg, J. and Schulzrinne, H. "The SIP Supported Header," work in progress.

Rosenberg, J. and Schulzrinne, H. "Reliability of Provisional Responses in SIP," work in progress.

Marshall, W. et al. "Integration of Resource Management and SIP for IP Telephony," work in progress.

Zimmerer, E. et al. "MIME media types for ISUP and QSIG Objects," work in progress.

Singh, K. and Schulzrinne, H. "Interworking Between SIP/SDP and H.323," work in progress.

Chapter 6

Greene, N. et al. April 2000. "Media Gateway Control Protocol Architecture and Requirements," RFC 2805.

Arango, M. et al. October 1999. "Media Gateway Control Protocol (MGCP) Version 1.0," RFC 2705.

Rosen, B. et al. "MEGACO Protocol," work in progress.

Chapter 7

ANSI. 1996. "Telecommunications-Signaling System Number 7 (SS7)—Message Transfer Part (MTP)," T1.111.

ANSI. 1995. "Telecommunications-Signaling System Number 7 (SS7)—Integrated Services Digital Network (ISDN) User Part," T1.113.

ANSI. 1996. "Telecommunications-Signaling System Number 7 (SS7)—Signaling Connection Control Part (SCCP)," T1.112.

ANSI. 1996. "Telecommunications-Signaling System Number 7 (SS7)—Transaction Capability Application Part (TCAP)," T1.114.

ITU-T. March 1993. "Functional Description of the Message Transfer Part (MTP) of Signaling System No. 7," Recommendation Q.701.

ITU-T. December 1999. "Signaling System No. 7—ISDN User Part Functional Description," Recommendation Q.761.

ITU-T. July 1996. "Functional description of the Signaling Connection Control Part," Recommendation Q.711.

ITU-T. June 1997. "Functional description of Transaction Capabilities," Recommendation Q.771.

Telcordia Technologies, December 1999. "Specification of Signaling System Number 7," GR-246-CORE.

Lin , H. "VoIP Signaling Performance Requirements and Expectations," work in progress.

Ong, L. et al. October 1999. "Framework Architecture for Signaling Transport," RFC 2719.

Stewart, R. "Stream Control Transmission Protocol," work in progress.

Morneault, K. et al. "SS7 MTP2-User Adaptation Layer," work in progress.

Sidebottom, G.et al. "SS7 MTP3-User Adaptation Layer (M3UA)," work in progress.

Loughney, J. "SS7 SCCP-User Adaptation Layer (SUA)," work in progress.

Xie, Q. et al. "Use SCTP as MEGACO Transport," work in progress.

Freed, N. and Borenstein, N. November 1996. "Multipurpose Internet Mail Extensions (MIME) Part Two: Media Types," RFC 2046.

Chapter 8

Braden, R. et al. September 1997 "Resource ReSerVation Protocol (RSVP)—Version 1 Functional Specification," RFC 2205.

Blake, S. et al. December 1998. "An Architecture for Differentiated Services," RFC 2475.

Jacobson, V., Nichols, K. and Poduri, K. June 1999. "An Expedited Forwarding PHB," RFC 2598.

Heinanen, J. et al. "Assured Forwarding PHB," RFC 2597.

Rosen, E., Viswanathan, A. and Callon, R. "Multiprotocol Label Switching Architecture," work in progress.

Durham, D. et al. January 2000. "The COPS (Common Open Policy Service) Protocol," RFC 2748.

Wroclawski, J. September 1997. "The Use of RSVP with IETF Integrated Services," RFC 2210.

Jacobson, V. et al. "The 'Virtual Wire' Behavior Aggregate," work in progress.

Shenker, S. and Wroclawski, J. September 1997. "General Characterization Parameters for Integrated Service Network Elements," RFC 2215.

Shenker, S., Partridge, C. and Guerin, R. September 1997. "Specification of Guaranteed Quality of Service," RFC 2212.

Wroclawski, J. September 1997. "Specification of Controlled-Load Network Element Service," RFC 2211.

Awduche, D. September 1999. "Requirements for Traffic Engineering Over MPLS," RFC 2702.

Callon, R. et al. "A Framework for Multiprotocol Label Switching," work in progress.

Andersson, L. et al. "LDP Specification," work in progress.

Jamoussi, B. "Constraint-Based LSP Setup using LDP," work in progress.

Chapter 9

ITU-T. February 1998. "A modem operating at data Signaling rates of up to 33 600 bits/s for use on the general switched telephone network and on leased point-to-point 2-wire telephone-type circuits," Recommendation V.34.

ITU-T. September 1998. "A digital modem and analogue modem pair for use on the Public Switched Telephone Network (PSTN) at data signaling rates of up to 56,000 bits/s downstream and up to 33,600 bits/s upstream," Recommendation V.90.

ITU-T. November, 1995. "Basic user-network interface—layer 1 specification," Recommendation I.430.

ANSI. 1996. "Integrated Services Digital Network (ISDN)—Minimal Set of Bearer Services for the Basic Rate Interface," T1.604.

ANSI. 1998. "Asymmetric Digital Subscriber Line (ADSL) Metallic Interface," T1.413.

ITU-T. June 1999. "Splitterless asymmetric digital subscriber line (ADSL) transceivers," Recommendation G.992.2.

Cable Television Laboratories, Inc. April 2000. "Data Over Cable Service Interface Specifications Radio frequency Interface Specification," SP-RFIv1.1-I04-000407.

Cable Television Laboratories, Inc. December 1999. "PacketCable™ 1.0 Architecture Framework Technical Report," PKT-TR-ARCH-V01-991201.

3rd Generation Partnership Project. October 1999. "Architecture for an All IP Network," 3GPP TR23.922.

Chapter 10

ITU-T. July 1996. "Standardization of Group 3 facsimile terminals for document transmission," Recommendation T.4.

ITU-T. July 1996. "Procedures for document facsimile transmission in the general switched telephone network," Recommendation T.30.

ITU-T. 1984. "300 bits per second duplex modem standardized for use in the general switched telephone network," Recommendation V.21.

ITU-T. June 1998. "Procedures for real-time Group 3 facsimile communication over IP networks," Recommendation T.38.

ITU-T. April 1999. "Procedures for real-time Group 3 facsimile communication over IP networks," Recommendation T.38, Amendment 1.

Masinter, L. March 1999. "Terminology and Goals for Internet Fax," RFC 2542.

Toyoda, K. March 1998. "A Simple Mode of Facsimile Using Internet Mail," RFC 2305.

Allocchio, C. March 1998. "Minimal PSTN address format in Internet Mail," RFC 2303.

Allocchio, C. March 1998. "Minimal Fax address format in Internet Mail," RFC 2304.

GLOSSARY OF ABBREVIATIONS

2B1Q	2 Binary, 1 Quaternary
3GPP	Third-Generation Partnership Project
3GPP2	Third-Generation Partnership Project 2
ABNF	Augmented Backus-Naur Form
ACELP	Algebraic Code-Excited Linear Prediction
ACF	Admission Confirm
ACM	Address Complete Message
ADPCM	Adaptive Differential Pulse-Code Modulation
ADSL	Asymmetric Digital Subscriber Line
AF	Assured Forwarding
AMI	Alternate-Mark Inversion
AMR	Adaptive Multi-Rate
ANM	Answer Message
ANSI	American National Standards Institute
API	Application Programming Interface
ARJ	Admission Reject
ARQ	Admission Request
AS	Application Server
ASN.1	Abstract Syntax Notation 1
ASP	Application Server Process
ATM	Asynchronous Transfer Mode
ATU-C	ADSL Transmission Unit—Central office
ATU-R	ADSL Transmission Unit—Remote
AUCX	Audit Connection (MGCP command)
AUEP	Audit Endpoint (MGCP command)
BCF	Bandwidth Confirm
BCP	Best Current Practice
BGP	Border Gateway Protocol
BRI	Basic Rate Interface
BRJ	Bandwidth Reject
BRQ	Bandwidth Request
CAP	Competitive Access Provider
CAP	Carrierless Amplitude Phase
CAS	Channel-Associated Signaling
CATV	Cable Television
CBR	Constant Bit Rate
CBS	Committed Burst Size
CCBS	Call Completion to Busy Subscriber
CCS	Common Channel Signaling

CDMA	Code-Division Multiple Access
CDR	Committed Data Rate
CED	Called terminal identification tone
CELP	Code-Excited Linear Prediction
CFR	Confirm to Receive
CIC	Circuit Identification Code
CLEC	Competitive Local Exchange Carrier
CMS	Call-Management Server
CMTS	Cable Modem Termination System
CNG	Comfort-Noise Generation
CNG	Calling tone
COPS	Common Open Policy Service
COT	Continuity Test
CPE	Customer-Premises Equipment
CPG	Call Progress
CRCX	Create Connection (MGCP command)
CR-LDP	Constraint-Based LDP
CRLF	Carriage Return and Line Feed
CR-LSP	Constraint-Based LSP
CS-1	Capability Set 1
CSCF	Call State Control Function
Cseq	Command Sequence
CSI	Called Subscriber Identification
CSRC	Contributing Source
DCF	Disengage Confirm
DCN	Disconnect
DCS	Digital Command Signal
DiffServ	Differentiated Services
DIS	Digital Identification Signal
DLCX	Delete Connection (MGCP command)
DMT	Discrete Multitone
DNS	Domain Name System
DOCSIS	Data-over-Cable Service Interface Specification
DPC	Destination Point Code
DPCM	Differential Pulse-Code Modulation
DRJ	Disengage Reject
DRQ	Disengage Request
DS	DiffServ
DSCP	Differentiated Services CodePoint
DSL	Digital Subscriber Line

DSLAM	Digital Subscriber Line Access Multiplexer
DSP	Digital Signal Processor
DTMF	Dual-Tone Multi-Frequency
DTX	Discontinuous Transmission
DWDM	Dense Wave-Division Multiplexing
EBS	Excess Burst Size
EC	Echo Cancellation
EDGE	Enhanced Data Rates for GSM Evolution
EF	Expedited Forwarding
EFR	Enhanced Full Rate
EOL	End Of Line
EOP	End Of Procedure
EPCF	Endpoint Configuration (MGCP command)
ER	Explicit Route
ESP	Enhanced Service Provider
ETSI	European Telecommunications Standards Institute
FCS	Frame-Check Sequence
FDDI	Fiber-Distributed Data Interface
FDM	Frequency-Division Multiplexing
FEC	Forwarding Equivalence Class
FEXT	Far-End Cross-Talk
FFT	Fast Fourier Transform
FISU	Fill-In Signal Unit
FoIP	Fax over IP
FR	Full Rate
FTP	File Transfer Protocol
FTTC	Fiber to the Curb
GCF	Gatekeeper Confirmation
GMSK	Gaussian Minimum Shift Key
GPRS	General Packet Radio Service
GRJ	Gatekeeper Reject
GRQ	Gatekeeper Request
GSM	Global System for Mobile communications
GSTN	General-Switched Telephone Network
GT	Global Title
HDLC	High Level Data Link Control Protocol
HDSL	High Bit-Rate Digital Subscriber Line
HFC	Hybrid Fiber Coax
HTTP	Hypertext Transfer Protocol
IAB	Internet Architecture Board

IACK	Information Request Acknowledgment
IAM	Initial Address Message
IANA	Internet Assigned Numbers Authority
ICMP	Internet Control Message Protocol
IESG	Internet Engineering Steering Group
IETF	Internet Engineering Task Force
IGMP	Internet Group Protocol
IGRP	Internet Gateway Routing Protocol
IN	Intelligent Network
INAK	Information Request Negative Acknowledgment
INAP	Intelligent Network Application Part
IP	Internet Protocol
IRQ	Information Request
IRR	Information Request Response
ISDN	Integrated Services Digital Network
ISDN-UP or ISUP	ISDN User Part
ISP	Internet Service Provider
ITSP	IP Telephony Service Provider
ITU	International Telecommunication Union
ITU-T	International Telecommunication Union-Telecommunications (formerly CCITT)
IUA	ISDN Q.921-User Adaptation Layer
IVR	Interactive Voice Response
JPEG	Joint Photographic Experts Group
LAN	Local Area Network
LCF	Location Confirm
LD-CELP	Low-Delay Code-Excited Linear Prediction
LDP	Label-Distribution Protocol
LEC	Local Exchange Carrier
LMDS	Local Multi-Point Distribution Service
LPC	Linear Prediction Coding
LRJ	Location Reject
LRQ	Location Request
LSP	Label-Switched Path
LSR	Label-Switching Router
LSSU	Link Status Signal Unit
M2UA	MTP2-User Adaptation Layer
M3UA	MTP3-User Adaptation Layer
MAP	Mobile Application Part
MC	Multi-Point Controller

MCF	Message Confirmation
MCU	Multi-Point Control Unit
MDCX	Modify Connection (MGCP command)
MF	Multi-Frequency
MFC	Multi-Frequency Compelled
MG	Media Gateway
MGC	Media Gateway Controller
MGCP	Media Gateway Control Protocol
MIME	Multi-Purpose Internet Mail Extension
MIPS	Million Instructions per Second
MMDS	Multi-Channel Multi-Point Distribution Service
MMUSIC	Multi-Party Multimedia Session Control
MOS	Mean Opinion Score
MP	Multi-Point Processor
MPE	Multi-Pulse Excited
MPLS	Multi-Protocol Label Switching
MP-MLQ	Multi-Pulse Maximum Likelihood Quantization
MPS	Multi-Page Signal
MRF	Multimedia Resource Function
MSU	Message Signal Unit
MTA	Media Terminal Adapter
MTP	Message Transfer Port
MTU	Maximum Transmission Unit
NEXT	Near-End Cross-Talk
NGW	Network Gateway
NHLFE	Next Hop Label Forwarding Entry
NIF	Nodal Interworking Function
NIST	National Institute of Standards and Technology
NSF	Non-Standard Facilities
NSM	Non-Standard Message
NT1	Network Termination 1
NTFY	Notify (MGCP command)
NTP	Network Time Protocol
OFDM	Orthogonal Frequency-Division Multiplexing
OPC	Originating Point Code
OSI	Open Systems Interconnection
OSPF	Open Shortest Path First
PAM	Pulse Amplitude Modulation
PBS	Peak Burst Size
PBX	Private Branch Exchange

PC	Personal Computer
PC	Point Code
PCM	Pulse Code Modulation
PDD	Post-Dial Delay
PDP	Policy Decision Point
PDR	Peak Data Rate
PEP	Policy-Enforcement Point
PHB	Per-Hop Behavior
PINT	PSTN and Internet Interworking
POTS	Plain Old Telephone Service
PPP	Point-to-Point Protocol
PQ	Priority Queue
PRACK	Provisional Response Acknowledgment
PRI	Primary Rate Interface
PSK	Phase-Shift Key
PSTN	Public-Switched Telephone Network
PT	Payload Type
PVC	Permanent Virtual Circuit
QCIF	Quarter Common Intermediate Format
QOS	Quality-of-Service
RAC	Resource Available Confirm
Rack	Response Acknowledgment
RADSL	Rate-Adaptive Digital Subscriber line
RAI	Resource Available Indicate
RAN	Radio Access Network
RAS	Registration, Admission, and Status
RCF	Registration Confirmation
REL	Release Message
RCF	Request for Comments
RIP	Routing Information Protocol
RIP	Request in Progress
RLC	Release Complete
RPE-LTP	Regular Pulse-Excited Long-Term Prediction
RQNT	Notification Request (MGCP command)
RRJ	Registration Reject
RRQ	Registration Request
Rseq	Response Sequence
RSIP	Restart in Progress (MGCP command)
RSVP	Resource-Reservation Protocol
RTC	Return to Control

RTCP	Real-Time Control Protocol
RTP	Real-Time Transport Protocol
RTSP	Real-Time Streaming Protocol
SAP	Session Announcement Protocol
SCCP	Signaling Connection Control Part
SCP	Service Control Point
SCTP	Stream Control Transmission Protocol
SDES	Source Description
SDP	Session Description Protocol
SG	Signaling Gateway
SID	Silence Insertion Descriptor
SIF	Signaling Information Field
SIO	Service Information Octet
SIP	Session Initiation Protocol
SLA	Service Level Agreement
SLS	Signaling Link Selection
SMS	Short Message Service
SNMP	Simple Network Management Protocol
SOHO	Small Office/Home Office
SONET	Synchronous Optical Network
SP	Signaling Point
SPC	Signaling Point Code
SS7	Signaling System 7
SSN	Subsystem Number
SSP	Service Switching Point
SSRC	Synchronization Source
STP	Signal Transfer Point
SUA	SCCP-User Adaptation layer
TA	Terminal Adapter
TCAP	Transaction Capabilities Application Part
TCF	Training Check
TCP	Transmission Control Protocol
TDM	Time-Division Multiplexing
TDMA	Time-Division Multiple Access
TE1	Terminal Endpoint 1
TE2	Terminal Endpoint 2
TFA	Transfer Allowed
TFP	Transfer Prohibited
TIFF	Tag Image File Format
TLV	Type-Length-Value

TOS	Type of Service
TSAP	Transport Service Access Point
TSI	Transmitting Subscriber Identification
TTL	Time to Live
UCF	Unregistration Confirm
UDP	User Datagram Protocol
URI	Uniform Resource Identifier
URJ	Unregistration Reject
URL	Uniform Resource Locator
URQ	Unregistration Request
UTF	Unicode Standard Transformation Format
VAD	Voice Activity Detection
VBR	Variable Bit Rate
VDSL	Very High Bit-Rate Digital Subscriber Line
VoATM	Voice over ATM
VoIP	Voice over IP
WAN	Wide-Area Network
WCDMA	Wide-Band Code-Division Multiple Access
WRR	Weighted Round Robin
XRS	Unknown Message Response

INDEX

ABOUT THE AUTHOR

Daniel Collins has worked in the telecommunications industry for over 13 years. He spent approximately nine years with Ericsson in various countries, including Ireland, Australia, the United Kingdom and the United States. During that time he worked extensively with both wireline and wireless network technologies. He helped to develop and deploy 2G wireless systems in Europe; he played a major role in the adaptation of GSM standards for use in the United States; and he was a major contributor to the launch of some of the earliest PCS networks.

Since leaving Ericsson, Mr. Collins has worked for a new telecommunications carrier and, more recently as a consultant. In a consultancy capacity, he has provided wireless and VoIP engineering expertise to numerous customers in the United States, Germany, Mexico, Portugal, and Spain. Mr. Collins holds a degree in Electrical and Electronic engineering from the National University of Ireland.